BEEF CATTLE FEEDING AND NUTRITION

ANIMAL FEEDING AND NUTRITION

A Series of Monographs and Treatises

Tony J. Cunha, Editor

Distinguished Service Professor Emeritus
University of Florida
Gainesville, Florida

and

Dean, School of Agriculture
Professor of Animal Science
California State Polytechnic University
Pomona, California

Tony J. Cunha, SWINE FEEDING AND NUTRITION, 1977

W. J. Miller, DAIRY CATTLE FEEDING AND NUTRITION, 1979

Tilden Wayne Perry, BEEF CATTLE FEEDING AND NUTRITION, 1980

in preparation

Tony J. Cunha, HORSE FEEDING AND NUTRITION, 1980

BEEF CATTLE FEEDING AND NUTRITION

Tilden Wayne Perry
Department of Animal Science
Purdue University
Lafayette, Indiana

1980

ACADEMIC PRESS
A Subsidiary of Harcourt Brace Jovanovich, Publishers
New York London Toronto Sydney San Francisco

ACADEMIC PRESS, INC.
111 Fifth Avenue, New York, New York 10003

United Kingdom Edition published by
ACADEMIC PRESS, INC. (LONDON) LTD.
24/28 Oval Road, London NW1 7DX

Library of Congress Cataloging in Publication Data

Perry, Tilden Wayne.
Beef cattle feeding and nutrition.

(Animal feeding and nutrition)
Includes bibliographies and index.
1. Beef cattle--Feeding and feeds. I. Title.
SF203.P46 636.2'13 79-8537
ISBN 0-12-552050-6

PRINTED IN THE UNITED STATES OF AMERICA

80 81 82 83 9 8 7 6 5 4 3 2 1

To Our Three Sons

Dr. James Tilden Perry
Harold Wayne Perry
Mark Randolph Remes

Contents

III THE BREEDING HERD

11 Breeding Herd Nutrition and Management

12 Forages and Environment Effect on Brood Cows

13 Milk Production and Calf Performance

IV FINISHING BEEF CATTLE

14 Starting Cattle on Feed

15 Feeding Stocker Cattle

16 Cattle Finishing Systems

17 Feedlot Disease

18 Economics of Cattle Feeding

19 Environmental and Housing Effect on Feedlot Cattle Performance

Foreword

This is the third in a series of books on animal feeding and nutrition. The first was "Swine Feeding and Nutrition," which appeared in 1977. The second was "Dairy Cattle Feeding and Nutrition," which was published in 1979. The fourth in the series is "Horse Feeding and Nutrition," which will appear in November 1980. These four books, and others to follow in this series, are designed to keep the reader abreast of the rapid developments in feeding and nutrition that have occurred in recent years. These new developments have resulted in changes in diets, the use of new feed processing methods, improved use of by-product feeds, and more supplementation with minerals, vitamins, amino acids, and nonprotein nitrogen compounds. New developments in the use of feed additives have made them even more essential. Farmers are also developing animals that produce and reproduce at a higher rate. Moreover, increasing attention is being given to improved carcasses with more lean meat, less fat, and even better eating quality. Since feed constitutes a major share of the cost of animal production, proper feeding and nutrition become increasingly important as farmers strive to decrease feed requirements per unit of animal product marketed. The volume of scientific literature is expanding rapidly and becomes increasingly larger. As this occurs, interpretation becomes more complex and requires a continuing need for summarization in up-to-date books. This necessitates that top scientists and authorities in the field collate all available information in one volume for each species of farm animal.

More attention is being paid to the seriousness of the world's food supply problem. A recent World Bank report indicates that over 1 billion people suffer from chronic malnutrition. Half of them are children under 5 years of age. Every 2½ years, the world's population increases by over 200 million people. This is the equivalent of another United States to feed. Animals provide excellent quality protein, plus many important minerals and vitamins. In the United States, for example, 44% of the food intake and 56% of the nutrients in the human diet come from animal products.

All of the developing countries have about 60% of the world's livestock and poultry, but produce only 22% of the world's meat, milk, and eggs. The United States, with about 8% of the world's animals, produces 19% of the world's meat, milk, and eggs. Better feeding, breeding, and management would help increase animal production in the developing countries. Many scientists feel they need to double their animal protein production in the next 20 years in order to improve the protein status of the world's population. It is hoped that this series of books on animal feeding and nutrition will be of some assistance in the United States and in world food production.

This third book in the series, "Beef Cattle Feeding and Nutrition," is exceptionally well-written by Dr. T. W. Perry, a distinguished scientist who is recognized worldwide for his outstanding work in animal nutrition. The book will be of great value to all who are concerned with beef production and the finishing of cattle throughout the world.

Tony J. Cunha

Preface

Beef Cattle production is indeed an exciting discipline wherever it can be followed in this world. In the United States, the sale of cattle for beef is the number one source of income to agriculture by a wide margin.

Type and style of beef cattle may change for a time, and thus several extreme types have appeared. Therefore, the author has no sections whatsoever on breeds or types because such sections get out of date too quickly, and thus they would appear to "date" such a text too rapidly.

Nutrition and management are based on the best interpretations of scientific facts and the appplication of such facts. Therefore, even though changes are occurring in the nutrition and management of beef cattle, in general they tend to be more gradual. Because this text contains the latest scientific findings, such information remains intact until newer and more refined knowledge has been obtained.

The area of beef cattle nutrition has advanced a long way, but most beef cattle nutritionists realize that so much remains unknown, compared to what we know, about this discipline that it behooves us to pursue the subject diligently. A good case in point has been the first real breakthrough in the nutrition of the rumen in the mid-1970s; the volatile fatty acid production of the rumen could be changed dramatically through the oral administration of a minute amount of monensin.

Beef Cattle Nutrition and Management is meant to be a compilation and interpretation of the latest on the subject. It is meant to represent (1) a literature review, (2) a text, and (3) a useful handbook.

Tilden Wayne Perry

I

Nutrient Requirements of Beef Cattle

1

Rumen Physiology* and Energy Requirements

The most abundant form of renewable energy in this world is cellulose and fiber. Unfortunately man is not able to utilize cellulose; fortunately ruminant animals are able to utilize cellulose. It is the purpose of this chapter to differentiate between ruminants and nonruminants and to identify the unique capability of ruminant animals through their rumen. Ruminants are found in nearly every environment where plants transform solar energy into chemical forms. This capability is a manifestation of the evolutionary processes which provided an alimentary apparatus capable of releasing, in useful form, the chemical energy from the structural carbohydrates of plants. Their nutritional value, however, is linked to a β-1,4-glucosidic bond.

Repeating Maltose Unit of Starch

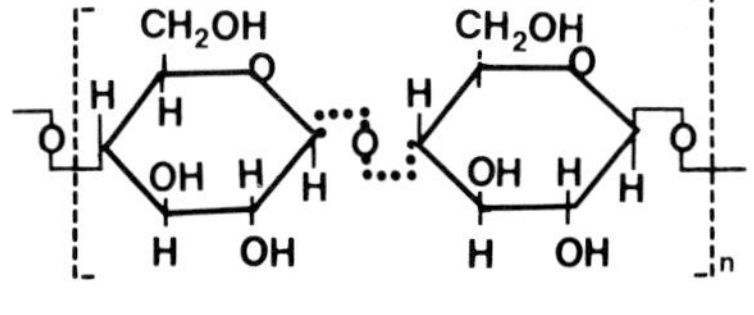

Repeating Cellobiose Unit of Cellulose

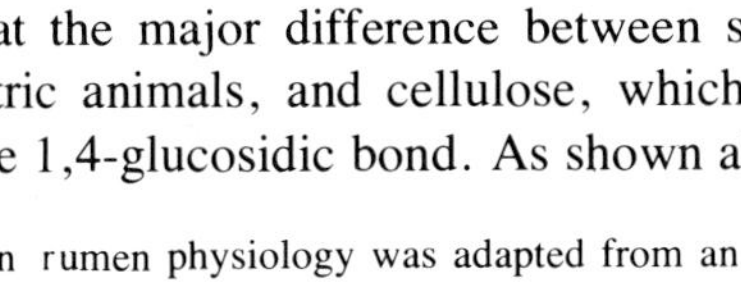

No mammalian degradative enzyme is capable of "unlocking" this bond. It is interesting to note that the major difference between starches, which can be digested by monogastric animals, and cellulose, which cannot, is the spatial configuration about the 1,4-glucosidic bond. As shown above, glucose units are

*Much of the material on rumen physiology was adapted from an article by Thompson (1976) (with permission).

joined in β-1,4 linkages in cellulose and in α-1,4 linkages in starch. This rather subtle difference perpetrated by Mother Nature has been one of the major factors which led to the evolutionary development of the ruminant animal. Ruminants and the herbivores have thus acquired symbiotic microbial populations which provide the required degradative enzymes. Because of the symbiotic relationship between the ruminant animal and its rumen microorganisms, this species possesses a uniqueness which is of great value to man. The microbial fermentation of ingested feedstuffs in the rumen also has significant nutritional and metabolic implications for the host animal, a subject which will be considered later.

I. STRUCTURE AND DEVELOPMENT OF THE RUMINANT STOMACH

Contrary to what students are usually instructed, the ruminant animal does not have four stomachs. The ruminant stomach is divided into four *compartments,* namely, reticulum, rumen, omasum, and abomasum. The reticulum and rumen are joined by a fold of tissue (reticulorumen fold) with the result that ingested materials may flow from one to the other. It is in the reticulorumen that most microbial activity takes place, and it is this area of the rumen that will receive most attention in this discussion (Fig. 1.1).

The function of the omasum is not understood. However, it does remove large quantities of water from the ingesta passing through this portion of the stomach; this may well be its sole function. The function of the abomasum is similar to that of the simple stomach of the monogastric group of animals.

The rumen is nonfunctional in newborn ruminants, but rumen fermentation starts within a few weeks after birth. Considerable growth of the rumen occurs during the first months of life with the main stimulus being solid food in the system. When the four compartments have attained their permanent relative sizes, the rumen constitutes approximately 80% of the total stomach volume. Very early in the life of the ruminant a mixed population of bacteria and protozoa become established in the rumen. The rumen then may be regarded as a large fermentation chamber providing a suitable environment for the continuous culture of the microbial population.

The pH of the rumen ranges between 5.5 and 7.0 and the temperature stays very close to 103° F, which is near optimum for the many enzyme systems contained therein. The food supply to the microorganisms is provided in a more or less continuous manner. Contractions of the rumen wall help to stir and mix intimately the microbes and the ingesta. The moist conditions are ideal for the reactions.

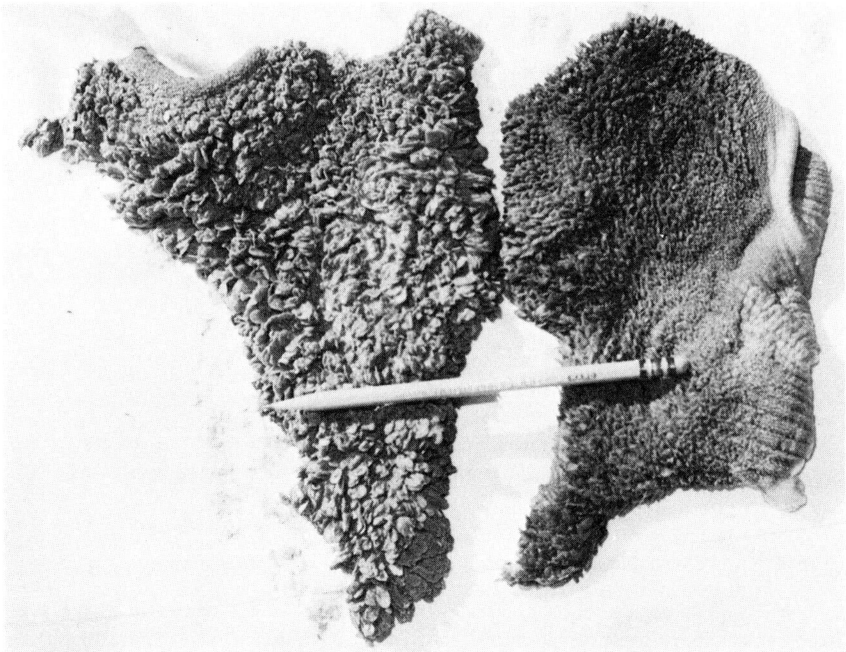

Fig. 1.1. Rumen section on the left is a typical example of rumen parakeratosis. Rumen section on the right is normal.

II. DIGESTION OF NUTRIENTS

A. Nitrogen Metabolism

Nitrogen metabolism in the rumen is a striking example of the influence of rumen microorganisms on the nutrition of the host animal. It has been recognized for more than a century that nonprotein nitrogen could be used—peculiar only to ruminant animals. Nitrogen metabolism in the ruminant is extremely complex. The many possible routes that ingested nitrogen may take in the ruminant animal are shown in Fig. 1.2.

The majority of the protein entering the rumen is degraded to the point where ammonia is produced, depending upon the solubility of the protein. In other words, less soluble protein is degraded less whereas more soluble protein is degraded almost totally. Casein, a highly soluble protein, is almost totally degraded. Fishmeal protein and blood albumin are much more resistant, and a large portion of these proteins will escape the rumen intact. Forage and grain proteins

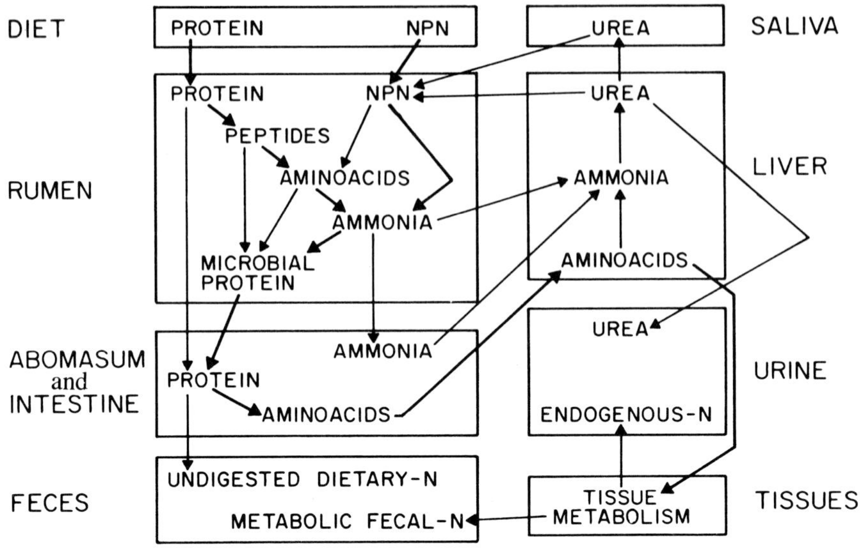

Fig. 1.2. Nitrogen metabolism in the ruminant.

are intermediate in their resistance to microbial breakdown. It appears that on the average, 55–75% of the nitrogen in the rumen is in microbial cells, the bulk of which is protein. Scientists have demonstrated that the solubility pattern of certain proteins may be altered to the extent that they are less soluble in the rumen and thus may bypass degradation to ammonia. Such proteins reach the site of intestinal absorption relatively unscathed. "Protected" proteins appear to have advantages under certain ruminant feeding conditions. Heating of soybean protein to approximately 300° F appears to make it less subject to rumen degradation. Complexing of proteins with casein and safflower oil is another technique which apparently decreases solubility of such protein in rumen fluids (Fig. 1.3).

Although some rumen bacteria need preformed amino acids as their nitrogen source, by far the majority of the rumen bacteria grow abundantly with ammonia as their sole nitrogen source. In fact, research has demonstrated rumen bacteria actually have an ammonia requirement.

In addition to consuming protein, all ruminants obtain some nonprotein nitrogen (NPN) in their natural diets, i.e., corn protein is 4% NPN, alfalfa protein is 10–20% NPN, and corn silage protein is 50% NPN. Even though no NPN is added to the diet, some NPN is certain to be consumed. Some of the NPN may be undigestible, but the majority of it is converted to ammonia in the environment of the rumen. At this point, it is interesting to point out that the addition of NPN to ruminant diets is not as drastic a change in nutrient composition as we are led to believe. It is apparent then that ammonia is a common and important intermediate in both protein and nonprotein nitrogen digestion in the rumen. It is important, therefore, to maximize the conversion of ammonia to microbial proteins and to minimize the loss of ammonia from the rumen by absorption.

Some ammonia is inevitably absorbed from the rumen and carried by the blood to the liver where it is converted to urea. In addition to the ammonia absorbed from the rumen, there is the ammonia resulting from normal protein metabolism in the body. Ammonia in more than token quantities is toxic to the cells. It must, therefore be converted to urea to render it harmless. When one talks in terms of "urea toxicity," it is thus a misnomer since it is actually ammonia toxicity that must be dealt with. Such a basic condition can arise when the concentration of ammonia in the rumen is so great that its rate of absorption into the bloodstream overwhelms the ability of the liver to convert it to urea.

Urea formed by the liver may take one of two possible routes: (1) it may be excreted from the body by dissolving in the urine or (2) it may be recycled into the rumen via saliva or directly through the rumen wall. The quantity of recycled urea is variable depending on the level of dietary nitrogen; 20–55% of dietary nitrogen may be recycled in this manner, perhaps eventually becoming synthesized microbial protein.

Microbial protein is passed to the lower gastrointestinal tract where it is digested and utilized by the host animal much the same as ingested intact protein. The quality of protein—or balance of essential amino acids—of microbial protein is quite good, but not exceptional. It is because of the extensive synthesis of fairly high quality microbial protein that we have not as yet overly concerned ourselves with the amino acid balance of ruminant protein supplements.

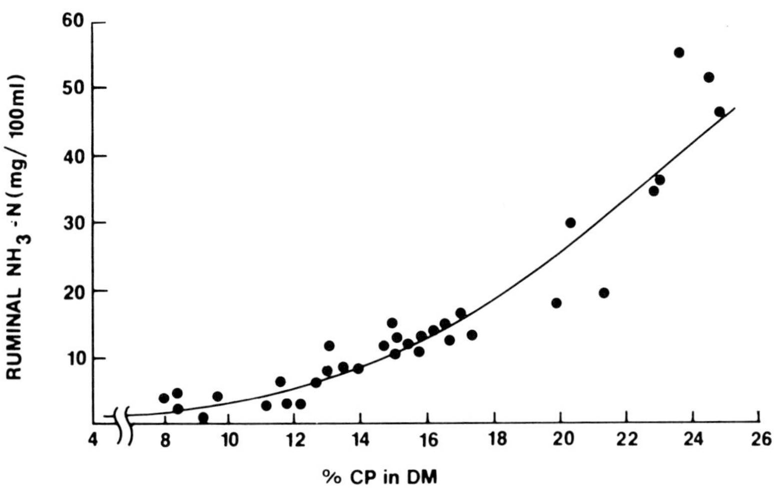

Fig. 1.3. Relationship between mean ruminal ammonia concentration and dietary crude protein.

Ammonia is not only formed from degradation of true protein, but also from breakdown of NPN in the feed. In addition, saliva contains urea which was formed in the liver. Three major sources, therefore, contribute to the ammonia produced in the rumen, with the bacteria utilizing ammonia equally well from all three sources. It is desirable, therefore, to understand maximum ammonia load that can be utilized efficiently by the rumen. Maintenance of ruminal ammonia concentrations in excess of the bacterial capability for utilization results in nitrogen waste.

Based on research at the University of Wisconsin, as dietary crude protein increases above 13% (on a dry matter basis), ruminal ammonia increases rapidly, in excess of 5 mg ammonia nitrogen per 100 ml rumen fluid. Research at Wisconsin also indicated that ruminal ammonia concentrations in excess of 5 mg per 100 ml resulted in no increased microbial protein synthesis and, thus, efficiency of nitrogen conversion to protein declined. Thus it would appear that when dietary protein exceeds 13%, more ruminal ammonia is produced than the microbial protein synthetic machinery can utilize.

The efficiency with which NPN supplements are utilized is not constant but varies depending upon the composition of the diet prior to supplementation and the amount of NPN added (Fig. 1.2). Efficiency of utilization of NPN decreases rapidly as the amount of NPN fed is increased, as the protein level in the supplemented diet is increased, and as the total digestible nutrient (TDN) content is decreased. In other words, most efficient utilization of NPN occurs when small amounts are added to low protein, high TDN diets (Table 1.1).

B. Carbohydrates

Quantitatively, carbohydrates are very important to the ruminant animal. Plant tissues contain about 75% carbohydrates. Cellulose is the most abundant organic compound in the world and comprises from 20 to 50% of the dry matter of most plants. Consequently, carbohydrates provide the primary source of energy for both the rumen organisms and the host ruminant. The carbohydrates found in plant tissues are primarily polysaccharides, including hemicellulose, cellulose, pectins, fructans, and starches. The energy of cellulose is available to the animal only through the intermediary microbial fermentation. The nutritive quality of all cellulose is not the same. Large amounts of cellulose exist in ruminant herbage in combination with hemicelluloses, lignin, cretin, and silica in the cell wall. All of these factors can vary the amount of cellulose available to the microorganisms.

The main end products of microbial carbohydrate metabolism in the rumen are short-chain organic acids, several of which are referred to as volatile fatty acids (VFA). These are made available to the host ruminant as metabolic substrates and become one of its prime energy sources. Because almost no carbohydrate

TABLE 1.1

Influence of Ration Composition on Mean Ruminal Ammonia Concentration and NPN Utilization

% CP[a] in DM	% TDN in DM (mg/100 ml)							NPN utilization (%)
	55	60	65	70	75	80	85	
8	6	5	4	3	2	2	1	
9	6	5	4	3	2	2	1	
10	6	5	4	3	2	2	1	>90
11	6	5	4	3	3	2	2	
12	7	6	5	4	4	3	3	0–90
13	8	7	6	6	5	4	4	
14	10	9	8	7	6	6	5	
15	12	11	10	9	8	8	7	
16	14	13	12	11	10	10	10	
17	17	16	15	14	13	13	12	0
18	20	19	18	17	16	16	15	
19	23	22	21	20	19	19	18	
20	27	26	25	24	23	23	22	

[a] CP, crude protein;
DM, dry matter;
TDN, total digestible nutrient.

escapes fermentation in the rumen, little if any sugar is made available for absorption from the lower gut of the ruminant animal. The absence of an exogenous supply of glucose is not fully compensated for by a decreased glucose requirement, but instead it has increased capability of gluconeogenesis, as compared to that of the monogastric animal.

1. Cellulose

The greatest activity—and probably the least understood—of the rumen is the reduction of cellulose to its constituent units. Some authors describe this as a "three-stage" process whereby (a) cellulose is broken down into smaller polysaccharides which are insoluble, (b) a second stage similar to hydrolysis of other polysaccharides to glucose and cellobiose, and (c) the hydrolysis of cellobiose to glucose. It is the initial stage that is not understood very well.

It appears that a number of enzymes are involved in the hydrolysis of cellulose and that the enzyme responsible for the initial attack is labile. The source of such enzymes appears to be bacterial, but some authors have suggested that protozoa may contribute to the pool of cellulases capable of breaking down cellulose.

2. *Starch*

Ruminal fermentation of starch is an inefficient process compared to intestinal breakdown which results in rapidly absorbed glucose. Ruminal fermentation of starch is relatively slow. Furthermore, methane production as a byproduct of starch fermentation in the rumen represents an energy loss and thus a lowering of the efficiency of carbohydrate utilization. The origin of starch affects its utilization by the rumen microorganisms. For example, corn starch is much more readily degraded than potato starch. In Europe where potato starch is utilized as a ruminant feed, it is readily obvious that some method of steaming or cooking of potatoes be provided to make such feed more digestible.

When diets rich in starch are fed to ruminants which are not accustomed to such diets, a radical change occurs in the acids present. Lactic acid content rises rapidly as does the proportion of propionic acid. Under these conditions, the pH falls (acidity rises) and marked changes in the microflora occur. The above conditions exist to a lesser extent when the introduction of higher levels of starch is made more gradually; thus the practice of "bringing cattle up to a full feed gradually" can be explained. Higher starch diets consistently result in the above conditions, but cattle fed such diets more gradually tend to adapt quite well.

Other polysaccharides, such as xylan and other pentosans, constituting one-fifth of grass or hay dry matter, are digested extensively in the rumen. The end products of such fermentation are volatile fatty acids.

3. *Sugar*

In the rumen fermentation of glucose, fructose, and sucrose results in the production of lactic, acetic, propionic and butyric acids. Maltose, lactose and galactose are fermented more slowly. Rate of fermentation of glucose, for example, is related to the diet. It is more slowly fermented when poor quality rather than high quality hay is fed. An interesting fact concerning carbohydrate metabolism in the rumen is that volatile fatty acids invariably arise as the end products.

C. Volatile Fatty Acids

The total concentration of volatile fatty acids (VFA) in the rumen and the proportions thereof are dependent on diet. More recently it has been demonstrated that certain agents, typified by monensin, can alter the pattern of VFA production in the rumen. Acetic acid tends to predominate under most conditions, with propionic acid and butyric acid following, respectively. Diets heavy in starch or sucrose favor propionic acid production. In general, feeds which are fermented rather rapidly, as are starch and sucrose, give rise to less acetic acid and more propionic acid.

Acetic acid is the major end product of the fermentation of carbohydrates by rumen microorganisms; also, the degradation of protein results primarily in acetic acid formation. The importance of acetate in ruminant nutrition cannot be overemphasized since it is a major energy source. In the lactating animal, acetate is used for milk fat production. Anything which disturbs acetate production results in lowered butterfat production. As an example, finely ground hay results in increased fermentation, but for a shorter time. Acetate which is continuously absorbed from the rumen is used most efficiently when fluctuations in its levels are minimized. Thus, fine grinding of hay results in higher levels of acetic acid than the animal can absorb from the rumen, but only for a short time. This is followed by a lowered level of acetic acid production.

Glucose is a key product formed in the pathway from dietary carbohydrates to VFA's. One molecule of glucose is converted by rumen microorganisms to two molecules of 3-carbon pyruvic acid. Pyruvic acid is a second key structure in carbohydrate metabolism in the rumen in that ultimately it can be converted to any of the VFA's.

Acetic acid is produced from pyruvic acid following the loss of one carbon as CO_2

$$1 \text{ pyruvic acid} + H_2O \rightarrow 1 \text{ acetic acid} + CO_2 + H_2$$

Propionic acid results from the addition of hydrogen to pyruvic acid

$$1 \text{ pyruvic acid} + H_2 \rightarrow 1 \text{ propionic acid} + H_2O$$

Butyric acid is formed by the condensation of two molecules of acetic acid

$$2 \text{ acetic acid} + 2\ H_2 \rightarrow 1 \text{ butyric acid} + H_2O$$

Methane is derived from the reduction of carbon dioxide by hydrogen

$$1 \text{ carbon dioxide} + 2\ H_2 \rightarrow 1 \text{ methane} + 2\ H_2O$$

Carbon dioxide and hydrogen are produced as a result of acetic acid formation. The production of both carbon dioxide and methane results in energy lost to the host ruminant since neither is a form of energy which can be utilized. The production of propionic acid in the rumen does not result in energy losses from gas production. Thus propionic acid production results in more efficient energy production in rumen fermentation than is true for either acetic acid or butyric acid production (Table 1.2).

It is not possible to generalize as to the relative proportions of the volatile fatty acids in the rumen since the type of diet has a profound effect. However, generally cattle fed high roughage diets will have a distribution ratio of 70% acetic, 20% propionic, and 10% butyric acids; those fed high concentrate diets tend to have a distribution ratio of 50% acetic, 40% propionic, and 10% butyric acids.

The volatile fatty acids are absorbed into the portal blood largely through the

TABLE 1.2

Efficiencies of Volatile Fatty Acid Production[a]

Volatile fatty acid	Moles derived per mole of glucose	Gross energy (kcal/mole)	Usable energy derived/mole glucose (kcal)	Relative efficiency[b]
Acetic	2	209.4	418.8	62.2
Butyric	1	524.3	524.3	77.9
Propionic	2	367.2	734.4	109.1

[a] Derived from Anonymous (1975).
[b] Relative to glucose (glucose = 673 kcal/mole).

rumen wall (about 76%); some are absorbed from the omasum and abomasum (19%) and a small amount is passed on to the intestine (5%).

Monensin sodium ($C_{36}H_{61}O_{11}Na$), introduced by Eli Lilly and Company, Indianapolis, Indiana, acts to increase the molar percentage of propionic acid produced during rumen fermentation with an accompanying decrease in acetic and butyric acid production. In their technical manual, Eli Lilly published the pooled results of 19 trials in which the effect of monensin on beef cattle feedlot performance was studied (Table 1.3). At a level of 20 or 30 gm monensin per ton of complete diet, cattle require approximately 10% less feed per unit of gain. In drylot, finishing cattle fed monensin tended to gain at approximately the same rate as those not fed monensin. However, on roughage-type diets or on pasture, cattle fed effective levels of monensin gained more rapidly than those not fed monensin. The net effect of monensin feeding is explained on the basis of altering volatile fatty acid patterns—more specifically, by increasing propionic acid production at the expense of acetic and butyric acid production.

D. Minerals

Another group of nutrients which play an important role in the rumen are the minerals. In addition to their requirement by the microorganisms, some of the

TABLE 1.3

Monensin Effect on Beef Cattle Performance (19 Trials)

Monensin (grams/ton complete feed)	Number of pens	Feed per lb gain (lb)	Improvement in feed efficiency (%)	Daily gain (lb)	Daily feed (lb)
0	63	9.46	—	2.29	21.5
5	31	8.80	6.98	2.39	20.8
10	56	8.83	6.66	2.37	20.8
20	61	8.57	9.41	2.33	19.8
30	60	8.46	10.57	2.28	19.2

mineral elements have other functions. For example, the physiological pH is maintained primarily by the buffering effect of the minerals; maintenance of osmotic pressure is attributed to the minerals.

Data from several investigators indicate that phosphorus is required by rumen microorganisms for cellulose digestion or cell growth. Iowa State researchers demonstrated there was an apparent preference for phosphorus sources by the rumen bacteria, particularly, dicalcium phosphate followed by defluorinated rock phosphate and steamed bonemeal; preference for colloidal clay phosphorus was rather low.

Sulfur is another mineral element which is critical to normal rumen function. The metabolism of sulfur parallels that of nitrogen rather closely. Because of the rumen microorganisms, the ruminant animal has the ability to utilize various forms of sulfur, including inorganic sulfur. With minor exceptions, sulfide is probably the central metabolite in rumen sulfur metabolism. Sulfur is incorporated into cystine, cysteine, and methionine by the rumen microorganisms. If sulfur levels are inadequate, total protein synthesis will be decreased. The known conditions of ruminal sulfur metabolism, allowing for losses from the system and the direct requirements for the microorganisms, give general support for the suggestion that the dietary nitrogen to sulfur ratio should be close to 10:1. This appears to be more critical when more than token quantities of the dietary nitrogen are not derived from amino acid nitrogen, as in the case of feeding high urea supplements.

E. Vitamins

Most discussions on vitamin requirements tend to be of short duration, and it is concluded generally that the rumen microorganisms have no vitamin needs they cannot meet by synthesis. Naturally it is assumed the ruminant animal will need supplemental vitamin A and that vitamin D will be supplied by exposure of subcutaneous materials to the sun's ultraviolet rays. However, that philosophy may need revision as we learn more about the subject. Data being collected at a few experimental centers currently suggest that supplemental niacin may be beneficial to beef cattle.

III. QUANTITATIVE REQUIREMENTS OF BEEF CATTLE*

The major nutrient requirements of beef cattle are listed in Appendix IV, Tables AIV.1 through AIV.8. They are expressed either as concentration in the ration dry matter or as amounts per animal per day. In order to interpolate to levels per unit air dry feed—let us say 10% H_2O—the respective nutrient re-

*Adapted from the Subcommittee on Beef Cattle Nutrient Requirements of the Committee on Animal Nutrition, Agricultural Board, National Research Council (Perry *et al.*, 1976).

quirements per unit of dry matter can be reduced by 10%. All results are based on published findings and are believed to be adequate under most feeding conditions for normal health, growth, finish, and reproduction. Feed manufacturers may find it desirable, however, to increase the concentration of nutrients and ingredients that are especially susceptible to deterioration upon being mixed and/or stored. In addition, certain ingredient levels may be altered under certain stress conditions, e.g., increasing vitamin A levels in times of shipping fever stress.

The nutrients required in smaller quantities, including calcium, phosphorus, and vitamin A, which are included in Appendix IV, Tables AIV.1 through AIV.5, are set forth in Table AIV.8.

IV. ENERGY

Energy is the first demand in all of animal nutrition. The energy need will be met first, and at the expense of all other nutrients, e.g., if the energy need is not satisfied and if protein is available, it will be broken down to satisfy the energy needs first before any of it will be used to meet the protein or amino acid needs. Furthermore, the total energy requirement of animals surpasses all other nutrient requirements, quantitatively. For example, a 650-pound finishing yearling steer has the following nutrient requirements (see tabulation below):

	Daily	Percentage of total requirement
TDN (energy)	13.2 lb	86.2
Protein	2.0 lb	13.0
Calcium	29 gm	0.4
Phosphorus	21 gm	0.3
Salt	9 gm	0.1

The utilization of modern day high urea supplements has held the cost of fortifying cattle rations with supplemental protein to a minimum. As a result, the large cost in cattle feeding is for the energy portion of the ration.

Based on current feed prices, the following tabulation presents the cost of supplying energy, protein, minerals, and vitamin A to a 650-lb finishing yearling steer:

Daily amount (lb)	Ingredient	Cost/unit	Cost[a]	Percentage of total cost
13	Corn (No. 2)	$2.50/bu	$0.58	66
18	Corn silage	$22.00/ton	0.20	23
1.5	48% Urea supplement (with minerals and vitamin A)	$130.00/ton	0.10	11
			$0.88	

From the above calculation, it is obvious the energy cost of the ration is the major cost, or 89% of the total. Furthermore, when one considers feed cost as representing approximately 70% of the total cost of a cattle finishing enterprise, then the cost of the energy portion represents 62% of the total cost of the enterprise.

Much of the energy consumed by beef cattle is wasted. Approximately 40% may be passed through the animal undigested in the feces; about 10% is lost in gases and urine; and as much as 30% is used in producing heat in the digestive process, leaving as little as 20% for body maintenance and weight gains.

There are at least two methods of defining the nutrient requirements for beef cattle. Therefore some definitions of the energy terms commonly used are in order.

A. Definitions of Terms

A *large calorie* or *kilocalorie* (kcal) is defined as the amount of heat necessary to raise the temperature of one liter of water from 15° to 16°C.

A *megacalorie* (Mcal) is equal to 1000 kcal; another term applicable here is the *Therm*. However, the latter term might be confused with the British thermal unit, or Btu (amount of heat required to raise the temperature of one pound of water 1°F).

Gross energy represents the heat of total combustion. This is also called the *heat of combustion*. The average heat of combustion for animals is listed below:

1 gm of average carbohydrate produces 4.1 kcal
1 gm of average fat produces 9.3 kcal
1 gm of average animal protein produces 4.25 kcal
1 gm of average vegetable protein produces 3.98 kcal
1 gm of average mixed protein produces 4.1 kcal

Digestible energy, as the name implies, is that portion which the animal can digest. It is determined by digestion trials by subtracting fecal dry matter loss from dietary intake of a nutrient or combination of nutrients.

Metabolizable energy, often referred to as available energy, is that left over after all the losses of energy in the feces, urine, and combustible gases. The so-called "work of digestion" energy is converted to heat and is not available for productive purposes. However, it is utilized to keep the animal warm and thus plays a critical role in cold weather. Metabolizable energy does not show the true potential of a feedstuff for productive purposes.

Net energy is calculated by subtracting the energy of heat increment loss from the metabolizable energy. This actual determination is tedious and expensive to determine because it involves respiration calorimetry in order to determine heat increment losses. It is used for maintenance only or for production plus maintenance.

Productive energy is that energy above maintenance requirement and is utilized for productive work, tissue gain, or for production of milk, eggs, wool, and fur. This measure is used mostly in poultry formulations.

Total digestible nutrients is a summation of all the energy potential digested by an animal. It is determined by the following equation:

[Digestible fat × 2.25) + digestible NFE + digestible fiber + digestible protein]/100

B. Use of Energy in Determining Nutrient Requirements

The tables in Appendix IV utilize both the total digestible nutrient (TDN) and the net energy system of calculating beef cattle nutrient requirements. The National Research Council (NRC) book on beef cattle nutrient requirements (Perry *et al.,* 1976) utilizes both systems.

The TDN system based on estimates and observations, coupled with limited performance data, has grown over the years. However, because of the years of experience on which it has been based, it has become quite accurate in establishing nutrient requirements for a given set of conditions.

The net energy system is relatively new, when compared to the TDN system. However, because of its extreme accuracy in predicting both nutrient requirements as well as anticipated performance under a specified system of feeding, it has become widely accepted and used.

The net energy system is based on two calculations, namely, net energy for maintenance (NE_m) and net energy for gain (NE_g). Perhaps the major advantage of separate net energy calculations for maintenance and for gain is that animal requirements do not vary when different roughage to concentrate ratios are fed.

The multiple net energy system can be used to calculate the quantity of feed necessary to meet an animal's energy needs and to formulate a diet to supply the needed concentration of energy per unit of dry matter.

C. The Net Energy Method of Predicting Rate of Gain of Beef Cattle

This method is illustrated by the following problem. A 550-lb yearling steer is to be fed primarily on dry shelled corn (1.9% of body weight or 10.5 lb), one pound of high urea supplement, plus whatever corn silage it will consume. The NRC table (see Appendix IV) indicates such an animal can consume a minimum of 13.5 dry matter per day. If both the shelled corn and the high urea supplement contain 88% dry matter (d.m.), then those two ingredients will represent 10.1 lb of the dietary dry matter, leaving 3.4 lb of dry matter equivalent to come from corn silage, or 11.3 lb of 30% dry matter corn silage. The calculations for the above problem follows below:

550 × 0.019 × 10.4 air dry corn × 88% dry matter	=	9.20 lb d.m. from corn
1 lb high urea supplement × 88% dry matter	=	0.88 lb
Total dry matter from corn plus supplement	=	10.08
Total d.m. intake 13.5 lb (approx.) − 10.1 lb (corn plus suppl.)	=	3.4 lb
d.m. to come from corn silage ÷ 30% (d.m. of silage) corn silage	=	11.3 lb

Calculations for net energy for maintenance (NE_m) and net energy for gain (NE_g) can be found in Appendix IV.

	Pounds 100% d.m. basis	Pounds, as is basis	Megcalories (Mcal) NE_m	NE_g
Corn, yellow, dent	9.2	10.4	9.53	6.18
Supplement	0.88	1.0	0.91[a]	0.59[a]
Corn silage (30% dry matter)	3.4	11.3	2.41	1.53
Total		22.7	12.95	8.30

[a] High urea supplement calculated at 60% the value of corn.
12.95 Mcal NE_m ÷ 22.7 lb ration = 0.57 Mcal NE_m/lb
8.30 Mcal NE_g ÷ 22.7 lb ration = 0.37 Mcal NE_g/lb
A 550-lb steer (Appendix Table AIV.7) requires 4.84 Mcal for maintenance
4.84 Mcal ÷ .57 Mcal NE_m/lb = 8.49 lb diet need for maintenance
22.7 lb total diet − 8.49 lb for maintenance = 14.2 lb diet remaining for gain
14.2 lb diet for gain × .37 Mcal NE_g = 5.25 Mcal NE_g
For a 550-lb yearling, each pound gain requires 1.7 Mcal NE_g
5.25 Mcal NE_g # 1.7 Mcal NE_g/lb gain × 3.1 lb daily gain.

D. Sources of Energy

Among the nutrients there are two main sources of energy, namely, the carbohydrate materials plus ether extract, or fat portions of the diet. Earlier in this chapter it was noted that energy has first "call" on the nutrients. Therefore if energy is deficient, the protein of the diet also will serve as a source of energy. The use of dietary protein primarily as an energy source is expensive and inefficient. Thus the energy needs should be met primarily from the carbohydrate and fat portions of the diet.

1. Carbohydrates

Carbohydrates are routinely classified as belonging to one of several categories including monosaccharide (glucose, fructose, glactose), disaccharides (sucrose, maltose, and lactose) and polysaccharides (starch, glycogen, and cellulose). This represents primarily a chemical classification of the carbohydrate family and is quite applicable to much of monogastric nutrition. However, in the case of ruminant animals, a more relevant classification of the carbohydrates is to divide most of the broad family into nitrogen-free extract (NFE) and crude fiber, with NFE including most of the more soluble carbohydrates (monosaccharides, disaccharides, and starch) and crude fiber including the more insoluble portion. Under this system, crude fiber includes both cellulose and lignin, even though the latter is not a true carbohydrate; it is included because it is almost always associated with cellulose.

a. Nitrogen-Free Extract. The relatively soluble carbohydrates are classified as the nitrogen-free extract (NFE) and include the mono- and disaccharides plus the starches and perhaps a part of the hemicelluloses, based on their relative solubility and digestibility. There is no practical method for exact determination of the NFE portion of feedstuffs. Rather, it can be determined mathematically by subtracting all the other determinations from 100:

$$100 - (\text{water} + \text{crude fiber} + \text{ether extract} + \text{crude protein} + \text{ash}) + \text{NFE}$$

For feeding purposes this calculation has proved satisfactory, although it is obvious that it is not too accurate and errors can be introduced with each variable.

NFE is of prime consideration in cattle feeding because it represents the most important energy source in the finishing of beef animals destined for slaughter. The primary source of NFE in finishing diets is the starch contained in feed grains such as corn, milo, barley, wheat, and oats.

Cattle saliva contains no enzyme capable of degrading starch so the starch of the dietary grains passes into the cows rumen unscathed. Rumen fluids do not break down starch to any great extent, except as needed in the synthesis of amino acids from ingested nonprotein nitrogen, such as urea. Thus, for the most part, ingested starches go almost directly to the small intestine before substantial breakdown occurs. The mono- and disaccharides, in contrast, are attached by the rumen microorganisms and are utilized as sources of energy.

In the small intestine the typical amylases break down starches, plus any disaccharides which escaped the rumen, into the 6-carbon sugars. These breakdown products are absorbed into the bloodstream, primarily as glucose. They probably combine with phosphorus for absorption across the villi of the small intestine. Not all of the sugars proceed across the villi at the same rate. Galactose is absorbed most rapidly, followed by glucose and fructose.

b. Crude Fiber. The second classification is crude fiber, which is defined as that nonmineral portion of the feedstuffs which is not soluble in weak acid or weak alkali.

After removal of the fat and water, the feed sample is boiled for 30 minutes with weak sulfuric acid (1.25%) and then for 30 minutes with weak sodium hydroxide (1.25%). This removes the proteins, soluble sugars, and starches, leaving lignin, cellulose and other complex carbohydrates along with the mineral matter. The loss on ignition of the remaining material is defined as crude fiber.

Cellulose, a principal constituent of crude fiber, has been shown to consist of a large number (200 to 2000) of β-glucose units joined together in a nonbranched chain of 1,4-glycoside bonds which are nonresponsive to any enzyme secreted by mammalian tissue. However, it is utilized, to a limited extent, by the microorganisms which inhabit the rumen portion of ruminant animals.

Lignin is not truly a carbohydrate, but because of its close association with cellulose in the plant kingdom, it is classified as a portion of the crude fiber fraction of feedstuffs. Generally, the presence of a higher percentage of lignin in feedstuffs results in lower digestibility of cellulose and other products. The grains and oil meals contain very little lignin while straw may contain as much as 8% or more. The percentage of lignin in a plant increases as the plant matures. Corn geneticists have selected for a lower lignin corn plant characterized by the name, "brown mid-rib." Metabolism trials with cattle have shown a higher digestibility for this type of corn silage. However, such plants do not appear to stand as well in the field after maturity, which confirms the conclusion that a major function of lignin is to provide rigidity to plant stems and stalks for a reasonable length of time after the plant matures.

Digestion of cellulose and crude fiber is confined largely to the ruminant animal. Essentially, no farm animal produces enzymes capable of degrading cellulose. However, the microscopic life which inhibit the first three sections of the ruminant's stomach and of the cecum and colon of the horse secrete enzymes capable of hydrolyzing the chemical bonds holding the cellulose molecule together. None of these enzymes have a marked effect on lignin. Thus cellulose, which is associated with higher levels of lignin, as in the case of wood, is not acted upon to any marked degree. The large intestines of many animals contain bacteria capable of aiding in cellulose breakdown, but this area is beyond the area of nutrient absorption so that such products are not used.

Typically, ruminal microorganisms hydrolyze cellulose to monosaccharides. Organic acids are then formed via fermentation, most of which are volatile fatty acids (VFA). Acetic acid is the principal fatty acid formed from fiber digestion (65–70%), followed by propionic (18 to 20%) and butyric acids (15–17%). However, the coccidiostat monensin, when administered at a level of 30 ppm of the diet, has been shown to be quite effective in altering the pattern of fatty and acid production. Its presence results in increased production of propionic acid.

Crude fiber digestion is accompanied by the production of relatively large amounts of methane gas (CH_4), CO_2, and heat. The heat thus produced is of importance in colder environments in aiding the host animals to maintain body temperature. Neither the methane nor the CO_2 produced has any economic value to the animal and production of methane represents a large loss of energy—perhaps as high as 10% or more of the total. For many years scientists have researched methods which might depress methane production in the rumen, and thus decrease the energy losses entailed with its production.

Some starch, but especially sugar, may be attacked by the rumen microorganisms and utilized by them as energy. On a high roughage-type diet, the addition of more than token quantities of sugar result in decreased digestion of the roughage material, probably merely because the microorganisms prefer sugar over fiber and thus utilize it first. Thus it is obviously advantageous that ruminant animals were designed to have virtually no ptyalin in their saliva to break starch down into lower sugars or much of the resulting lower sugars might be utilized in the rumen by the microorganisms.

Many researchers believe the products of crude fiber digestion—the volatile fatty acids—are absorbed at their point of production, or across the wall of the rumen into the blood stream. This suggestion has been opposed by many who have felt that these acids traveled to the small intestine for absorption.

Of the three major fatty acids formed, propionic acid is of major significance in the formation of glucose. This transformation takes place primarily in the liver. This reaction is critical to ruminant animals, which consume no starch, as a source of blood glucose. Butyric acid also is metabolized primarily in the liver, while acetic acid may bypass hepatic conversion and enter directly into the peripheral circulation to be oxidized as an energy source. Acetic acid can be used by the mammary gland in producing milk fat. The three fatty acids also may be used as direct energy sources via the carboxylic acid cycle.

c. Carbohydrate Metabolism. Absorbed carbohydrates move rapidly to the liver where they are stored, at least temporarily. During rapid absorption, the glucose level of the portal blood may reach 200 mg% while the remaining blood contains no more than 100 mg%, indicating the liver is removing glucose from the blood. The carbohydrate thus stored in the liver cells is deposited in the form of glycogen, commonly called animal starch. During such times liver glycogen storage may equal 10% of the weight of the liver, but the level may fall to practically zero when carbohydrates are being absorbed. The muscles also store glycogen and their content may reach 2% of their weight.

The immediate fate of glucose is governed by the level of at least two hormones, namely, insulin and epinephrine, more commonly called adrenaline. Most researchers believe insulin accelerates conversion of glucose to glycogen and carbohydrate oxidation in the liver and muscles, and that adrenaline increases the rate of hydrolysis of liver glycogen to glucose and increases the conversion of muscle glycogen to hexose phosphate.

The most important step in carbohydrate metabolism is the oxidation to CO_2

and H_2O with the release of energy. The following equation offers the net reaction in an oversimplified form:

$$C_6H_{12}O_6 + 60_2 + 6CO_2 + 6H_2O + 673 \text{ cal}$$

Glucose is catabolized to pyruvate which in turn enters the tricarboxylic cycle. By a series of reactions in which as many as a dozen enzymes participate, a molecule of glucose yields two molecules of 3-carbon pyruvate and eventually two molecules of lactic acid. These final steps occur under anaerobic conditions and take place in the muscle.

d. Conversion of Carbohydrates to Fat. The nutritional basis of fattening all livestock for slaughter is that carbohydrate is readily converted to fat in the animal body. This necessitates the formation of two types of compounds, namely, glycerol and fatty acids. Glycerol may arise from glucose metabolism, possibly from phosphoglyceraldehyde, a product in carbohydrate metabolism.

Acetyl coenzyme A, from pyruvate, probably provides the starting point for fatty acid synthesis. First, carboxylation of acetyl coenzyme A takes place, and then through a series of complicated condensations, 2-carbon fragments are attached to form the typically even-numbered carbon chain fatty acids. It is suggested that the so-called essential fatty acids cannot be synthesized from carbohydrates.

2. *Fats*

Fats or lipids are a group of naturally occurring substances characterized by their insolubility in water and their solubility in fat solvents such as ether, chloroform, boiling alcohol, and benzene. The lipid group includes not only the true fats but also materials which are related chemically (lecithin) and materials which have comparable solubility properties (cholesterol, waxes). The true fats are of interest not only because of their concentrated source of energy (2.25 times that of carbohydrates or proteins) but because a number of the vitamins are associated with fat (fat-soluble vitamins, A, D, E, and K). In addition, even though fat is not considered as an indispensable nutrient per se most nutritionists recognize the need for certain so-called essential fatty acids (linoleic and linolenic).

Of the nutrients in beef cattle nutrition, fat is generally found in only small quantities, except where fat is routinely added to the diet. Even when fat is added to the diet, it usually represents no more than from 3 to 5% of the total. Common feedstuffs contain fairly low levels of fat, ranging from practically none up to 2% in hays, 4 or 5% in grains, 7 to 10% in distillery byproducts, 13% in rice bran, and up to 98% in oils and tallows. Thus a diet based on hay, corn, and oil meal contains less than 4% fat. However, 4% of digestible fat will contain the energy equivalent of 9% of digestible carbohydrate or protein due to its concentrated energy.

The actual digestion process for fats is not definitely known. It has been rather well accepted that hydrolysis of fat to glycerol and constituent fatty acids takes place. Bile plays a role in preparing fats for absorption. It has been demonstrated that neutral fat is found in the intestinal lymph. Some believe fat hydrolysis in the digestion process is limited to the formation of tri-, di-, and monoglycerides plus perhaps limited quantities of free fatty acids. Diets high in fats tend to be digested more slowly.

Water-soluble products of lipid digestion are absorbed directly into intestinal mucosa; insoluble fatty acids and monoglycerides probably are emulsified and then pass into epithelial cells; short-chain fatty acids enter the portal system. Resynthesis and recombination of glucerol and three fatty acids into triglycerides appears to take place during the absorption process. Most of these products move to the liver where they are removed from the circulatory system.

Although fat may be deposited in various portions of the animal body, it is stored primarily (1) in intramuscular connective tissue, (2) in the abdominal cavity, and (3) in subcutaneous connective tissue.

When fat is to be used as a source of energy, the first reaction is hydrolysis to glycerol and the three constituent fatty acids. Glycerol eventually enters the tricarboxylic acid cycle; the fatty acids are oxidized to CO_2 and water by a series of beta oxidation sequences in which oxygen enters the molecule at the beta carbon, eventually resulting in the release of a 2-carbon fragment and, at the same time, reducing the length of the chain by two carbon atoms.

REFERENCES

Anonymous (1975). "Rumensin Technical Manual." Eli Lilly & Company, Indianapolis, Indiana.

Perry, T. W., Garrett, W. N., Klosterman, E. W., Matsushima, J. K., Nelson, A. B., and Woods, W. R. (1976). "Nutrient Requirements of Beef Cattle," 5th rev. ed. Natl. Acad. Sci., Washington, D.C.

Thompson, D. J. (1976). International Minerals and Chemicals Corp., Libertyville, Illinois.

2

Vitamin Requirements of Beef Cattle

Perhaps the most universally accepted definition of vitamins is that suggested by Rosenberg (1945).

> Vitamins are organic compounds which are required for normal growth and maintenance of life of animals, including man, who, as a rule, are unable to synthesize these compounds by anabolic processes that are independent of environment other than air, and which compounds are effective in small amounts, do not furnish energy and are not utilized as building units for structure of the organism, but are essential for the transformation of energy and for the regulation of the metabolism of structural units.

There is a biblical reference to the fact "cattle were not able to see because there was no green grass for them to eat," and so possibly an awareness of this fact may represent some of the very earliest knowledge of the science of nutrition. However, there is a long period from that biblical beginning until the four decades between 1910 and 1950 when so much knowledge concerning the vitamins was uncovered. The last accepted vitamin to be discovered was vitamin B_{12}, in 1948.

Generally, the vitamins are classified into "fat-soluble" and "water-soluble" groupings. This method of classification is largely a physical one and does not tell much more about the vitamins than just their solubility characteristics. Fat-soluble vitamins contain only carbon, hydrogen, and oxygen while the water-soluble vitamins, except inositol and vitamin C, contain, in addition, nitrogen, sulfur, or cobalt.

It is generally accepted that animals with a developed rumen can synthesize all of the recognized water-soluble vitamins. However, at times of apparent rumen dysfunction the rumen microorganisms may experience some impaired capability for B-vitamin synthesis. Veterinarians often administer supplemental B vitamins as a part of a "stress pack" when ruminant animals need special attention.

I. FAT-SOLUBLE VITAMINS

A. Vitamin A

The ultimate functions of vitamin A in animals are brought about by several natural and synthesized compounds.

One grouping of nine compounds is the carotenoids each of which contain 40 carbon atoms, and which are the forms occurring in plants. The carotenoids possess no vitamin A activity, but can be converted to physiologically active vitamin A in varying degrees in accordance with the type of carotene and the type of animal converting it. Since the carotenes must be converted to vitamin A before they have vitamin A activity, they are called *provitamins A*. The carotenes occur in association with chlorophyll in plants. They do not occur in more than token quantities in animals, although some are found in fat deposits, milk and butter, blood, liver, corpus luteum of cows, and testes of bulls. The carotenes occurring in green and yellow plant materials are a most important source of vitamin A for animals. Typically for cattle, green pasture, silages, hay, yellow corn, carrots, and pumpkin represent feed sources richest in carotenes.

The carotenes are destroyed by exposure to oxygen, once the plant producing it has been harvested. Therefore techniques to limit the exposure of feedstuffs to oxygen, such as ensiling, are often employed to limit carotene destruction. It has been demonstrated with swine (McDonald *et al.*, 1966), for example, that yellow corn stored in conventional bins loses much of its vitamin A activity (carotenes) over a period of one to three years. Green hay curing in the swath may lose one-half its vitamin A activity in one day's exposure to the sunlight and perhaps it may lose practically all of it if it is exposed to rain as well as sunlight. Mature, overly ripe plant materials have greatly decreased quantities of carotene.

Species vary in their ability to convert carotene to vitamin A. The rat quite efficiently converts carotene to vitamin A, whereas this process is limited in the cat. Furthermore, there is a within-species variation in that capability, i.e., Holsteins convert carotene efficiently, while Guernseys are much less efficient. Research at the Purdue Experiment Station (Perry *et al.*, 1957) first demonstrated that finishing cattle receiving 20 mg of carotene per day suffered typical vitamin A deficiency symptoms. While cattle fed a similar diet plus 20,000 to 30,000 international units (IU) of crystalline vitamin A per head, daily, gained weight approximately 25% more rapidly and had no vitamin A deficiency symptoms. These data would indicate finishing beef cattle do not convert carotene to vitamin A very efficiently.

Vitamin A has not been identified in plants, and occurs only in animals. The most potent natural source of vitamin A are the various liver oils. Among fish, for example, vitamin A potency per gram of liver oil ranges from 65 IU for haddock to 600,000 for black sea bass. Cod liver oil, perhaps the most commonly used fish liver source of vitamin A, contains about 600 IU per gram. Because the liver oils are so highly unsaturated, they are subject to oxidation which, in turn, rapidly destroys their vitamin A potency. Therefore when stabilized synthetic sources of vitamin A became available, the use of fish liver oils as sources of vitamin A declined rapidly.

Vitamin A plays many critical roles in the animal body. Vitamin A is involved in a generalized maintenance of epithelial tissue. This role is not understood, but, in the absence of vitamin A the epithelial tissues of the body tend to keratinize (Wolbach and Howe, 1925) and thus lose most of their useful function. This effect will occur in the alimentary, genital, reproductive, respiratory, and urinary tracts. Such altered characteristics make the affected tissues more susceptible to infection. Thus, colds and pneumonia are typical secondary effects of a vitamin A deficiency. However, greater than optimal intakes of vitamin A will not aid in preventing infections. A vitamin A deficiency also results in reproductive problems, less rapid weight gain, and impaired vision. Vitamin A is concerned in normal bone development through the osteoblasts of the epithelial cartilage. Thus, if a vitamin A deficiency exists, abnormal bone formation will result.

The vitamin A requirements of beef cattle may be met in part by administration of provitamin A (carotene) plus supplementary vitamin A either by intramuscular injection or orally (Fig. 2.1). International standards for vitamin A are based upon

Fig. 2.1. Intramuscularly injected vitamin A ensures uniform dosage for an extended period of time.

TABLE 2.1

Injectable versus Oral Vitamin A for Fattening Steer Calves[a]

Item	Control, none	20,000 IU per day orally	1 million IU injected initially	6 million IU injected initially
		Method of administering vitamin A		
Daily gain (lb)	1.76	2.07[b]	2.00[b]	1.94[b]
Feed per lb gain (lb)	9.4	8.5	8.5	8.8
Blood serum data				
Vitamin A (μg per 100 ml)				
Initial	38	42	32	41
167 Days	24	59	27	48
210 Days	21	49[c]	16	40[b]
Carotene (μg per 100 ml)				
Initial	105	154	133	94
167 Days	134	100	124	105
210 Days	104	85	111	107
Final liver data				
Vitamin A (μg per gm)	3	15[c]	2	12[c]
Carotene (μg per gm)	10	19[b]	13	19[b]

[a] Average initial weight, 540 lb, 210 days. Data from Perry *et al.* (1962).
[b] Significantly different ($P < 0.05$).
[c] Significantly different ($P < 0.01$).

the capability of the rat to convert β-carotene to vitamin A. However, beef cattle convert carotene to vitamin A much less efficiently than do rats. For rats, 1 mg of β-carotene is equivalent to 1667 IU of vitamin A, but for beef cattle it is proposed that 1 mg of β-carotene is equivalent to only 400 IU of vitamin A.

Under practical feeding conditions it is desirable for cattlemen to recognize the following: (1) In drought years, with prolonged consumption of bleached grasses or hay, or when the background of purchased cattle is unknown and they appear unthrifty, body stores of vitamin A might be suboptimal. (2) The carotene content of dried or sun-cured forages decreases upon storage with the rate of such destruction depending upon factors such as temperature, exposure to air and sunlight, and length of storage. (3) Vitamin A and carotene destruction occurs due to processing of feeds with steam and pressure, or when mixed with certain oxidizing materials such as minerals.

Growing and finishing cattle require 1000 IU of vitamin A per pound of dry diet, pregnant heifers and cows 1270 IU per pound of dry diet, and lactating cows and bulls 1770 IU per pound of dry diet. Intramuscular injection of emulsified vitamin A at a level of one million IU will provide sufficient vitamin A to prevent deficiency symptoms for about 3 months in growing or breeding beef cattle

(Table 2.1). Such injected vitamin A is transferred to the liver for storage and parcelled out only as needed by the body.

Vitamin deficiency symptoms for all species of animals are almost impossible to identify except when deficiencies are very marked. The first observable deficiency symptom is decreased growth which could be the result of a number of factors including genetics and energy or protein deficiency. The role of vitamin A in promoting growth (Table 2.2) is not understood except that it is involved in the activity of bone cell formation at the epiphyseal cartilage.

In vitamin A deficiency of rather long duration, several visual impairment conditions may result. Night blindness due to inability to regenerate visual purple is the result of a vitamin A deficiency. Inflammation of the eyes due to a drying up of the tear ducts and thus inability to wash the eyes is also due to vitamin A deficiency.

Probably the most accurate indicator of a borderline vitamin A deficiency is an analysis of the blood plasma of several animals from the group. A level of less

TABLE 2.2

Effect of Levels of Supplemental Vitamin A on Growth Rate of Beef Cattle[a]

Vitamin A per steer per day (IU)	Daily[b] gain (lb)	Daily feed (lb)	Feed per lb gain (lb)
No alfalfa[c]			
0	1.82	16.8	9.2
10,000	2.18	18.6	8.6
20,000	2.39	20.3	8.5
30,000	2.21	19.0	8.6
40,000	2.33	19.7	8.5
50,000	2.36	19.5	8.3
10% Alfalfa[c]			
0	2.15	18.6	8.6
10,000	2.43	19.4	8.0
20,000	2.52	20.6	8.2
30,000	2.46	20.5	8.3
40,000	2.46	20.0	8.1
50,000	2.47	19.7	8.0

[a] Average initial weight, 470 lb, 256-day trial. Data from Perry *et al.* (1962).

[b] Feeding of all levels of supplemental vitamin A resulted in increases ($p < 0.01$) in gain. Cattle fed 10% alfalfa gained more rapidly ($p < 0.01$) than cattle not fed alfalfa.

[c] Control diet (high corn) contributed 16.8 mg carotene per day; those containing 10% sun-cured alfalfa contributed 45 mg carotene per day.

than 40 μg of vitamin A per 100 ml of blood serum indicates vitamin A deficiency (Table 2.3).

B. Vitamin D

It is assumed that beef cattle usually receive sufficient vitamin D from exposure to direct sunlight or from sun-cured feedstuffs. However, because of the relatively large numbers of cattle finished in confinement and thus not having exposure to sunlight, a discussion of the role of vitamin D is included.

The specific role of vitamin D is in the prevention and curing of rickets (Table 2.4). It is involved in the metabolism of calcium and phosphorus, its function related to their mobilization. Vitamin D is critical for normal absorption of calcium and phosphorus from the gut. Apparently it is not essential for the deposition of calcium and phosphorus in bone.

The body has some vitamin D storage capability, mainly in the liver, and to a lesser extent in the lungs and kidneys.

Deficiency of vitamin D in growing animals results ultimately in external symptoms characterized by deformed bones and excess deposit of cartilage in the usual areas of bone growth. The blood level of calcium and/or phosphorus is

TABLE 2.3

Effect of Vitamin A and Carotene Intake on Blood Plasma Levels and Liver Storage[a]

	Blood plasma						
	Vitamin A (μg/100 ml)			Carotene (μg/100 ml)			
Vitamin A per steer per day (IU)	Initial	107 days	256 days	Initial	107 days	256 days	Liver vitamin A (IU/gm)
No alfalfa[b]							
0	56	26	11	47	64	40	4
10,000	60	49	37	50	68	57	7
20,000	57	59	45	48	56	45	22
30,000	50	59	53	40	44	32	49
40,000	58	69	52	57	43	34	64
50,000	48	68	59	54	51	35	120
10% Alfalfa							
0	54	39	16	51	108	51	3
10,000	49	65	34	45	92	59	8
20,000	54	68	49	37	66	51	23
30,000	50	76	64	46	54	47	55
40,000	50	77	65	38	60	39	80
50,000	53	72	61	46	53	38	146

[a] From Perry *et al.* (1962).

[b] High energy diet contributing 16.8 mg carotene per day; those containing 10% sun-cured alfalfa contributed 45 mg carotene per day.

TABLE 2.4

Composition of Normal and Rachitic Bones[a]

	Water	Ash	Organic matter	Calcium	Phosphorus
Normal rib	14–33	40–47	27–39	16–18	5–8
Rachitic rib	42–66	8–32	21–22	3–12	1–6

[a] Data in percentage of dry fat-free matter. From Hess (1929).

lowered but the level phosphatase is increased. There is a widening of the epiphyseal junction. In severe or prolonged vitamin D deficiency, the tension of the muscles will cause a bending and twisting of the long bones to give the characteristic deformity of bone. There is enlargement at the ends of the bones due to the deposit of excess cartilage, giving the characteristic "beading" effect along the sternum at the point of attachment of the rib bones. Rickets is fairly common in calves and is characterized by decreased growth, stiffness, enlarged joints, and arching of the back.

In mature animals a condition known as osteomalacia or, literally, a wearing away of the bones characterizes a prolonged vitamin D deficiency. If the deficiency is prolonged, the bones will become sufficiently depleted of calcium and phosphorus that they will fracture.

There are essentially two sources of vitamin D. Liver oils represent one source whereas products which have been exposed to irradiation by ultraviolet rays represent another. Most plants contain a sterol known as ergosterol, which, when irradiated by ultraviolet light is converted to calciferol, an active form of vitamin. Several years after this had been established, subsequent research demonstrated calciferol was not very active for poultry, but that irradiated 7-dehydrocholesterol (animal source) was much more active. This resulted in categorizing vitamin D into D_2, the plant source, and D_3, the animal source. It was noted that vitamin D_2 was about one-seventh as active for poultry as was D_3 In the case of beef cattle there is essentially no difference in efficacy between vitamins D_2 and D_3.

The National Research Council (Perry *et al.*, 1976) lists the vitamin D requirements for all types of beef cattle at 124 IU per pound of dry diet.

C. Vitamin E

Vitamin E has a wide distribution in nature and especially rich sources are the germs of seeds. The 1976 NRC booklet on beef cattle states, "Under most conditions natural feedstuffs appear to supply adequate quantities of alpha tocopherol (vitamin E) for adult cattle." There appears to be a partial overlap in the requirement for vitamin E and selenium, but most researchers agree there are some requirements for each that cannot be met by the other.

Since vitamin E is especially susceptible to oxidative destruction, various methods of processing and/or storage may result in varying degrees of destruction of naturally occurring vitamin E in feedstuffs. Its role for beef cattle will be discussed below.

It is perhaps unfortunate that vitamin E has been related so closely to reproduction, since apparently this aspect of vitamin E is critical only to the rat. Proof has not been presented to indicate that a vitamin E deficiency affects reproduction per se in any species with the exception of the rat. Yet many "shy-breeding" and sterile bulls, rams, and stallions have been administed copious amounts of vitamin E in the hope that somehow this would cause them to become potent. The most common lesion in a vitamin E deficiency in some animals is a deterioration of the striated musculature. Thus, muscular dystrophy is the most common manifestation of a vitamin E deficiency in farm animals. This condition is known as "white muscle disease" in young calves. Vitamin E is an effective antioxidant because it is oxidized so readily itself. This capability of vitamin E has been overplayed in an attempt to assess its function. However, as an antioxidant, it prevents peroxidation of unsaturated fatty acids to form free radicals and hyperperoxides which destroy a part of the cell.

The body has tremendous potential for storage of vitamin E. Therefore efforts to study deficiency symptoms have been impeded by such body stores which can prevent true symptoms from appearing for long periods of time.

Vitamin E is one of the fat-soluble vitamins. The richest natural source of vitamin E is wheat germ oil. All cereal grains, legumes, and nuts are excellent sources of vitamin E. There are at least three forms of vitamin E, namely, α, β, and γ-tocopherol. α-Tocopherol is the most potent of the three. The tocopherols are resistant to heat but they are readily oxidized.

The National Research Council makes no recommendation for vitamin E requirements for beef cattle except for, "Estimates of the requirement for younger calves range from 15 to 60 IU (mg) of *dl* α-tocopherol acetate per kilogram of dry diet (7 to 27 IU per pound of dry diet)." The NRC bulletin goes on to point out that under most conditions natural feedstuffs appear to supply adequate α-tocopherol for adult cattle.

D. Vitamin K

The fourth fat-soluble vitamin is vitamin K, which derived its letter name from the Danish word *Koagulation*, and was discovered by Henrik Dam.

Vitamin K is concerned primarily with blood coagulation, and in certain situations where the action of vitamin K is inhibited, increased clotting time of the blood is encountered. This should not be confused with hemophilia which is a condition in males, inherited through the mother, characterized by a tendency to bleed excessively.

Vitamin K is required for the formation of prothrombin, plus possibly other related proteins, in the liver. In the mechanics of blood clotting it is suggested that thromboplastin, released from broken tissue, in the presence of calcium ions, converts prothrombin into thrombin. Once thrombin has been formed it converts soluble fibrinogen of the blood plasma into insoluble fibrin, which is the clot. Thus any interference with thrombin formation, for example, will in turn slow down blood clotting.

Vitamin K is quite prevalent in nature, especially in green leaves; it occurs abundantly in seeds, and much less abundantly in fruits and roots. Vitamin K is synthesized in the rumen. However, the problem that might exist relative to vitamin K adequacy in cattle diets is the matter of an anti-vitamin K substance known as dicumarol. Spoiled sweet clover contains dicumarol which, when consumed serves as an anti-metabolite, blocking the action of vitamin K in forming prothrombin.

Since vitamin K is a normal dietary constituent as well as a normal rumen synthesis product, it is generally accepted there is no dietary need for vitamin K for cattle except at such times as when they consume dicumarol. The NRC bulletin does not list a vitamin K requirement for beef cattle.

II. WATER-SOLUBLE VITAMINS

Water-soluble vitamins are assumed to be synthesized in sufficient quantities by cattle with a developed rumen. Beef calves normally nurse until an average weaning age of 6 or 7 months. The milk received by calves in nursing is an excellent source of all the water-soluble vitamins. Rumen development begins very early in the calf's life due to its consumption of such roughage materials as hay or grass and by the time it is weaned, usually its rumen is sufficiently developed to take care of all the B-vitamin (water-soluble) synthesis it needs. Naturally, if the calf is to be weaned early, before sufficient rumen development takes place, B-vitamins should be added as a dietary supplement.

Once a calf has reached a typical weaning age of 6 or 7 months, it has rather far advanced rumen development. Such animals as well as even more mature animals—including young herd replacement stock, breeding stock and feeder cattle—will manufacture sufficient water-soluble, or B-vitamins, to meet their physiological needs.

During times of stress there may be an interference in the normal rumen synthesis of some of the water-soluble vitamins. At such time veterinarians will often prescribe rather large quantities of certain of the water-soluble vitamins. For example, the injection of rather large dosages of thiamin (vitamin B_1) is used by veterinarians in certain diseases which affect the nervous system, such as listerosis, also often called "circling disease."

REFERENCES

Hess, T. (1929). "Rickets, Osteomalacia and Tetany." Lea & Febiger, Philadelphia, Pennsylvania.

McDonald, T. A., Smith, W. H., Pickett, R. A., and Beeson, W. M. (1966). Influences of aged corn and supplemental vitamin A on growing finishing swine. *J. Anim. Sci.* **25,** 1024.

Perry, T. W., Burns, R. L., Mohler, M. T., and Beeson, W. M. (1957). Mixed supplements versus soybean meal for fattening cattle self-fed ground ear corn. *Purdue Univ. Agric. Exp. Stn., Mimeo* **A.S. 201.**

Perry, T. W., Beeson, W. M., Mohler, M. T., and Smith, W. H. (1962). Levels of supplemental vitamin A with or without sun-cured alfalfa meal for fattening steer calves. *J. Anim. Sci.* **21,** 333.

Perry, T. W., Garrett, W. N., Klosterman, E. W., Matsushima, J. K., Nelson, A. B., and Woods, W. R. (1976). "Nutrient Requirements of Beef Cattle," 5th rev. ed. Natl. Acad. Sci., Washington, D.C.

Rosenberg, H. R. (1945). "Chemistry and Physiology of Vitamins." Wiley (Interscience), New York.

Wolbach, G. B., and Howe, P. R. (1925). Tissue changes following deprivation of fat-soluble A vitamin. *J. Exp. Med.* **42,** 753.

3

Mineral Requirements of Beef Cattle

I. INTRODUCTION

The total mineral, or ash, content of the animal body represents a very small percentage of the total dry matter. Furthermore, several factors such as age of the animal and percentage of fat carried by the body affect the percentage composition. Reid and co-workers (1955) expressed body composition of the bovine on a "fat-free" basis, minus the contents of the digestive tract. On this basis, the composition was 72.9% water, 21.6% protein and 5.3% ash. In addition, there is a very small amount of carbohydrate, which is found primarily in the liver, muscles, and blood.

Missouri researchers (Hogan and Nierman, 1927) analyzed the bodies of 18 steers of varying ages and characterized the mineral composition of the body for several mineral elements as follows: calcium, 1.33%; phosphorus, 0.74%; sodium, 0.16%; potassium, 0.19%; chlorine, 0.11%; magnesium, 0.04%; sulfur, 0.15%. In addition to those listed, there are a host of other mineral elements found in the bovine body in very minute quantities. For example, the author (Perry *et al.*, 1976a) analyzed hair for selenium content from a number of beef cattle and showed a level of 0.38 ppm selenium, which would be equivalent to less than one ten-thousandth of 1% of selenium.

Approximately 99% of the calcium, 70% of the phosphorus, plus magnesium, sulfur, sodium, and chlorine are employed in the bone and its cartilaginous organic matrix. Reasonably large quantities of phosphorus, potassium, and sulfur are involved in muscle and gland tissues; sodium, potassium, and chlorine, plus other elements, are in solution in the body fluids which establish pH and are involved in secretions, osmotic pressure, and irritability of the nerves and contractility of the muscles.

II. ESSENTIAL MINERAL ELEMENTS

A. Calcium*

Calcium is the most abundant mineral element in the body; 99% is found in the bones and teeth and 1% in various soft tissues. Normal blood plasma contains 10–12 mg per 100 ml, but this level may fall as low as 5–7 mg per 100 ml in conditions of severe deficiency. Calcium is involved in a number of roles in the body. It is required for (a) normal bone and teeth formation and maintenance, (b) normal blood clotting, (c) muscular contraction, (d) the regulation of the heartbeat (e) secretion of certain hormones, (f) and milk production.

The most characteristic symptoms of a calcium deficiency are rickets (in the growing animal) and osteomalacia, or "wearing away of the bones" (in mature animals). However both of these symptoms represent advanced acute deficiency symptoms and are rarely observed. Milk diets for the young and typical high roughage diets fed to brood cows normally contain sufficient calcium to meet at least minimal needs. Most feeding grains are extremely low in calcium (corn, 0.02%; barley, 0.05%; milo, 0.03%; wheat, 0.05% compared to grass–legume hay, 0.47%). Therefore, a calcium deficiency for beef cattle may be most often anticipated with finishing beef cattle on a high energy ration. Unless supplemental calcium is provided, one may expect poorer gains, poorer digestibility of feedstuffs, and even occasional tetany. Blood assay will show lowered blood calcium in the range of 5–7 mg per 100 ml.

In addition to its structural role in bone and tooth formation, calcium plays several metabolic roles. Furthermore the role of both calcium and phosphorus in the bone is not strictly structural. The bones serve as a metabolic pool for these elements which may be drawn upon by the soft tissues of the body as needed. Thus in times of late gestation or in lactation, calcium and phosphorus may be drawn from the bones to meet needs not satisfied by dietary intake.

The chemistry of bone calcification is not well understood but the composition does not vary greatly, and consists of a mixture of 85% $Ca_3(PO_4)_2$, 10% $CaCO_3$, and 1.5% $Mg_3(PO_4)_2$. However, this is not the true crystalline structure of the bone. X rays indicate bone is similar to hydroxyapatite, which has the approximate formula $Ca(OH)_2 \cdot 3Ca_3(PO_4)_2$.

The bone serves as a metabolic pool for calcium and phosphorus even in old age and thus if dietary levels of calcium and phosphorus are not adequate, and this state is prolonged, a condition known as osteomalacia develops due to a greater withdrawal of calcium and phosphorus than is deposited.

Calcium plays a role in blood coagulation. The calcium in the blood is found in the plasma. When blood clots, calcium apparently forms a complex with pro-

*Much of the material used in this section is based on data from Beeson and Perry (1975).

thrombin which is acted upon by thromboplastin to form thrombin; thrombin then acts on soluble fibrinogen to form fibrin, which is the blood clot. Without the presence of calcium, blood will not clot.

Calcium is involved in the control of muscle activity. This may be by direct action on the muscle itself or by action on the portion of the nervous system which controls such muscles. Nevertheless, in marked calcium deficiency, manifestation of lack of muscle control is in the form of tetanic spasms.

The need for calcium in milk secretion is obvious. In times of dietary calcium deficiency, the bone storehouse is called upon to supply the calcium needed for milk production.

The calcium balance within the body is affected by dietary intake and by factors which affect its absorption from the intestinal tract into the bloodstream. Recommended dietary levels for the various beef cattle situations are given in the Appendix IV and also in abbreviated form in Table 3.1. It is hypothesized that most calcium absorption takes place in the upper portion of the small intestine because of the possible greater acid concentration there. This is the region at which the gastric juice empties into the small intestine. Absorption of calcium is aided by the presence of vitamin D. Although the proper ratio of Ca:P is critical in affecting absorption of both in most animals, perhaps it is less critical in beef cattle than in most species. Research has shown that beef cattle can tolerate calcium:phosphorus ratios of 7:1 without detrimental effect. However, the most optimum ratio is probably from 2:1 to a 1:1 ratio of each (Wise *et al.*, 1963). Although cattle apparently grow normally when much higher calcium levels are fed in the diet, Smith *et al.* (1964) presented data which indicated doubling the

TABLE 3.1

Calcium and Phosphorus Requirements of Beef Cattle[a,b]

Feeding situation	Day gain (lb)	Weight range (lb)	Dry matter calcium level (%)	Dry matter phosphorus level (%)
Finishing calves	2.4–2.7	400–600	0.55–0.45	0.40–0.35
Finishing yearlings	2.6–3.0	650–900	0.40–0.30	0.34–0.26
Finishing yearlings	2.6–3.0	900–1100	0.29–0.22	0.27–0.22
Growing heifers	2.2–2.5	450–900	0.45–0.23	0.35–0.18
Dry pregnant beef cows	—	800–1300	0.18	0.18
Cows nursing calves, average milking	—	800–1200	0.28	0.27
Cows nursing calves, superior milking ability	—	800–1200	0.41	0.38

[a] For more detailed requirements see Appendix IV.
[b] Perry *et al.* (1976c).

calcium level of the diet from 0.25 to 0.50% interfered with zinc utilization (Table 3.2).

The availability of calcium from various sources was quantitated by Hansard *et al.* (1957) using the radioisotope technique in 109 balance studies with young and mature steers. True digestibility of calcium was greater in young than mature cows. This effect was more pronounced than that due to source. Calcium from mineral sources has a greater availability than calcium from alfalfa. When a biological value of 100% is assigned to calcium carbonate, mineral sources of calcium ranged from 95 to 138 and the hays from 78 to 100. When feedstuffs are grouped according to availability for calcium, bonemeal, monocalcium phosphate, and dicalcium phosphate are the highest, ground limestone, calcium carbonate, and defluorinated rock phosphate are intermediate, and the hays had the lowest calcium availability.

B. Phosphorus

Since phosphorus is so intimately involved with calcium in bone and tooth formation, it is easy to lose sight of its many other metabolic involvements. Phosphorus is found in many soft tissues and thus plays a major role in many reactions, including fat metabolism. Through the complex creatine phosphates, it plays a primary role in carbohydrate metabolism and as an integral part of the buffering system, it is concerned with neutrality regulation. In areas such as South Africa, and under conditions of unthriftiness and emaciation, death of cattle is common unless supplemental phosphorus is provided. Such cows have very poor reproductive performance due to a disturbance of the estrous cycle.

Phosphorus is found as a constituent of many parts of the body, in addition to the bones and teeth. Thus functions in which phosphorus plays a major role are

TABLE 3.2

Effect of Dietary Calcium Levels on Serum and Hair Zinc[a]

Item	Dietary level of calcium	
	0.25%	0.50%
Number of lots	4	4
Number of cattle	24	24
Serum zinc		
Initial (μg/100 ml)	130	135
Final (182 days) (μg/100 ml)	213	188
Hair zinc (112th day) (μg/gm)	212	174

[a] Data from Smith *et al.* (1964).

many. In fact, Forbes and Keith (1914) suggested that phosphorus compounds enter into practically all physiological functions.

1. Phospholipids

These are important compounds involved in the transport of fat materials in the body. They probably are intermediates in the utilization of fats and participate in the oxidation–reduction reactions involved in the release of energy. Choline is derived from phospholipids.

2. Nucleoproteins

Nucleoproteins occur as components of cell nuclei. In nucleic acid, phosphorus in the form of phosphoric acid is combined with a number of compounds including purines, pyrimidines, and carbohydrates. The nucleic acids not only play a role in cell activity as such, but also are involved in the enzymatic control of intermediate metabolism and tissue respiration. Nucleic acid cooperates with phosphorus—combinations with three B-vitamins (thiamin, riboflavin, and niacin) in the metabolism of carbohydrates and the chemistry of muscular work and tissue respiration.

3. Phosphoric Acid Esters of Carbohydrates

Phosphoric acid esters are critical in the release of energy from carbohydrates. When glucose is converted to the storage form of glycogen, or when glucose is utilized for energy, phosphorus compounds are involved in the reaction. The breakdown of glucose and the release of energy is characterized by the oversimplified equation: glucose + adenosine triphosphate → glucose 6-phosphate + adenosine diphosphate + release of heat energy.

4. Riboflavin–Phosphate–Enzyme Complex

This complex is a part of the tissue respiration enzymes, starting with "Warburg's yellow enzyme" and encompassing a number of such enzymes. This group functions not only to release energy for work, but also in maintaining muscle tone and resilience.

5. Diphosphothiamin

Also known as cocarboxylase, diphosphothiamin is active in the breakdown of carbohydrates through pyruvic acid and lactic acid stages.

6. Pyridoxal Phosphate

Pyridoxal phosphate, also known as codecarboxylase, is utilized especially in the removal of the COOH radical from amino acids which are destined to go through the energy release cycle rather than the protein building cycle.

Research concerning the site of absorption of phosphorus from the intestine is sparse. However, it is well established that both calcium and phosphorus are absorbed and excreted through the intestinal wall, making balance studies of the two mineral elements almost impossible. Absorption of phosphorus from the intestinal tract is very rapid, perhaps reaching a peak within 30 min. The vehicle for absorption of ingested phosphorus is mainly that of phospholipids formed in the intestinal wall. It has been generally accepted that vitamin D does not play a major role in phosphorus absorption from the gut, but that abnormally high calcium levels may depress absorption. It seems the cow is affected less by this than other animals. In an earlier discussion it was pointed out organically bound phosphorus in the phytin form poses little availability problems for cattle.

Generalized phosphorus requirements for beef cattle are listed in Table 3.1 and requirements for a variety of conditions are listed in the Appendix.

Phosphorus may be provided to beef cattle from a number of supplemental sources when the content of the diet of typical feedstuffs is inadequate. The availability of phosphorus to beef cattle from most sources is relatively high and so it is mostly a matter of economy in selecting which source to use. The one exception is the use of raw rock phosphate which contains toxic levels of fluorine and thus should not be used. However, it is a common practice to remove practically all of the fluorine of raw rock phosphate by heating it to very high temperature.

Phosphorus and calcium contents of several sources of the two elements are listed in Table 3.3.

C. Sodium and Chloride

Sodium and chloride or salt have been recognized as necessary constituents of the diet of man and animals for centuries. The value of salt is referred to in the Bible. The word "salary" is derived from the Latin word for salt. Salt in

TABLE 3.3

Calcium and Phosphorus Content of Several Supplementary Sources

Product	Calcium (%)	Phosphorus (%)
Steamed bonemeal	31.3	14.4
Dicalcium phosphate	22	18.5
Diammonium phosphate	0.5	20
Phosphoric acid, feed grade	—	23.5
Defluorinated rock phosphate	32	18
Limestone	38	—
Oyster shell	38	—

minimal quantities serves to enhance the palatability of foods and feeds while in larger quantities it limits food and feed intake. This latter effect has been used in cattle feeding to control intake of more expensive and more concentrated feedstuffs such as either protein supplements or corn, when they are fed on a free-choice basis. Although sodium and chloride have widely differing functions, they will be discussed together since that is the way in which they are found most commonly.

The most common deficiency symptom for salt can be seen in the "salt-seeking" behavior of animals. They appear to go to great lengths to satisfy their salt hunger. Animals in the wild state will travel miles to gain access to natural salt licks. Confined cattle will lick at the earth or wood, and especially at the bodies of other cattle, in search of salt. If a marked salt deficiency persists, cattle will show signs of lack of thrift such as a roughened hair coat and decline in body weight. Lactating cows will show a decline in milk production. The latter symptoms are rarely seen because salt is quite easy to supply. Cattle should never be without salt; it should be supplied on a free-choice basis at all times.

The physiological functions of salt must be broken into that of the individual elements, sodium and chloride, since their functions are unrelated.

Sodium is present in the body primarily as the sodium ion and its function appears to be independent of whatever ion it happens to be associated with, like bicarbonate, phosphate, chloride. A major function of sodium is in the regulation of osmotic pressure within the body, or the pressure which affects the passage of water, nutrients, and waste products across membranes. Another obvious function of sodium is in the regulation of the acid–base relationship within the body. The ion is the chief cation of blood plasma, and this is true also for the other extracellular fluids of the body. The sodium ion has an effect on irritable tissues, such as muscles. In fact, the rate at which the heart beats is regulated by the proportions of the sodium, potassium, and calcium present.

Chloride functions as a part of the gastric juice, in accompaniment with the hydrogen ion (hydrochloric acid). When gastric juice—and the hydrochloric acid accompanying it—is lost by vomiting, alkalosis may briefly persist due to a relative excess of bicarbonate, which had been neutralized by the HCl. Chloride is involved in regulation of osmotic pressure. The majority of the anions of blood plasma and extracellular fluids are comprised of chloride. Chloride is involved in the "chloride shift" which aids in the regulation of the acid–base balance of the blood. The chloride ion can exert its "base effect" in the blood plasma, thereby maintaining the desired acid–base relationship. When bases such as bicarbonate enter the blood, chloride can shift from the plasma into the blood cells, no longer exerting its base effect. In contrast, when the lungs remove bicarbonate from the blood or when acid enters the bloodstream, chloride shifts from the inactive state, within the blood cells, into the active state (base effect) by entering the blood plasma.

Salt is often fed at elevated levels in feedlots where urinary calculi, or "water belly," is a problem. Its function here is not understood. It may be that the additional salt causes increased consumption of water, and, subsequently, increased urination. Thus its role herein may be merely that of flushing out the urinary calculi-predisposing materials.

Sodium is absorbed readily from the small intestine, apparently requiring no special conditions. However, a possible exception is that excessive consumption of potassium results in excessive excretion of sodium, and vice versa. This apparent relationship might be due to poorer absorption of each when the other is present in excessive quantities. Since forages are especially rich in potassium, it is assumed ruminants have a relatively larger sodium requirement than nonruminants. However, this is due to the effect mentioned above—that the increased intake of potassium of forage eaters results in increased excretion of sodium. Sodium is absorbed, to a limited degree, from the stomach. Chloride, on the other hand, is absorbed primarily from the small intestine.

Practically all of the excreted sodium chloride exits via the urine and tends to reflect ingested levels. Hagsten and Perry (1975) demonstrated that lambs excrete large quantities of salt in the urine when large quantities are consumed, but on markedly salt-deficient diets, the excretion of salt is extremely low. The same researchers showed that plasma levels of sodium remained quite constant (33% ppm) over an 11-week period in which extremely low levels of sodium were fed (0.01% of the dry matter). This indicated that the role of sodium in the blood is so critical that it is maintained at a constant level, when possible. However, the potassium level of the plasma declined 23% (211 to 170 ppm) over the same period. Apparently the adrenal glands produce a hormone, other than adrenaline, which regulates sodium levels in the blood, because animals suffering from Addison's disease excrete greatly increased quantities of sodium in the urine. (Addison's disease is a disease of the adrenal gland and affects the physiology of the adrenal cortex, which in turn apparently regulates sodium metabolism.)

Hagsten *et al.* (1975) established the supplemental salt requirements of growing and finishing lambs at 0.20% of the "air dry" diet, when it was shown most diets contain 0.20% salt, to give a total requirement of 0.40%. These recommendations should be quite applicable to beef cattle. Thus, a level of 0.20–0.25% supplemental salt for beef cattle is adequate.

Theoretically, salt "toxicity" could occur and would be characterized primarily by excessive dehydration as the result of the body excreting extra water to compensate for excessive intakes of salt by the urine. Earlier effects would be an edema (or excessive accumulation of extracellular water) in an attempt to maintain the proper ionic concentration in the body fluids.

Because cattle will not consume excessive levels of salt, unless they are given no alternative, it can be used to regulate intake by cattle of relatively more palatable feedstuffs. For example, the incorporation of salt into free-choice protein supplements for beef cows grazing low-quality roughages has been practiced for decades; the incorporation of 5% of salt in free-choice ground shelled corn fed on pasture caused a 22% decrease in corn consumption [18.4 versus 14.3 lb per head per day for cattle averaging 835 lb, over a 195-day grazing period (Perry *et al.*, 1976b)].

III. TRACE MINERAL ELEMENTS

Several trace mineral elements are required (Table 3.4). Normally, natural feedstuffs meet most of the trace mineral requirements of beef cattle, but under some conditions this may not be the case. Furthermore, as the virgin deposits of certain trace elements in the soil tend to become depleted, more frequent occurrence of deficiency symptoms of several of the trace mineral elements have appeared. Relevant discussions of several trace mineral elements are included in this discussion.

A. Iodine

The only known physiological function of iodine is through the function of thyroxine, of which iodine is a part. Ingested iodine is transported rapidly to

TABLE 3.4

Mineral Recommendations for Beef Cattle[a]

Element	Growth	Finishing	Lactation	Maintenance
Calcium[b] (%)	0.3–0.15	0.4–0.2	0.3	0.20
Phosphorus[b] (%)	0.25–0.15	0.3–0.15	0.25	0.20
Chlorine (%)	0.20	0.20	0.20	0.20
Sodium (%)	0.15	0.15	0.15	0.15
Potassium (%)	0.5	0.3	0.3	0.3
Sulfur (%)	0.1	0.1	0.1	0.1
Magnesium (ppm)	440	440	440	440
Iron (ppm)	40	40	40	40
Zinc (ppm)	40	40	40	40
Manganese (ppm)	20	20	20	20
Copper (ppm)	5	5	5	5
Cobalt (ppm)	0.1	0.1	0.1	0.1
Iodine (ppm)	0.1	0.1	0.1	0.1
Molybdenum (ppm)	0.1	0.1	0.1	0.1
Selenium (ppm)	0.1	0.1	0.1	0.1
Fluorine (ppm)	1–5	1–5	1–5	1–5

[a] Expressed on units per unit dry matter.
[b] Decreases with relative growth rate.

the thyroid gland where it is incorporated as protein-bound iodine into thyroxine. The broad function of thyroxine is that of controlling the rate of metabolism of the body. Thus, a decreased level of thyroxine results in a lower metabolic rate. Conversely, excess thyroxine results in an increased metabolic rate. In addition to its effect on the rate of metabolism, thyroxine affects other processes such as differentiation of cells, body growth, and tonus of the muscles. When an iodine deficiency exists (and thus of thyroxine), the thyroid gland greatly enlarges in an apparent attempt to compensate for lowered thyroxine production. The enlarged thyroid condition is known as "big neck" or goiter in animals whose thyroid gland is outside of the chest cavity, thus becoming visible in the intact animal (cattle and sheep).

The only iodine deficiency usually detected in cattle can be seen in young calves at the time of birth. They are weak, have enlarged thyroids ("big neck"), may be hairless, and quite often stillborn.

Iodine deficiency can be anticipated when feedstuffs grown inland, away from the ocean, are fed. Crops grown near the ocean probably contain adequate quantities of iodine.

Iodine requirements for pregnant and lactating cows have been estimated at 400 to 800 μg per day. This can be supplied readily by feeding salt containing 0.007% of stabilized iodine.

B. Magnesium

Magnesium deficiency symptoms in beef cattle are rarely encountered. However, it appears that the incidence of "grass tetany" or "grass staggers" is increasing, perhaps reflecting a depletion of the earth's natural store of the element. Magnesium deficiency, then, is characterized by hyperirritability, tetany, and convulsions. In its final stages before death, cattle so affected thrash about with uncontrolled muscular spasms. Lowered blood magnesium seemingly is the underlying cause. Grass tetany can be considered a magnesium-deficient disease, in the strictest sense. It may, however, be a "relative" deficiency disease since a lowering of the bone levels of magnesium cannot be demonstrated.

A high percentage of cases of grass tetany occur in the spring when grass is young and lush. It has been postulated that the availability of magnesium is decreased greatly in the springtime grass due to accumulation of ammonium ions, which then interferes with magnesium absorption. More recently, it has been demonstrated that the symptoms of grass tetany may occur under drylot feeding conditions with cattle who do not have access to young grass.

The physiological effect of increased magnesium levels is that of a tranquilizing effect. Sleeping, hibernating animals maintain a much higher serum level of magnesium than to active animals. The serum magnesium content of nonhibernating animals increases when the body temperature is artificially lowered. The intravenous injection of magnesium results in muscular paralysis similar to that following an injection of curare. Magnesium in excess, then, depresses the central nervous system whereas a deficiency results in the opposite effect.

The requirement of magnesium for beef cattle can be met by feeding 15 gm of magnesium per head per day. Magnesium oxide is 60% magnesium whereas magnesium sulfate is approximately 10% magnesium. So 1 ounce of MgO or 5 ounces of $MgSO_4 \cdot 7H_2O$ per cow daily will meet magnesium requirements. However, magnesium compounds generally are not very palatable and, therefore, it is sometimes difficult to get a cow to voluntarily consume the designated amounts of the above-mentioned magnesium compounds unless they are somehow masked. A free-choice mineral mixture consisting of equal parts dicalcium phosphate, plain salt, MgO, and dried cane molasses has been used with some success with brood cows on pasture.

C. Cobalt

Cobalt requirement of beef cattle is approximately 0.1 ppm of the dry matter of the diet. Cattle feeding pastures with levels of cobalt in this range are free from deficiency signs. Cobalt toxicity has been produced in calves by feeding 0.5 mg cobalt per pound of body weight. Cobalt is an integral part of the vitamin B_{12} molecule, which can be synthesized by the microorganisms of the reticulorumen. Cobalt deficiency is not observed under typical feedlot situations, but cattle fed forages grown in areas such as Florida, Michigan, Wisconsin, Massachusetts, New Hampshire, Pennsylvania, and New York have developed deficiency symptoms. Such cattle appear starved. Cobalt may be supplied as a part of trace mineralized salt.

D. Manganese

Manganese requirement for beef cattle is low, usually in the range of 10 to 20 ppm of the dietary dry matter. Most feedstuffs are adequate in manganese and thus deficiency is relatively rare. Deficiency symptoms in the cow are characterized by reproductive disorders. These include delayed estrus, reduced fertility, abortions, and deformed young. Calves born to manganese-deficient cows have exhibited deformed legs (enlarged joints, stiffness, twisted legs, "overknuckling"), weakened shortened bones, and poor growth.

E. Zinc

Zinc requirements of beef cattle appear to be about 30 ppm of the diet. It is highly tolerated and it is not until a level of 900 ppm is fed that toxicity symptoms occur. At this level a slight decrease in feed consumption and slower gains are seen. Severe zinc deficiency results in rough scaly skin, which itches and causes much discomfort. The nose and mouth become inflamed, and submucosal hemorrhage occurs. The animal develops an unthrifty appearance, roughened hair coat, and stiffness of joints. Finishing cattle show lowered weight gains in less severe deficiencies.

F. Potassium

The potassium requirements for beef cattle have been reported to be about 0.75% of the dietary dry matter. Since forages generally are quite high in potassium, cattle fed a medium to high forage diet should suffer no potassium deficiency. Finishing cattle on a high grain diet conceivably could be borderline deficient in potassium. Since potassium is required for a number of body functions, a deficiency could result in nonspecific symptoms or delayed growth, reduced feed consumption and efficiency, stiffness, and emaciation.

Excessive levels of potassium may interfere with magnesium absorption and, also, in high levels of phorphorus, increasing the incidence of urinary calculi. Also, excessive intakes of potassium results in greatly increased excretion of sodium.

G. Sulfur

Sulfur requirements for beef cattle are primarily those involving the sulfur-containing amino acids (mainly cystine and methionine). Sulfur may be supplied in either organic or inorganic forms. The organic forms are most readily utilized; elemental sulfur is utilized least and sulfates are intermediate in this respect.

The sulfur requirements of ruminants have attracted attention as a result of the greatly increased utilization of nonprotein nitrogen in cattle feeding. An ample supplement for diets low in sulfur is 3 gm of inorganic sulfur to each 100 gm of urea, which is calculated at 1 part inorganic sulfur to 15 parts of NPN.

H. Copper

Copper requirement of beef cattle is met with 4 ppm of dietary dry matter when the diet is not excessively high in molybdenum and sulfate. In areas where Mo and So_4 are high, the copper requirement may need to be increased 2- to3-fold. Most feedstuffs supply adequate copper, but they do reflect the copper content of the soil on which they were grown. For example, the soils of parts of Florida and of the Coastal Plain region of the southeastern states are quite low in copper and thus produce feedstuffs which reflect that deficiency. Copper deficiency may occur in calves fed milk diets for long periods of time or in older cattle subsisting on forage produced on copper-deficient soil. The signs of a copper deficiency in cattle include a depraved appetite, loss of condition, stunted growth, rough hair coat, anemia, diarrhea, depigmentation of the hair, and sudden death.

I. Selenium

Selenium plays a unique role in beef cattle nutrition in that supplemental selenium is required in some areas of the country whereas in others toxic levels exist in natural feedstuffs. In fact, within the state boundaries of South Dakota, for example, both selenium-deficient and selenium-toxic situations exist.

The selenium requirements of beef cattle have been widely accepted as 0.1 ppm of the dry diet. Cows grazing forages containing less than this amount tend to produce calves with nutritional muscular dystrophy, or "white muscle disease." Apparently selenium and vitamin E have somewhat overlapping nutritional functions in that proper use of the vitamin can prevent or cure white muscle disease that would be caused by a selenium deficiency. However, there appears to be functions attributable to each which cannot be met by the other.

Little is known of the mechanisms of selenium toxicity. It has been postulated, however, that selenium can substitute for sulfur in the amino acids, cystine and methionine, causing them to lose their physiological specificity.

Four techniques have been utilized in supplying selenium: (1) administering selenium supplements as a drench, (2) subcutaneous or intramuscular injections, (3) placing selenium in fertilizers applied to pastures, and (4) using selenium as a feed additive.

J. Fluorides

Specific evidence is lacking to show an essential function for fluorides in beef cattle. However, the nutritional interest in fluoride is its toxic aspect. Research on fluorine has been concerned primarily with toxic levels that affect teeth and bone structure and the utilization of other nutrients (Fig. 3.1). Certain rock phosphates must be defluorinated to make them safe for cattle feeding. Raw rock phosphate often contains 3.5–4.0% of fluoride, which is toxic when such rock phosphate constitutes 1% of the diet.

The harmful effects of fluoride over prolonged periods of feeding are due to fluoride accumulation in the tissues; the bones become thickened and soft and their breaking strength decreases. The teeth may erode and the enamel may become mottled. Appetite decreases and slow growth results. Fluoride is a cumulative poison and the toxic effects may not be noticed for some time. Because of this, maximum care should be exercised with breeding animals to be

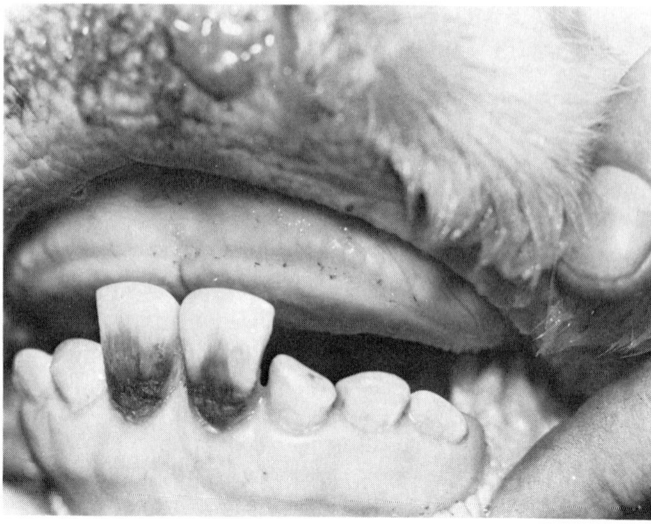

Fig. 3.1. Tooth mottling typical of fluorine toxicity.

retained in the herd. Safe levels are as follows: no more than 100 ppm of fluoride in the diet of finishing cattle and not more than 40 ppm in the diet of cattle to be kept in the breeding herd.

IV. ROLE OF BUFFERS IN BEEF CATTLE NUTRITION

Many of the so-called buffers are composed of mineral elements. Purdue (Wheeler, 1976) research has demonstrated an interesting response of growing heifers by doubling the calcium level using a high magnesium limestone. The diet used consisted of approximately equal parts forage and concentrate; corn silage was used as the forage. Corn ear silage plus a supplement constituted the concentrate with soybean meal being the principal source of supplemental protein. The comparative diets contained (dry matter basis) 12% protein, 0.35% phosphorus, 0.50% trace mineralized salt plus either 0.35% calcium or p.72% calcium with the latter being considered the buffered diet. Magnesium limestone (21% Ca, 13% Mg and 0.18% P) was the source of added calcium in the second diet (Table 3.5).

Weight gains by the heifers fed the buffered diet were greater (3.78 versus 2.71 lb per day). Feces from heifers fed the buffered diet had less starch (1.14 versus 5.6%) and higher pH values (6.76 versus 7.37). Possibly the most remarkable result was a 44% decline in feed required per unit gain by heifers fed the buffered diet (5.0 versus 9.0 lb feed per pound gain).

TABLE 3.5

Buffer Effect on Performance of Heifers

Item	Diet type: Control, 0.32% Ca	Diet type: Buffered, 0.72% Ca
Feed intake dry matter (lb)	18.8	18.3
Feed intake (dry matter, %)	2.2	2.1
Weight gain per day (lb)	2.71[a]	3.78[b]
Feed dry matter per lb gain (lb)	9.0[a]	5.0[b]
Fecal starch (%)	5.6[a]	1.1[b]
Fecal pH	6.4[a]	6.8[b]

[a,b] Values in the same row with differing superscripts differ by $p < 0.01$.

In contrast, Iowa State University (Rouse, 1978) research showed no benefit from added calcium carbonate to a diet also fortified with rumensin. At the same Station (Trenkle, 1978), the addition of 0.4% sodium bicarbonate resulted in an 8% increase in daily gain (2.79 versus 2.58 lb per day) and an 8% improvement (7.78 versus 8.44 lb feed per lb gain) in feed efficiency. A higher level of sodium bicarbonate decreased performance. In contrast, wheat diets or different types of silage diets may give a positive response to higher levels of buffers, possibly because of increased levels of lactic acid production. Several studies have indicated a more marked response to sodium bicarbonate supplementation during the first month or two in the feedlot. This is a period of adjustment wherein rumen lactic acid levels may be quite erratic, even approaching a condition of lactic acidosis. Purdue research has indicated that supplemental sodium bicarbonate may be especially useful in helping new cattle adjust to a rumensin regimen.

REFERENCES

Beeson, W. M., and Perry, T. W. (1975). Calcium for beef cattle. In "Calcium in Beef and Dairy Nutrition," Natl Feed Ingred. Assoc., Des Moines, Iowa.

Forbes, E. B., and Keith, M. H. (1914). A review of literature of phosphorus compounds in animal metabolism. *Ohio, Agric. Exp. Stn., Tech. Bull.*

Hagsten, Ib, and Perry, T. W. (1975). Effect of dietary sodium levels on blood levels, urinary excretion and adrenal histology of lambs. *J. Anim. Sci.* **40,** 1205.

Hagsten, Ib, Perry, T. W., and Outhouse, J. B. (1975). Salt requirements of lambs. *J. Anim. Sci.* **40,** 329.

Hansard, S. L., Crowder, H. M., and Lyke, W. A. (1957). The biological availability of calcium in feeds for cattle. *J. Anim. Sci.* **16,** 437.

Hogan, A. G., and Nierman, J. L. (1927). Studies in animal nutrition. VI. The distribution of the mineral elements in the animal body as influenced by age and condition. *Mo. Agric. Exp. Stn., Bull.* **107.**

Perry, T. W., Caldwell, D. M., and Peterson, R. C. (1976a). Selenium content of feeds and effect of dietary selenium on hair and blood serum. *J. Dairy Sci.* **59,** 760.

Perry, T. W., Beeson, W. M., and Mohler, M. T. (1976b). "Value of Salt in Controlling Free Choice Corn Intake on Pasture and Value of Supplemental Magnesium," Cattle Feeders Rep., p. 53. Purdue University, Agric. Exp. Stn., Lafayette, Indiana.

Perry, T. W., Garrett, W. N., Klosterman, E. W., Matsushima, J. K., Nelson, A. B., and Woods, W. R. (1976). "Nutrient Requirements of Beef Cattle," 5th rev. ed. Natl. Acad. Sci., Washington, D.C.

Reid, J. T., Wellington, G. H., and Dunn, H. O. (1955). Some relationships among the major chemical components of the bovine body and their application to nutritional investigations. *J. Dairy Sci.* **38,** 1344.

Rouse, G. H., (1978). *In* "Animal Industry Week," Iowa State University, Ames. A.S. Leafl. R-271.

Smith, W. H., Perry, T. W., Mohler, M. T., Harrington, R. B., and Beeson, W. M. (1964). Performance of steers fed three levels of zinc and two levels of calcium. *Proc. Annu. Indiana Cattle Feeders Day. Proc.*

Trenkle, A. (1978). *In* "Animal Industry Week," A.S. Leafl. R-277. Iowa State University, Ames.

Wheeler, W. E. (1976). Ph.D. Thesis, Purdue University, West Lafayette, Indiana.

Wise, M. B., Ordoveza, A. L., and Barrick, E. R. (1963). Influence of variations in dietary calcium:phosphorus ratio on performance and blood constituents of calves. *J. Nutr.* **79,** 79.

4

Protein Requirements of Beef Cattle

I. INTRODUCTION

Specific protein needs for various feeding situations for beef cattle are listed in Appendix IV and represent a modification of those presented by the National Research Council (Perry *et al.*, 1976). The requirements are expressed on the basis of total protein requirement, either as a percentage of the dietary dry matter or as an absolute figure. They can be exceeded without toxicity by protein from natural sources but it should be borne in mind that excessive levels of nonprotein nitrogen are highly undesirable and may actually place a burden on the kidneys or liver.

The protein requirements presented are based on the sum of (1) endogenous nitrogen (grams of N = 0.12 times weight of the animal to the 0.75 power, or $W^{0.75}$), (2) nitrogen loss through hair and surface cells (grams N = 0.02 $W^{0.75}$), (3) metabolic fecal nitrogen (4 gm nitrogen per kg feed dry matter), and (4) composition of gain. Composition of gain varies from 18% protein for steers weighing 220 pounds to 9% for steers weighing 1100 pounds. Corresponding values for heifers vary from 18 to 7%. Protein requirements for dry pregnant cows and for nursing cows calculated as above tend to be relatively high.

II. AMINO ACIDS

The proteins are a very complex group of compounds characterized primarily by the fact that all proteins contain nitrogen. Furthermore, all "true" proteins contain amino acids. In addition to the element nitrogen, all proteins contain carbon, hydrogen, and oxygen, many contain sulfur and phosphorus, and some contain iron, copper, manganese, or iodine. If a protein is hydrolyzed, as by treatment with acid, free amino acids will result. An amino acid is an ordinary organic acid in which an NH_2^- is found in the molecule. For example, the structure of propionic acid, commonly used as a preservative for high moisture grains is

$$CH_2\text{-}CH_2\text{-}COOH$$

propionic acid

With the addition of NH_2^- at the alpha carbon, the compound now becomes the amino acid alanine (α-aminopropionic acid)

$$\underset{\displaystyle NH_2}{CH_2\text{-}\overset{}{CH}\text{-}COOH}$$

alanine

There are many types and kinds of amino acids. Most nutritionists agree there are a total of 23 amino acids constituting the proteins of animal nutrition. Of that number approximately 10 or 11 are so-called "essential amino acids," which are defined as those which an animal cannot synthesize in sufficient quantities.

It is fairly well established that ruminant animals which have a functioning rumen can synthesize their total amino acid requirements through the microscopic life that inhabits their rumen if the proper raw materials are supplied. Therefore, the supplementation of preformed amino acids to beef cattle becomes much less important than it does to such monogastric animals as the pig and the chicken, with the exception of the young beef animal, whose rumen has not become functional.

It is obvious that the beef animal has similar tissue requirements for amino acids as do poultry and swine. However, such amino acids appear to be synthesized by the functioning rumen and then passed to the small intestine for absorption into the bloodstream.

III. THE ROLE OF PROTEIN

The name "protein" was suggested many years ago because of its basic constituent role in protoplasmic materials and refers to a product of primary importance. It would be most difficult to pinpoint specific functions of proteins—or amino acids—since almost every function of the body utilizes proteins or amino acids. In addition to its obvious role as a part of protoplasm, protein is involved as a principal constituent of all hormones and enzymes in the body; it constitutes the principal dry matter content of muscle, skin, blood, and of bodily secretions. The role of protein, and of amino acids, in animal physiology is so basic and so far reaching that it is within the realm of reason to say merely that protein is "basic" to life.

IV. PROTEIN DIGESTION

Protein digestibility in the ruminant prior to entry into the small intestine differs greatly from that in the monogastric animal. Although some small quantities of undegraded protein may pass through the rumen, most protein is broken down by the microorganisms to the constituent amino acids. Some of the amino acids thus released may pass down the digestive tract to the small intestine where they may be absorbed into the bloodstream. However, the majority of the amino acids will be deaminated in the rumen to form free ammonia plus the carbon–hydrogen–oxygen fraction from which the ammonia was removed. The free ammonia so released may either combine with an available carbohydrate fraction provided by such feedstuffs as molasses or corn or barley, or it may pass directly across the wall of the rumen into the bloodstream. Any absorbed ammonia is transported to the liver where it is transformed into urea. Urea may then (1) go to the kidneys for excretion in the urine, (2) pass on into the saliva and then back into the rumen, or (3) pass back into the bloodstream and then back into the rumen. The fate of urea in the rumen is similar to that of amino acids (referred to above). Urea in the rumen is deaminated to release free ammonia which then attaches to a carbohydrate fraction or passes through the wall of the rumen into the bloodstream. The rumen microorganisms attack any ammonia–carbohydrate complexes to synthesize amino acids and bacterial protein which, in turn, pass down the digestive tract for absorption in the small intestine.

The capability of the rumen in synthesizing amino acids from a diet devoid of protein was demonstrated by Oltjen and Putnam (1966) (Table 4.1).

The fate of absorbed amino acids may be threefold. A primary requirement is for the resynthesis of tissue protein and other nitrogen-containing constituents, such as enzymes, hormones, and milk. Apparently the functioning rumen can synthesize all the amino acid requirements of the tissues. Theoretically, formation of tissue protein is the reversal of the hydrolysis observed in the digestion process. It constitutes the building up of protein from amino acids. Blood plasma proteins are primarily manufactured in the liver.

A second fate of absorbed amino acids is that of deamination. Both the kidneys and the liver are capable of deaminizing the amino acids. The enzyme involved is amino acid oxidase, which is further subdivided into (+)- and (−)-amino acid oxidases which act on either the (+)-amino acids or the (−)-amino acids. The latter form [(−)-amino acids] is the most common form found in feedstuffs. In this process the amino acid is dehydrogenated to the corresponding amino acid, which, in turn, is hydrolyzed into the keto acid plus free ammonia. The keto acid thus formed can be (1) converted to fat, (2) converted to carbohydrate, (3) resynthesized into an amino acid, or (4) oxidized to carbon dioxide and water.

A third fate for digested amino acids is that of storage. Although this capacity is quite limited, compared to that of storage of fat, at least a limited capability for protein storage does exist. For example, heavy exercise can result in increased muscular bulk. This represents protein storage. A small amount of protein storage occurs briefly in the blood.

The fate of depleted amino nitrogen is through the urea cycle. Urea is the principal nitrogen excretory product of beef cattle. The liver represents the chief

TABLE 4.1

Plasma Amino Acids of Steers Fed No Natural Protein[a]

Amino acid	Sources of soluble carbohydrate (μmoles/100 ml)	
	Glucose plus starch	Starch
Lysine	5.3	7.2
Histidine	3.5	3.1
Tryptophan	0.4	0.8
Arginine	4.4	5.4
Aspartic acid	0.3	0.4
Threonine	2.4	3.4
Serine	3.7	3.5
Glutamic acid	5.6	4.1
Glycine	30.5	27.0
Alanine	14.0	11.5
Valine	11.6	12.3
Cystine	1.0	1.1
Methionine	1.0	1.2
Isoleucine	5.3	6.0
Leucine	5.5	5.1
Tyrosine	1.4	1.4
Phenylalanine	2.0	1.0
Total	103.8	99.1

[a] From Olfjen and Putnam (1966).

site of urea formation through what is called the "urea cycle." The theory of the urea cycle proposes that ornithine combines with carbamylphosphate (ammonia plus carbon plus phosphorylation) to form citrulline, which, when energy is added, is converted to argininosuccinate; this is split into fumarate, which leaves the cycle, plus arginine. Arginine is hydrolyzed into ornithine and urea, with the latter being excreted in the urine. Ornithine then remains to go through the cycle again.

Ingestion of all nutrients elevates the metabolic rate, but ingestion of protein has a more pronounced effect than that for carbohydrates or fats. This tendency to increase the release of energy (increased metabolic rate) is called the specific dynamic action of nutrients. Materials which have no calories, such as cellulose, cannot cause this effect. Ingested amino acids are not stored, to any great extent. Therefore, ingestion of relatively large quantities of amino acids may temporarily overwhelm the need of the tissues for amino acids, thus releasing the excess for deamination and subsequent oxidation. This effect plus the formation of urea in the liver—and its excretion by the kidneys—probably accounts for at least one-half of the specific dynamic action effect attributed to protein ingestion.

The specific dynamic action effect should be taken into account in calculating a calorically adequate diet. This effect may account for as much as 6–10% of the total calories needed.

V. NONPROTEIN NITROGEN (NPN)

Characterized primarily in the form of urea and ammonia, NPN is now one of the most critical and typical ingredients utilized in formulating cattle diets. Its place in cattle feeding is justified for only one reason—a matter of economy. Since one pound of urea contains the crude protein equivalent of over six pounds of soybean meal, the price advantage of urea as a source of protein for beef cattle is obvious.

The adoption of urea as a feed ingredient did not occur overnight. In many respects urea has been a difficult one to adapt to cattle feeding—it has been the "Peck's bad boy" of feed formulation for several reasons. However, several decades of research with urea have brought it into line as an integral part of almost all beef cattle feeds except those for the very young nursing calf. However, many adjustments had to be made in the formulations before the incorporation of large quantities of NPN became such a common practice.

Urea contains no available energy. Researchers recognized this and pointed out that when one pound of urea replaces slightly over six pounds of an oil meal, on a crude protein basis, a loss of six pounds of high-energy feed also occurred. Therefore, a more nearly correct replacement value for urea is that one pound of urea *plus six pounds* of corn, or other high-energy feedstuff, will replace six pounds of soybean meal. Therefore the use of urea in formulating protein supplements must be evaluated from two aspects. Its use by the feed manufacturer is extremely advantageous because he can replace six pounds of soybean meal or cottonseed meal with one pound of urea without compensating for the loss of the energy from six pounds of soybean meal or of cottonseed meal. Thus, a trade of one pound of urea for six pounds of one of the oil meals is very attractive. However, the cattle feeder purchasing such supplements must add extra corn, milo, or barley to compensate for the energy lost by substituting urea for oil meals in the supplement formulation.

It is only through the symbiotic relationship existing between a steer and the billions of microscopic life that inhabit his rumen that make it possible for cattle to utilize urea for energy. Once urea reaches the rumen, it is broken down into ammonia and carbon dioxide. If the appropriate conditions exist in the rumen, the microorganisms will utilize the ammonia liberated from urea and manufacture high-quality protein. However, if ideal conditions do not exist, such ammonia will be absorbed through the wall of the rumen into the bloodstream; the liver will remove this ammonia from the bloodstream and thus the absorbed ammonia becomes a burden on the liver.

Research has revealed a great deal concerning factors which assist the rumen microorganisms in utilizing urea nitrogen more effectively. Naturally, it is critical that the diet contain a good balance of the well-recognized nutrients, discussed earlier, in addition to urea. Additional factors, however, are needed and are elaborated below.

First, some natural protein is required for optimum performance, although cattle can survive if they recieve all their protein from NPN. However, they will do much better if at least a part of their dietary protein is natural protein, such as is derived from corn. In fact it is a pretty good "rule of thumb" that not more than one-third of their total protein be derived from NPN. This is not a hard and fast rule and often pregnant brood cows fed rather low-quality roughages may derive two-thirds of their total protein from NPN, and perform quite well.

Second, some source of readily available carbohydrate is almost essential for optimum utilization of NPN. When NPN breaks down to ammonia and CO_2 in the rumen, a readily available sugar is essential for complexing the ammonia until the rumen microorganisms can synthesize amino acids and protein from it. Thus, if carbohydrate is not available, the ammonia released from the dietary urea would be absorbed into the bloodstream and its effective use would be greatly diminished. An excellent source of carbohydrates for use in manufacturing protein supplements containing higher levels of urea is cane molasses. This urea supplement lends itself to manufacture of either liquid or dry formulations. Cattle fed high-energy diets containing higher levels of corn, milo, or barley also provide good carbohydrate substrate for improved utilization of urea. In contrast, low-quality roughages do not provide a readily available source of carbohydrates. Therefore it is necessary that the protein supplement to be fed in such a program should contain its own built-in source of carbohydrates, for example, 10–15% molasses in dry supplements and up to 50% molasses in liquid supplements.

Both basic and applied research have shown that the rumen microorganisms can synthesize protein from urea more efficiently when some source of urea protein factors (UPF) is present. Purdue research demonstrated that dehydrated alfalfa meal is an essential component for formulating high urea supplements which are comparable to a natural protein supplement for beef cattle. Later, at Purdue, it was shown that nitrogen retention by cattle was increased 66% by the incorporation of 5% dehydrated alfalfa meal in a diet in which a high percentage of the nitrogen is derived from NPN. Also, the incorporation of 5% distillers dried grains with solubles with a high urea diet increased nitrogen retention by 79%. Since neither of the above ingredients is soluble, new sources of UPF for liquid supplements needed to be developed (Fig. 4.1). It has been demonstrated that the incorporation of from 9 to 10% of 27% dry matter distillers solubles into a liquid supplement provides an excellent source of UPF. Undoubtedly there are many other sources of UPF, but unfortunately they have not as yet been identified and reported.

Fig. 4.1. Liquid supplements provided from a lick tank may be used to supplement many cattle feeding situations. (Photo by J. C. Allen and Son.)

Numerous feedlot tests and university research have shown rather clearly that there is no significant difference between the nutritional value of liquid or dry high-urea beef cattle supplements when both contain the same balance of known essential nutrients plus sources of UPF. In other words, cattle response cannot distinguish between the same nutrients supplied in the dry and liquid forms. Thus it is a matter of choice by the cattle feeder whether he feeds a protein supplement in the dry versus the liquid form.

A "condensation" product of two molecules of urea is biuret. This product gained a lot of attention because of its apparent nontoxic aspect. It would appear that biuret has an almost 100% record of nontoxicity for beef cattle. Therefore it seemed to be ideal for self-feeding situations, as for supplementing the diets of pregnant beef cows with access to low-quality roughages. However, since the manufacture of biuret is significantly more expensive than urea, cattle feeders tend to utilize urea in preference to biuret and thus suffer a bit more risk. However, with the advent of the slower ammonia release from urea in both liquid and dry forms, the use of urea has become increasingly more safe. Thousands of pregnant and lactating brood cows now have access to free-choice high-urea liquid supplements offered via a lick wheel. Biuret as a source of NPN has two drawbacks: (1) it is much more expensive per unit nitrogen than urea, and (2) it is not very soluble in aqueous solutions, making it impractical for liquid supplement formulation.

Michigan State University developed a formulation for applying a combination of ammonia and minerals suspended in water to corn silage at time of ensiling. The material contains 85% crude protein equivalent and is added to corn silage at a level of 6.43% of the total dry matter. In comparative research conducted at Purdue University in 1976 with silage made with or without the added ammonia–minerals–suspension (AMS), cattle fed only the corn silage treated with AMS gained slightly faster (2.27 versus 2.12 lb per day) and required 11% less dry matter per pound gain (8.1 versus 9.2 lb) than cattle fed regular silage plus 1.5 lb of protein supplement per day. Regular corn silage made without added AMS contained 7.3% protein, on a dry matter basis, whereas that treated with AMS at time of ensiling contained 12.7% protein, making it completely adequate for meeting the protein needs of growing–finishing beef cattle. Cattle fed the AMS-treated corn silage derived 42.5% of their total protein from non-protein nitrogen.

The nitrogen excretion product of poultry is uric acid. Therefore poultry droppings offer a source of NPN for cattle feeding provided they are not too contaminated with bedding materials or dirt. Furthermore, there is some question as to the legality of its use in feed formulation. Admittedly, several states have approved the use of poultry litter as a feed ingredient, but the Food and Drug Administration has not yet ruled on this subject. Research from the University of Arkansas and from Virginia Polytechnic Institute (V.P.I.), and other places, has shown that poultry litter can provide a large percentage of the total nitrogen in cattle diets. In one case, the use of poultry manure in cattle formulations resulted in an unfortunate situation. A large cage layer operation was saving the droppings and adding sawdust to absorb the excess high moisture. The resultant mixture was then ensiled, along with corn silage, to feed fattening cattle. Very shortly thereafter, several cattle in the feedlot receiving this mixture died. It was soon learned that the sawdust used to absorb the moisture from the cage layer droppings was walnut sawdust; this combination was toxic to beef cattle. It would be a much better combination to use a feed product such as ground corn cobs, straw, or cornstalks to absorb such excess moisture.

VI. EFFECT OF PROTEIN SUPPLEMENTATION ON THE PERFORMANCE OF FINISHING CATTLE

According to the National Research Council Committee on Nutrient Requirements of Beef Cattle, growing–finishing beef cattle require rather definite minimal levels of protein, based on liveweight and rate of gain (Table 4.2). Diets formulated on corn and corn silage will probably require supplemental protein if optimal growth performance is to be obtained. Even though a growing–finishing beef calf's total protein, as a percentage of the diet, is reduced as it increases in body size and age, very few corn–corn silage–type diets will contain adequate protein. Only where supplemental leguminous roughage is used will the total protein needs of the diet be met withoug supplemental protein.

A. Withdrawal Effects of Supplemental Protein

Several years' research were conducted at Purdue University to study the effects of removing supplemental protein from the diets of growing–finishing beef cattle. The diet studied was a typical corn–corn silage diet containing 7.5% protein on a 100% dry matter basis. By referring to Table 4.2, it is obvious that is below the NRC requirement for optimum performance. The diet contained 14 pounds of corn silage plus a full feed of corn; and, in addition, a high-urea 64% protein supplement was used to add supplemental protein where the experimental design called for it (Table 4.3). It is obvious from the data that at no time was it satisfactory to remove the supplemental protein from the corn–corn silage diet of growing–finishing beef cattle.

TABLE 4.2

Protein Requirements for Growing–Finishing Beef Steers and Heifers[a]

			Total protein (%)			
Average weight (lb)	Daily gain (lb)		100% dry matter basis		85% dry matter basis	
	Steers	Heifers	Steers	Heifers	Steers	Heifers
331	2.4	2.4	15.6	15.0	13.3	12.7
441	2.4	2.4	13.6	12.8	11.6	10.9
551	2.9	2.6	12.7	11.9	10.8	10.1
661	3.1	2.6	11.9	10.9	10.1	9.3
772	3.1	2.6	10.9	10.0	9.3	8.5
882	3.1	2.4	10.5	9.7	8.9	8.2
992	3.1	2.4	10.0	9.5	8.5	8.1
1102	2.9	—	9.7	—	8.3	—

[a] Modification of National Research Council (Perry *et al.*, 1976).

TABLE 4.3

Effect of Removal or Addition of Supplemental Protein on Beef Cattle Performance (210-Day Trial)[a]

Item	Length of feeding supplemental protein, initially			
	1 None	2 70 days	3 140 days	4 210 days
0–70 Days				
Initial weight (lb)	601	602	599	598
Final weight (lb)	654	809	797	800
Daily gain (lb)	0.77	2.97	2.82	2.90
Supplemental protein removed from group 2 after 70 days				
71–105 Days				
Final weight (lb)	665	812	869	876
Daily gain (lb)	0.42	0.08	2.05	2.17
Supplemental protein added to group 1 after 105 days				
106–140 Days				
Final weight (lb)	789	837	947	962
Daily gain (lb)	3.54	0.71	2.22	2.45
Supplemental protein removed from group 3 after 140 days				
141–196 Days				
Final weight (lb)	949	992	1000	1090
Daily gain (lb)	2.85	2.76	0.95	2.29
Supplemental protein added back to group 2 after 154 days				

[a] Beeson *et al.* (1974).

Another aspect of the research was to determine the effect of adding supplemental protein back into the diet after it had been withheld from the cattle. Note from Table 4.3 the effect of adding supplemental protein to group 1 cattle starting on the 105th day. Their gains increased by over 3 pounds (500%). Notice also the response of group 2 cattle in which supplemental protein was removed after they weighed 812 pounds. After 7 weeks without protein, in which gains had dropped from near 3 pounds per day to less than 1 pound per day, adding supplemental protein at 154 days resulted in a jump back to 2.76 pounds daily gain.

In another Purdue experiment, finishing steer calves were self-fed a finishing diet containing a constant level of protein over a 180-day period (Table 4.4). A constant level of 10.3% protein gave best results.

B. Pasture

Several years' research at Purdue has indicated protein supplements on pasture is not very critical—at least not when a high-urea supplement is fed. However,

TABLE 4.4

Effect of Protein on Gain and Feed Efficiency at Various Weights[a]

Item	Level of protein in ration (%)			
	9.0	9.6	10.3	11.0
0–56 Days				
Initial weight (lb)	474	475	472	475
Daily gain (lb)	2.19	2.40	2.48	2.54
Feed per lb gain (lb)	6.7	6.4	6.0	5.8
57–84 Days				
Initial weight (lb)	596	610	612	617
Daily gain (lb)	2.63	2.34	2.84	2.50
Feed per lb gain (lb)	6.5	6.6	6.2	6.1
85–126 Days				
Initial weight (lb)	670	675	691	687
Daily gain (lb)	2.10	2.34	2.45	2.53
Feed per lb gain (lb)	7.1	7.0	6.6	6.4
127–182 Days				
Initial weight (lb)	871	895	923	931
Daily gain (lb)	2.02	2.16	2.31	2.45
Feed per lb gain (lb)	9.8	9.0	8.6	8.0
0–182 Days				
Final weight (lb)	982	1013	1050	1066
Daily gain (lb)	2.18	2.30	2.48	2.51
Feed per lb gain (lb)	7.9	7.6	7.2	6.9

[a] Beeson *et al.* (1974).

the use of an all-natural protein supplement such as Purdue Supplement A has given somewhat different results (Table 4.5).

Even though cattle fed Supplement A gained an average of 11% faster than those fed no supplemental protein, such cattle required the same amount of concentrates per pound of gain as those fed no supplemental protein, indicating those with access to supplemental protein did not have to graze as much to meet their total protein requirements. Those fed Purdue 64 (a high-urea supplement) gained 6% more rapidly than those fed no supplement, but they required even a bit more feed per pound of gain. The pasture available was a mixture of brome and canarygrass, devoid of legumes.

VII. PROTECTED OR SLOWLY DEGRADED PROTEIN

In a previous section it was pointed out that protein entering the rumen is degraded to ammonia and later synthesized to microbial protein. However, if in

TABLE 4.5

Corn Alone versus Supplement A[a] versus Purdue 64[b] for Finishing Cattle on Pasture (Purdue)

	Daily gain (lb)			Concentrates/lb gain (lb)		
Year	No supplement	Supplement A	Purdue 64	No supplement	Supplement A	Purdue 64
1967	2.30	2.60	2.44	6.2	6.3	6.6
1968	2.62	2.94	2.79	6.6	6.6	6.6
1969	2.41	2.61	2.50	6.8	6.6	6.9
Average	2.44	2.72	2.58	6.5	6.5	6.7

[a] Natural protein.
[b] High-urea protein.

the rumen the release of ammonia is too rapid, a relatively large portion may be absorbed directly through the wall of the rumen into the bloodstream. Recent research has shown protein sources vary in their ''rumen solubility'' or the rate at which they may be degraded in the rumen. Thus, those more slowly degraded may move through the rumen with less total destruction than those which are degraded more rapidly. Those which are degraded rapidly, such as soybean meal, may supply the rumen microorganisms with more than enough ammoniacal nitrogen with a loss of ammonia across the rumen wall. A major part of slowly degraded protein passes out of the rumen unscathed and can be utilized more efficiently further down the gastrointestinal tract.

Recent work with ''slowly degraded''proteins such as corn gluten meal, dehydrated alfalfa, brewer's grains, distiller's grains, and blood meal showed that these products, when fed in combination with urea, can provide performance superior to soybean meal. Why is urea necessary with the less soluble proteins? If nearly all of the ingested protein passes out of the rumen intact, little if any ammonia nitrogen would be available to the rumen microbes for the manufacture of microbial protein. Thus, it is important that some highly degradable source of nitrogen, such as urea, be fed in combination with slowly degradable proteins to meet the ammonia requirement of the rumen microbes. Otherwise, breakdown of fiber in the rumen would be decreased greatly, as would overall performance of the host bovine, and the results would be biased. Many research stations have investigated the role of ''protected protein'' in cattle feeding, and the results are not in complete agreement. However, it appears this is a relatively new concept in cattle feeding which is apt to become an accepted practice. The University of Nebraska has conducted interesting research on beef cattle with ''protected'' or slowly degraded protein (Rock *et al.*, 1979). In two experiments the performance of growing Angus and Hereford 440-pound calves was studied for protein source effect (Table 4.6). The basal diet consisted of 60% corn cobs and 20% corn to which the various protein supplements were added to make a 61% total digestible nutrients (TDN) and 11.5% protein equivalent diet. The trials were 112 days in length. The results indicated that all natural proteins were superior to urea alone as a supplement. In comparing natural proteins, distillers dried grains alone or in combination with soybean meal, dehydrated alfalfa meal, corn gluten meal, and dehydrated alfalfa meal plus corn gluten meal were superior to soybean meal alone. These results suggested that more protein is reaching the abomasum and small intestine intact for more efficient utilization from more slowly degraded proteins. Based on this research, a "relative protein efficiency" table was designed (Table 4.7).

VIII. PROTEIN ADJUSTMENT DURING TEMPERATURE STRESS

Cattle feeders should be aware of the thermal stress effect on daily gain of cattle and thus take this into account in prescribing the total protein requirements, based on data presented at the 1979 Kansas State University Cattle Feeders Day Program (Ames, 1979). The most common environmental factor that alters nutrient requirements is temperature. Both heat and cold stress have a direct influence

TABLE 4.6

Protein Source Effect on Performance of Growing Calves[a]

	Protein sources[b]						
Item	Urea	Soybean meal	Dehydrated alfalfa	Corn gluten meal	Distillers dried grains	Dehy[c] + CGM	DDG[c] + soy
Experiment I							
Number of cattle	10	9			10		10
Daily gain (lb)	1.39	1.54			1.69		1.63
Daily feed (lb)	12.5	12.6			12.6		12.6
Feed/gain (lb)	9.1	8.4			7.6		7.9
Experiment II							
Number of cattle	9	9	20	10		10	
Daily gain (lb)	0.81	0.94	1.15	1.01		1.12	
Daily feed (lb)	13.4	13.1	13.4	12.8		13.4	
Feed/gain (lb)	16.6	14.0	11.7	12.7		12.0	

[a] Rock *et al.* (1979).
[b] Test proteins were 40% and urea was 60% of supplement.
[c] Equal amounts.

TABLE 4.7

Relative Cattle Values of Slowly Degraded Proteins Compared to Soybean Meal[a]

Source	Relative protein value (%)
Soybean meal	100
Distillers dried grains	203
Distillers dried grains + soybean meal	113
Dehydrated alfalfa meal	226
Corn gluten meal	189
Dehydrated alfalfa meal + corn gluten meal	242

[a] Rock *et al.* (1979).

on the energy requirement for maintenance. Therefore, the relative amounts of the other feedstuffs in the total diet justify alteration in relation to energy.

It is important to understand a number of terms which help describe this relationship. The first is "thermal neutral zone," the temperature at which an animal is comfortable and maintains only a basal heat production. This is sometimes referred to as the "comfort zone," but it should be borne in mind the comfort zone for a steer may greatly differ than that for a human being. The comfort zone for a fattening beef animal should *not* be gauged to human comfort. "Critical temperature" is defined as the lower limits of the thermal neutral zone. This is the temperature where performance begins to decline as temperature drops. Naturally the reason for the accompanying decline in performance, as the temperature drops below the critical temperature, is that a greater percentage of the ingested feed is required just to keep the body warm. In like manner, when the upper limit of the comfort zone is reached and the temperature continues to rise, then the animal uses energy in panting, in an attempt to dissipate heat. Temperatures at the top of the comfort zone—and higher—are called heat; those at the bottom of the comfort zone—and lower—are called cold.

Table 4.8 can be used as a guide to determine where the critical temperature

TABLE 4.8

Estimated Critical Temperature for Finishing Beef Cattle

Hair coat description	Critical temperature (°F)
Summer coat, or wet	59
Fall coat	45
Winter coat	32
Heavy winter coat	18

may be. For instance, beef cattle with short hair and a dry coat have a critical temperature of about 59°F; for cattle wearing their heavy winter coat, the critical temperature is about 18°F. However, for an animal with a heavy winter coat, exposed to rain (wet), the critical temperature is near 59°F, indicating the wetted long hair has lost much of its insulating capability.

Note from Table 4.9 the protein requirement for maintenance of a 900-pound steer at a constant 251.8 gm per head daily. However, as the environmental temperature varies away from the comfort zone—or critical temperature—the animal grows less and thus requires less protein. A typical example can be used: A 900-pound steer has a "fall coat" of hair, but an unseasonably warm spell of 75°F sets in for about one week. In other words, the temperature is 30° above the critical temperature for a steer wearing a fall coat of hair. Thus, in Table 4.9, a 30° too-warm deviation means the rate of gain will drop by 18.2%. Proceeding across the table, we see the protein can be dropped to 10.49%, on a dry matter equivalent basis, without affecting the rate of gain of the steer. Normally, one would be utilizing a 12% level of protein. Removing protein from the diet during thermal stress (either too hot or too cold) can improve protein efficiency with no penalty on performance. Furthermore, removing sufficient protein supplement to

TABLE 4.9

Sample Ration Adjustment for a 900-Pound Steer Exposed to Thermal Stress

	Deviation °F from critical temperatures	Decline in ADG (%)	Protein for maintenance (gm)	Protein for growth (gm)	Protein in ration (gm)	Crude protein in ration (%)
Hot	45	52.3	251.8	269.7	521.5	7.66
	40	39.1	251.8	344.3	596.1	8.75
	35	27.7	251.8	408.8	660.6	9.70
	30	18.2	251.8	462.5	714.3	10.49
	25	10.5	251.8	506.0	757.8	11.13
	20	4.8	251.8	530.3	790.1	11.60
	15	0.7	251.8	561.4	813.2	11.94
Critical temperature	10	—	251.8	565.4	817.2	12.0
	5	—	251.8	565.5	817.2	12.0
	0	—	251.8	565.4	817.2	12.0
Cold	5	2.3	251.8	552.4	804.2	11.81
	10	4.5	251.8	540.0	791.8	11.63
	15	6.8	251.8	527.0	778.8	11.44
	20	9.0	251.8	514.5	766.3	11.26
	25	11.3	251.8	501.5	753.3	11.06
	30	13.5	251.8	489.1	740.9	10.88
	35	15.8	251.8	476.1	727.9	10.69
	40	18.0	251.8	463.6	715.4	10.51
	45	20.3	251.8	450.6	702.4	10.31

[a] ADG, average daily gain.

drop the total dietary protein from 12.0 to 10.49%, and substituting corn in its place, might mean a feed cost saving per day of several cents.

REFERENCES

Ames, D. R. (1979). Effect of environment on nutrient requirements of livestock. *Kans. State Univ. 1979 Cattle Feeders Day Program* p. A-1.

Beeson, W. M., Perry, T. W., Thomas, V. M., and Mohler, M. T. (1974). "Annual Indiana Cattle Feeders Day Program." Purdue University, Lafayette, Indiana, p. 15.

Oltjen, R. R., and Putnam, P. A. (1966). Plasma amino acids and nitrogen retention by steers fed purified diets containing urea or isolated soy protein. *J. Nutr.* **89**, 385–391.

Perry, T. W., Garrett, W. N., Klosterman, E. W., Matsushima, J. K., Nelson, A. B., and Woods, W. R. (1976). "Nutrient Requirements of Beef Cattle," 5th rev. ed. Natl. Acad. Sci., Washington, D. C.

Rock, D., Waller, J., Klopfenstein, T. and Brittan, R. (1979). Beef Cattle Rep., p. 11. University of Nebraska, Coop. Ext. Serv., Lincoln.

5

Computer Programming of Beef Cattle Diets

I. COMPUTER DIETS

The use of a computer to formulate "least-cost" diets for beef cattle has become a common practice in the feedlots in the western part of the United States, but is not used extensively east of the Missouri River. There are essentially two reasons for this geographical difference in procedure. The primary reason is that feedlots west of the Missouri River tend to be much larger than those in the eastern United States and thus perhaps more attention need be paid to very slight formula cost differentials. Second, because of size and location of the feedlots, cattle feeders in the western United States purchase more of the feeds they utilize whereas the eastern United States cattle feeder relies more on home-grown feeds such as corn silage and thus really has very little flexibility in choice of feeds in formulating diets. The western cattle feeder has a greater array of feed ingredients available from which to compute a diet and can be much more selective.

Supplying the feedstuffs component in a cattle feeding program represents at least 70% of the total cost—perhaps even more. Therefore accurate judgment decisions in formulating diets is critical and any assistance from the computer may be vital in selecting the makeup of the of the diet.

Energy in nutrition is equally as critical as it is in driving machinery—it is the *number one* cost. When all of the dietary requirements for most beef cattle feeding situations are considered on a percentage basis, the requirement for energy in the formula may approach 85% of the total.

A. Interchangeability of Nutrient Sources

The total basis for linear programming least-cost diets is the premise that source of nutrient is not critical. In other words, one must assume one pound of protein from cottonseed meal is equally as good as one pound of protein from linseed meal or from soybean meal or from a host of other sources. The example given in the preceding sentence is essentially true for most beef cattle feeding situations, and yet there are certain peculiarities about each of the sources of listed nutrients that cannot be given a numerical value. For example, certain types of linseed meal impart a hair-conditioning effect and thus a cattle feeder fitting cattle for show might gladly pay a premium for linseed meal in order to obtain the desired hair coat effect. In the case of phosphorus, availability to the animal may vary between inorganic and organic sources, and even within inorganic sources.

Urea is a prime example in which interchangeability of nutrient sources must be used with caution. There are many limitations regarding the substitution of urea protein for natural proteins which must be considered. Admittedly, restrictions, or maximums may be put into a program which would limit the extent to which any nutrient may be found in a formulation, but this often requires judgment decisions which can only be made by a qualified person, and not by a computer program. As was indicated above, such limitations or restrictions on certain feed ingredients are put into the specific program as tolerances. Tolerances represent either maximums or minimums, or both, at which a specific nutrient may be found in the formulation.

B. Human Judgment

Human judgment is required Linear programming of diets is a technique in which the least-cost combination of ingredients is formulated to meet a predetermined set of needs. Because of the complexity of the calculations necessary to determine this one solution, it is necessary to utilize a complex calculator. Such calculator proceeds with the information with which it is provided by the programming technician. In other words, an electronic computer capable of making thousands of calculations in a very short time takes the information given it and determines the solution. The answers will be no better than the information provided because the computer is incapable of making judgments. The computer output is the result of information input and the resulting formulations obtained from linear programming will be no more reliable than the reliability of the information provided by the technician operating it.

C. Dual Approach Calculations

Calculations require a dual approach. With the sophistication now available in linear programming of formulations, nutritionists may impose limitations of from a dozen up to double that figure of literally dozens of ingredients which are selling for different prices at different shipping points. Therefore, it is obvious the calculation of the least-cost combination by anything but a most complicated electronic calculator is beyond comprehension. The inputs, therefore, for calculating least-cost formulations designed to do a predetermined job require the dual capability of a mathematician capable of designing a specific program for a machine as well as that of a nutritionist who has the judgment to make dietary restrictions and to make evaluations of the final solution. In this day of specialization it is indeed rare to find one person who has the capability for both. Admittedly many nutritionists can follow the directions for implementing a "canned" program which has been designed and set up for him for a specific computer.

D. Constraints and Restrictions

Constraints, or restrictions, must be known, since the machine has no way of knowing that extremely high levels of calcium will upset the metabolism of several other nutrients. Yet calcium carbonate, per unit weight, is undoubtedly the least expensive feed ingredient available on the market. Therefore, if some maximum restriction were not placed upon calcium, it is conceivable that the computer would load the formula with ground limestone to make the formulation total 100%, after the other nutrient requirements were met. This perhaps represents an extreme case and not one which one would encounter but once—after which an upper restriction on calcium would be imposed. For a set of typical restraints for growing and finishing beef cattle, see Table 5.1.

Both minimum and maximum restrictions, or tolerances, will be utilized. However, both limits need only be utilized for a few ingredients, such as calcium. Generally, at most, only one constraint is needed. The more expensive ingredients including energy sources, protein, and phosphorus usually require minimum constraints, meaning that the computer is required to supply at least minimal levels at any cost. Obviously, then, it is usually the lower cost specifications such as crude fiber, calcium, and salt, on which maximum constraints must be imposed. It is seldom necessary to impose range (both minimum and maximum) constraints on nutrients. A typical exception to this is in the case of calcium wherein a minimum level is critical but because of the low price of calcium a maximum level must be imposed to prevent the computer from overloading the formulation with calcium carbonate.

E. Feedstuffs Composition

Feedstuffs composition must be known. There are many feed ingredient tables available which give average feed ingredient analyses. However, there is little value in linear programming a least-cost formula if only "average" or feed ingredient table values are used. The author analyzed the protein content of corn

TABLE 5.1

Beef Cattle Formulation Restraints for Growing and Finishing Beef Cattle[a]

	Feeding situation						
	Calf	Calf	Calf	Calf	Yearling	Yearling	Yearling–2-yr-old
Starting weight (lb)	450	450	550	550	650	800	1000
Daily gain (lb)	1.50	2.50	1.70	2.50	2.75	2.85	2.80
Restraints							
Protein (min) (%)	11.5	13.6	11.0	12.3	11.0	10.5	10.5
Urea (max) (%)	1	1.5	1	1.5	1.5	1.5	1.5
Fiber (max) (%)	25	20	25	20	20	15	15
Roughage (max) (%)	50	25	60	25	20	15	15
Calcium (min) (%)	0.32	0.59	0.33	0.43	0.35	0.35	0.26
Calcium (max) (%)	0.50	0.85	0.70	0.75	0.60	0.60	0.50
Phosphorus (min) (%)	0.28	0.43	0.28	0.35	0.30	0.30	0.25
Potassium (min) (%)	0.70	0.70	0.70	0.70	0.70	0.70	0.70
Sulfur (min) (%)	0.10	0.10	0.10	0.10	0.10	0.10	0.10
Magnesium (min) (%)	0.10	0.10	0.10	0.10	0.10	0.10	0.10
Zinc (min) (%)	0.002	0.002	0.002	0.002	0.002	0.002	0.002
Vitamin A (abs) (IU/lb)	2000	2000	2000	2000	2000	2000	2000
Salt (abs) (%)	0.20	0.20	0.20	0.20	0.20	0.20	0.20
NE_m (min) (Mcal/lb)	0.64	0.94	0.72	0.82	0.82	0.98	0.98
NE_g (min) (Mcal/lb)	0.35	0.62	0.43	0.54	0.54	0.62	0.62
TDN (min) (%)	68	86	71	77	77	80	82

[a] Listed on basis of 100% dry matter.

selected from several counties in the State of Indiana. The "average" protein content of those many corn samples, on a 100% dry matter basis, was 8.3%; the range of protein values was from 6.7 to 10.0%, with the largest proportion of the samples averaging near 8%. In many cattle finishing situations corn may represent well over one-half of the total dietary dry matter. So it would be ridiculous to calculate a formulation to two or three places behind the decimal point, utilizing a chart value of corn protein when the actual value might vary at least by 25%.

Some misunderstandings concerning the need for supplemental protein for heavy finishing cattle arose in the early 1970's because research based on corn with a higher than average protein was used. Later research utilizing corn with a lower protein content demonstrated it would depress finishing beef cattle performance drastically if supplemental protein were removed from the diet formulation at any stage of the finishing program. Barley and milo are two other typical major beef cattle ration ingredients whose protein levels will vary over rather large ranges.

Knowledge of the nutrient composition of the major ingredients is especially critical in establishing a mathematical program for formulating. Perhaps knowledge concerning the trace mineral elements of cobalt, copper, zinc, iodine, and manganese contained in the feedstuffs is not so critical since such so-called trace mineral elements probably will be supplied in chemical form in the premix.

F. Cost of the Ingredients

The most variable constituent in linear programming—and the only real justification for linear programming—is that prices fluctuate. Such price changes may occur abruptly such as might be triggered by the announcement of a new contract for grain shipment exports. Price changes, on the other hand, may be quite gradual or almost stable and thus no need might arise to calculate a new formula for quite a period of time. The computer will print out how much any one ingredient will need to change in price before a new calculation should be made.

G. Constant Ingredients

Certain ingredients are locked into the formula at a constant level. For example, the salt will remain constant; vitamin A level will remain constant, no matter what the cost. Where there may be a number of constant ingredients in a formulation, such may be grouped together and literally removed from consideration by the machine so that time may be conserved on the computer.

H. The Role of the Animal Nutritionist

Even a very accurate chemical analysis of feed ingredients will not necessarily be indicative of the usefulness of such ingredients in a formulation. There is no known chemical assay for palatability, for example. A good case in point here is the relative palatabilities of corn and wheat for beef cattle. Most feed tables will indicate wheat and corn have a comparable TDN value for beef cattle, about 80%. Yet knowledgeable nutritionists know that finishing beef cattle tend to tire of predominantly wheat diets after a few weeks. So, even though wheat might be more economical per unit weight than corn, a competent nutritionist would limit the wheat content of a cattle finishing diet to a maximum of 35–40% of the total concentrates. The nutritionist may not accept the least-cost formula a computer may design. The nutritionist must have the final word even though his demands on the formulation may increase its cost.

I. Avoiding Sudden Changes

Even though balance of nutrients is generally much more critical than source of nutrients in a beef cattle diet, sudden changes should not be made in order to take advantage of certain steep price fluctuations. However, there is nothing wrong in changing nutrient source gradually to reflect price fluctuations. Apparently rapid changes in nutrient sources may affect palatability and therefore consumption levels will be altered.

II. ROLE OF THE PROGRAMMING MACHINE

It is not the purpose or intent of this text to instruct in the mechanics of setting up a program to formulate least-cost beef cattle diets. This is primarily the function of a computer technician working in cooperation with the nutritionist. However, it is desirable to explain a bit of the principle of least-cost formulating with an electronic computor.

A. Storage of Vital Data

Generally, there are two types of data required for computation. One type is a relatively permanent set which does not change greatly, such as the nutrient requirements for the several feeding situations and chemical analyses for most of the feedstuffs utilized. The less permanent data are represented primarily by ingredient prices, which will change from time to time and which changes justify recalculations when the price change of ingredients is more than token amounts. The program is designed such that revision of either the more permanent or the less permanent data is easily possible.

The stored data will include (a) the ingredients to be considered for formulation, complete with numerical values of their characteristics to be considered, (b) the names of the characteristics to be restrained, (c) constraints on the amount of any specific ingredient which may be entered into the formula, if such are necessary, and (d) prices of the ingredients. All of these data are recorded on coded cards by a special type punch machine.

It is important that data concerning each feed source is compatible with that for other feed sources. For example, all feedstuffs must be listed on the same weights and measures system and not be a mixture of two or more systems, such as metric and English.

B. Diet Specifications

The diet specifications are the data which characterize the limitations of the formulation. For example, a minimum level is indicated for all of the more expensive items such as energy, protein, and vitamins. Normally, no maximum level is necessary or practical since because of higher costs such ingredient characteristics tend to stay at a near minimum. The less expensive items tend to require no minimums, but do require maximums. A good example is calcium.

Admittedly, because of the critical nature of minimal calcium, a minimum tolerance is required. However, since calcium carbonate is the least expensive of all feed ingredients, the computer is "tempted" to overload the diet with calcium unless a maximum restraint is imposed. There is still a third set of restraints known as "absolutes" (ABS) in which only a given quantity—no more or no less—is introduced. Salt is a typical example; potassium, sulfur and vitamin A are other examples where absolute values are imposed.

C. The Mathematical Basis of Linear Programming

A linear programming problem differs from the general type in that a mathematical model can be stated using relationships which are called straight line, or linear. Mathematically, these relationships are in the form:

$$a_1x_1 + a_2x_2 + a_3x_3 + a_4x_4 + \ldots + a_jx_j + \ldots + a_nx_n = \text{the solution}$$

The a's are known coefficients and the x's are unknown variables. Then the complete mathematical statement of a linear programming problem involves a set of simultaneous linear equations representing the conditions of the problem and a linear function which expresses the objectives. Mathematically, linear programming deals with nonnegative answers.

These calculations are solved by the mechanical computer and either projected on a television screen for observation or printed on paper for a permanent record.

6

The Effect of Processing on the Nutritive Value of Feedstuffs for Beef Cattle

It is only within the past two decades that the effect of feed processing on beef cattle performance has been recognized. Prior to that time there was only limited feed processing including: grinding of grains and ensiling of certain hay crops and the whole corn plant. However, in the 1960's, interest in feed processing increased dramatically and it is now common practice to process most feed grains as well as many roughages for cattle feeding. The most dramatic effect from processing is that on the feed grains for beef cattle. A variety of feed grain processing techniques are now in common usage.

I. PROCESSING OF FEED GRAINS

Several methods of grain processing have been developed. All but two of them require heat energy to complete the process—only grinding and moisturizing require no supplemental heat. Feed grain processing techniques in common practice today include extrusion, gelatinization, grinding, micronizing, moisturizing, popping, roasting, and steam flaking. Much of the discussion on grain processing, except for moisturizing, is derived from a National Academy of Sciences review (1973); the discussion on high-moisture grains is based on a review by Merrill (1971).

A. Extrusion

This is accomplished by forcing dry grain through an orifice, utilizing an augerlike rotor which crushes the grain prior to the time it reaches the orifice. The resulting ribbonlike strip from the orifice breaks up into flakes of different shapes and lengths. Much of the research with extruded grain was conducted at Colorado State University.

In one experiment (Matsushima *et al.*, 1969) finishing steers were fed a diet containing processed grain, limited corn silage (19 lb per day), 2.4 lb alfalfa hay, 1 lb dried beet pulp, and 0.71 lb protein supplement. Cattle fed extruded grain gained more slowly and required more feed per unit gain than cattle fed flaked corn; those fed high-moisture corn gained similarly to those fed flaked corn (Table 6.1). In subsequent research, McLaren and Matsushima (1971) reported that cattle fed extruded corn at the 85% concentrate level made gains comparable with those of cattle fed flaked corn (2.79 versus 2.76 lb per day). Efficiency of gain and carcass quality was comparable for the two groups.

Digestibility studies with extruded grain were compared with flaked (Matsushima, 1970). Both thin-flaking and dry extrusion improved digestibility of dry matter, crude fiber, and crude protein over that of whole corn (Table 6.2). The wet-extrusion process was not as effective in improving digestion as was dry extrusion. Feedlot tests on the four types of corn showed comparable daily gains (2.97–3.17 lb) but feed efficiencies were (lb feed per lb gain); whole corn, 7.6, thin flaked, 7.0, dry extruded, 7.1, and wet extruded, 6.8.

B. Gelatinization

Gelatinization is accomplished by subjecting ground grain to heating with steam to soften the grain, followed by forcing the resultant product with an auger through cone-shaped holes in an expander head. The holes are smaller where the material enters and gradually enlarge until the feed is expelled, which causes a release of pressure as the grain moves through the die. This causes expansion of grain. This technique is employed in the manufacture of expanded pet foods. Mudd and Perry (1969) conducted three metabolism and one feeding experiment with corn in which the starch was 100% gelatinized. In two of the three

TABLE 6.1

The Comparative Value of Flaked Raw Corn, Extruded Corn or Milo, and Ensiled High-Moisture Corn for Beef Cattle[a]

Treatment	Daily gain (lb)	Daily feed (lb)	Feed per lb gain (lb)
Flaked corn	2.85	22.7	8.0
Extruded corn	2.69	23.0	8.6
Extruded milo	2.74	23.6	8.6
High moisture corn	2.73	21.9	8.0

[a] Matsushima *et al.* (1969).

TABLE 6.2

Flaking versus Extrusion of Corn on Cattle Digestibility[a]

	Digestibility (%)		
	Dry matter	Crude protein	Crude fiber
Whole	65	41	17
Thin-flaked	74	55	23
Dry-extruded	71	52	21
Wet-extruded	68	48	20

[a] From Matsushima (1970).

metabolism studies, the substitution of gelatinized corn for raw corn as a major constituent of the diet resulted in a significant and linear depression in the digestibility of nutrients; in the feeding trial, gelatinized corn decreased feed intake and cattle gains.

Nebraska researchers (Wilson and Woods, 1966) reported that substitution of 15, 30, or 45% gelatinized corn for raw corn tended to increase gains over those of cattle fed raw corn; efficiency of feed conversion also tended to improve.

C. Grinding

Undoubtedly grinding of corn is the oldest processing technique applied to feedstuffs for cattle, especially for the feed grains. More recent research has demonstrated that grinding does not improve the nutrient value of corn for beef cattle. However, there may be situations wherein grinding is almost a necessity in order to obtain a relatively homogenous mixture of the ingredients.

It has been demonstrated that on extremely high corn diets, with little or no roughage present, it is a distinct advantage not to grind the corn. At time of slaughter, cattle fed high corn diets in which the corn is not ground have a lower incidence of rumen parakeratosis than do those whose corn was ground. An explanation for this difference is that the hard and sharp tip cap of unground corn may serve in part for the "scratch" effect provided by roughages in less concentrated diets.

D. Micronizing

Micronizing consists of heating grain to 300°F by gas-fired infrared generators. The term *micronization* was coined to describe this dry heat treatment since microwaves are emitted from the infrared burners during the process. Texas research (Schake *et al.*, 1970) utilized a field trial involving two lots of

100 steers each to evaluate micronized milo against steam-flaked milo. Feedlot gains favored steers fed micronized milo, but efficiency of feed conversion favored steam-flaked milo. However, the report indicated micronizing was a more economical process than was steam flaking.

E. Moisturizing

Moisturized, or high-moisture grain, was perhaps the processing technique which really gave impetus to the whole area of grain processing. The earliest reported data on the feeding value of ensiled high-moisture grains was in 1958 (Beeson and Perry, 1958), in which it was reported ensiled high-moisture ground ear corn had a 12–15% greater feeding value per unit of dry matter, based on comparable rates of gain and decreased dry matter intake. Unfortunately, the corn ear picker has been replaced by the picker–sheller combine and thus ear corn is no longer available for ensiling.

The storage of high-moisture grains is dependent upon either anaerobic fermentation, as in a silo, or the prevention of mold formation by the use of such materials as organic acids, e.g., propionic acid. The matter of choosing which technique to employ is a matter of which method fits best into a given system of cattle feeding and which is most economical.

One of the real advantages of high-moisture grain systems of feeding is that the harvest may be initiated two to three weeks ahead of that normally possible for harvesting grain which is to be artificially dried and stored in bins—the exception being barley. An equally substantial advantage to such programs is that expen-

Fig. 6.1 Ensiled high-moisture corn is a most excellent cattle feedstuff. (Photo courtesy of BEEF Magazine.)

sive artificial drying is not necessary. Perhaps the greatest drawback to the high-moisture grain system is that such grain cannot be sold in commercial channels once it has been stored as ensiled high-moisture grain. Acid-treated high-moisture grains may be dried to an acceptable moisture level and subsequently introduced into commercial trade.

Moisture level of high-moisture corn is important. Burroughs *et al.* (1971) summarized the data from 17 reports involving different moisture levels of corn stored in sealed storage. An average improved feeding value of 10% for high-moisture ear corn showed a low of +7% when the corn had from 23 to 32% moisture and a high of 13% for moisture levels of 33–44% (there was one high level of 23% improvement for 44% moisture corn in the higher moisture group). The higher moisture levels may give some harvesting problems and a realistic optimum for high-moisture ear corn is from 30 to 35% moisture.

For high-moisture shelled corn, the average improved feeding value of 6% was partitioned into 7% for moisture levels of 23–27% and 5% for 28–35% moisture levels. The recommended moisture level for high-moisture shelled corn is 25–30%.

The feeding results from "reconstituted" (dry corn treated with water to bring the moisture level to 25–30%) corn has been quite erratic. Generally, in the case of corn, it is recommended that original moisture high-moisture corn will give more consistent benefits than "reconstituted" high-moisture corn (Fig. 6.1).

Table 6.3 summarizes several experiments in which high-moisture corn was compared with dry corn as feed for beef cattle. From this research it is obvious that comparable gains were obtained from both types of corn but cattle consumed

TABLE 6.3

Dry versus High-Moisture Harvested Corn Stored Whole[a]

	Dry rolled corn			Ground ensiled corn			
Station	Daily gain (lb)	Corn dry matter per day (lb)	Corn cwt gain (lb)	Moisture content (%)	Daily gain (lb)	Corn dry matter per day (lb)	Corn/cwt[b] gain
Illinois, 1972	3.21	13.4	417	29.0	2.91	11.9	409
Iowa, 1970	2.00	11.7	585	24.3	2.09	10.9	522
Iowa, 1971	2.56	15.5	605	24.0	2.79	15.5	555
Iowa, 1972	2.32	13.3	573	27.3	2.40	12.6	525
Michigan, 1970	3.54	18.5	523	32.0	3.33	15.3	459
Minnesota, 1971	2.81	16.9	602	26.0	2.90	16.9	585
Minnesota, 1971	2.69	11.7	436	30.0	2.56	10.9	430
Average	2.73	14.4	534		2.71	13.4	498
Change (%)					0	−6.9	−6.7

[a] Annual Beef Cattle Research Reports.
[b] Cwt, Hundred weight.

an average of 6.9% less dry matter and thus required 6.7% less feed per pound of gain.

Embry (1971) presented data to show that rolled high-moisture shelled corn has greater advantage over rolled dry corn when fed with corn silage than when fed in an all-concentrate diet. With corn silage, rolled high-moisture shelled corn produced equal or greater gains and had greater feed efficiency, whereas in all-concentrate diets, all comparative data were nearly the same for both types of corn. When hay represented about 65% of daily dry matter, high-moisture shelled corn increased gain 8% and improved feed efficiency 7%; with haylage at 65% of daily dry matter, rolled high-moisture shelled corn resulted in 12% faster gains and 10% less feed per pound gain than for rolled dry corn.

It has been concluded generally by research data on the subject that cattle respond to high-moisture shelled corn treated with organic acids as well as they do ensiled high-moisture shelled corn.

1. High-Moisture Sorghum Grain

Since milo is the basic energy grain of the sprawling cattle feeding industry of the Southwest, it is appropriate that Texas, Oklahoma, and Kansas have conducted the majority of the research concerned with the nutritive value of high-moisture sorghum grains for beef cattle.

The benefits from high-moisture sorghum grain over comparable dry sorghum is much more dramatic than it is for corn. Their feeding value in the dry form is lower than their chemical composition would predict. Basic research into this discrepancy indicates the starch of dry sorghum grain is less available and the protein is not utilized as well as that of corn or barley. Therefore, almost anything that can be done to sorghum grain probably will improve its nutritive value for livestock.

Cattle fed ground moist sorghum grain have required less grain dry matter per pound of gain than did those fed ground dry grain, ranging from 15 to 25% with an overall average of 20%. In practically all comparisons high-moisture sorghum grains have resulted in a lowered feed intake compared to other processing methods such as steam flaked or ground or rolled dry grain. With similar weight gains, then, feed efficiency is increased.

Further processing of high-moisture sorghum grain is important. For example, beef cattle fed ground high-moisture sorghum grain gained 11% faster and required 37% less grain dry matter per unit of gain than cattle fed the same high-moisture grain in whole form. Rolling either high-moisture grain sorghum or dry sorghum is superior to fine grinding for increasing efficiency of feed conversion.

Some investigators feel the change that takes place in reconstituting grain sorghum is similar to that which occurs during germination in which the starch of the endosperm is liquefied to an extent for use by the growing seedling.

The ideal average moisture content for high-moisture sorghum grain is 30% with a range of 25 to 35%. Similarly, the ideal reconstituted level is 30% moisture. However, it is most difficult to add more than ten points of water from the starting point.

It is critical in the reconstituting of sorghum grain that the grain remain in the reconstituting process a minimum of approximately 21 days.

Increased dry matter digestibility has been proposed as the primary factor causing increased feed efficiency from using high-moisture sorghum grain. It has been shown, for example, that reconstituted sorghum grain increases in digestibility of dry matter, organic matter, and nonprotein organic matter of a magnitude of from 12 to 29%.

2. *High-Moisture Barley*

Barley kernels are physiologically mature when the moisture content drops below 40%. The ideal average moisture content for high-moisture barley is 30%, similar to that for high-moisture sorghum grain. All the physical advantages related to earlier harvesting for barley add to the increased feeding value of high-moisture barley. Research indicates that high-moisture barley has a place in cattle feeding, but not because of increased gain or because of improved feed efficiency. The chief advantage of high-moisture barley appears to be its high acceptability, with cattle going on feed faster, resulting in better early gains. Cattle stay on feed easier on high-moisture barley than do those on dry-rolled barley.

High-moisture barley should be rolled for beef cattle. In comparative studies, cattle fed whole high-moisture barley gained 0.3 lb less per day and required 63 pounds more feed per 100-lb gain than cattle whose high-moisture barley was rolled prior to feeding.

F. Popping

Popping is primarily restricted to sorghum grain and is accomplished by the use of gas-fired infrared generators, rated at about 50,000 Btu per hour each, suspended above the table to heat the grain as it passes beneath. The percentage of milo that will actually pop will range from 13 to 45% and appears to be influenced by moisture content, temperature, and rate of flow through the machine. As moisture content of the grain increases, the percentage of popping increases.

Riggs *et al*. (1970) made in depth studies of the effect of popping milo on its nutritional value. Four types of popping were compared: rolled (unpopped); normal run popped (13–45% popped); 100% popped fraction; and partially popped, or that fraction left over from screening out the 100% popped fraction. The processed milos were self-fed in a mixture composed of 92% milo, 7% cottonseed meal and minerals.

The most striking feature of these data is the great reduction in dry matter consumption of all three groups fed popped milo (Table 6.4). Steers fed rolled dry milo consumed 19–37% more dry matter per day than those fed popped milo. The reduction in feed intake was accompanied by improved feed utilization but also decreased daily gain.

Digestibility studies explained some of the reasons for the improvement of popped milo. It should be noted from Table 6.4 that digestibility of the nitrogen-free extract (NFE) was approximately 26% greater (61% for rolled versus an average of 77% for popped). Since NFE represents nearly 73% of milo, a 25% increase in digestibility would tremendously enhance its utilization.

The decline in dry matter consumption of popped milo is quite easily explained on the basis of Table 6.5. From the data in that table it may be observed that the feeding of popped milo resulted in greatly increased levels of propionic acid in the rumen. Research data (discussed in a later chapter), demonstrates rumensin results in (1) greatly increased production of propionic acid in the rumen, (2) a 10% depression in feed intake, and (3) approximately a 10% increase in feed efficiency.

G. Pelleting

Pelleting of beef cattle diets attracted a great deal of interest several years ago. However, very little pelleting of beef cattle diets is practiced anywhere today. Research from the Dixon Springs (Illinois) experiment station showed that pelleting of lower quality hay resulted in greater acceptability and also digestibility. However, research has not shown any benefit from pelleting either high-quality hay or high concentrate diets. Furthermore, the cost of pelleting would require quite large nutritional or economic benefits from such practice in order to justify using that technique.

TABLE 6.4

Comparative Feeding Value of Rolled and Popped Milo[a]

Milo treatment	Daily gain (lb)	Daily feed (lb)	Feed efficiency (lb)	Nutrient digestibility (%)		
				Dry matter	Crude protein	NFE[b]
Rolled	3.10	21.2	6.9	57	39	61
Normal popped	2.73	14.9	5.5	75	39	75
100% popped	2.55	15.1	5.9	79	38	80
Partially popped	2.75	17.5	6.4	76	41	77

[a] Riggs *et al.* (1970).
[b] NFE, nitrogen-free extract.

TABLE 6.5

Effect of Popped Milo on Average Molar Percentage of Volatile Fatty Acid Production in the Rumen[a]

Fatty acids	Rolled milo	Normal popped	100% popped	Partially popped
Acetic	54.9[b]	41.9[c]	44.6[c]	45.5[b,c]
Propionic	30.2[b]	47.6[c]	44.3[c]	43.0[c]
Butyric	8.8	7.8	8.3	8.9
Isovaleric	4.3[b]	1.0[c]	0.5[c]	.8[c]
Valeric	1.2[b]	1.6[b]	2.4[c]	1.6[b]

[a] Riggs *et al.* (1970).

[b,c] Values on same line with differing superscripts differ for acetic, propionic, and isovaleric by $p < 0.01$ and for valeric by $p < 0.05$.

H. Roasting

Corn roasting has perhaps offered more nutritional benefits as a processing technique than any other. However, because of the cost of heat energy, all processing techniques which require heat (steam flaking, popping, micronizing, roasting, gelatinizing) probably will be more expensive to use than the moisturizing technique of either ensiling high-moisture grains or else treating high-moisture grains with mold inhibitors.

The grain roasting technique, discovered by the author in the late 1960's has been shown to result in an average of 8% more rapid gain, plus nearly a 10% feed saving (see Table 6.6).

Roasting of corn is accomplished by exposing the grain to open flame heat. A common model for roasting grain consists of a cylinder housed within a jacket. The inside cylinder has fins on the outside which lift the grain through jets of flame pointing downward from the top of the jacket. Many revolutions of the inside drum take the grain through the flames many times until it is heated to 275°F. Roasted corn has a pleasant "nutty" aroma and a puffed, carmelized appearance. Very few of the kernels are actually popped. While raw corn weighs 45 lb/ft^3, roasted corn weighs only 39 lb/ft^3, indicating expansion during the roasting process. The moisture content of roasted corn is about 5–9% less than that of raw corn.

Cattle like roasted corn and tend to "go on feed" more readily than they do on unroasted, or raw corn. However, although cattle do not eat any more dry matter on a roasted corn diet, they gain more rapidly, thus causing such cattle to convert dietary dry matter to beef more efficiently.

Another effect from feeding roasted corn to finishing beef cattle is that the

TABLE 6.6

Comparative Performance of Cattle Fed Raw and Roasted Corn, 1970–1975, 6 Trials[a]

	1970		1971		1972		1973		1974		1975	
	Raw	300°F	Raw	300°F	Raw	300°F	Raw	300°F	Raw	300°F	Raw	300°F
Number of cattle	91	91	75	75	61	61	25	25	25	25	27	28
Initial weight (lb)	509	513	550	551	507	507	516	518	550	552	524	524
Length of study (days)	112	112	127	127	106	106	191	191	170	170	189	189
Daily gain	2.33	2.66	2.33	2.47	2.23	2.37	2.24	2.42	2.60	2.73	2.19	2.40
Percentage change		+14		+6		+6		+8		+5		+10
Daily corn (lb)	11.9	12.4	15.1	13.5	13.8	12.9	15.3	15.4	10.4	9.2	15.2	13.8
Dry feed/lb gain (lb)	6.8	5.2	7.5	6.4	6.7	6.1	7.6	7.0	7.9	7.4	8.4	7.3
Percentage change		−7.4		−14.7		−9.0		−8		−6		−13

[a] Author's summary of Purdue University research data. Average increase in gain = 8.2%. Average decrease in feed required per lb gain = 9.7%.

carcasses from cattle fed roasted corn consistently grade approximately one-half grade higher than do those from cattle fed regular corn.

I. Steam Flaking and Steam Rolling

These two processes are somewhat similar except the former technique is more specific and a longer time is given to the cooking or steaming process. In addition, the flaking process results in elevated moisture content of the grain. In the flaking of corn, a cooking or steaming time of 12 minutes at a temperature of 200°F will elevate the moisture content from 15 to 18%; holding milo at 200°F for 14 minutes will elevate the moisture content from 14 to 20%. Following the cooking process, the grain is passed through rollers set to produce flakes 1/32-inch in thickness. As soon as the grain is rolled, it should be dried to approximately 15% moisture. Colorado research (Matsushima and Montgomery, 1967) has demonstrated the importance of producing thin flakes in the rolling process (Table 6.7, Fig. 6.2).

Cattle fed raw ground corn did not gain as rapidly (Table 6.7) as those fed either of the two thicknesses of flaked corn; those fed corn of the thinner flake grew most rapidly and required least dry matter per unit of gain.

Flaking of corn increased efficiency of gain over pelleted corn by 5% and by 10–15% over grinding or cracking in Florida (Hentges *et al.*, 1966) research (Table 6.8).

Steam flaking of milo (Totusek *et al.*, 1967) resulted in increased intake (11.3 versus 10.6 lb/day) and increased rate of gain (2.63 versus 2.43 lb/day), but no improvement in efficiency of gain, in Oklahoma research. Hale *et al.* (1966) reported steam processing and flaking of milo resulted in increased gain (3.10 versus 2.83 lb/day) and improvement in efficiency of feed conversion (7.6 versus 8.0 lb feed/lb gain). The same investigators reported that steam processing and

TABLE 6.7

Comparative Feeding Value of Corn Steam-Flaked to Different Thicknesses[a]

	Thickness of flakes		
	1/32 inch	1/12 inch	Finely ground ¼-inch screen
Number of cattle	14	14	14
Initial weight (lb)	485	483	490
Daily gain, 163 days (lb)	2.82	2.70	2.65
Daily corn consumption (lb)	12.4	12.7	12.8
Feed per pound gain (lb)	6.1	6.7	6.9

[a] Matsushima and Montgomery (1967).

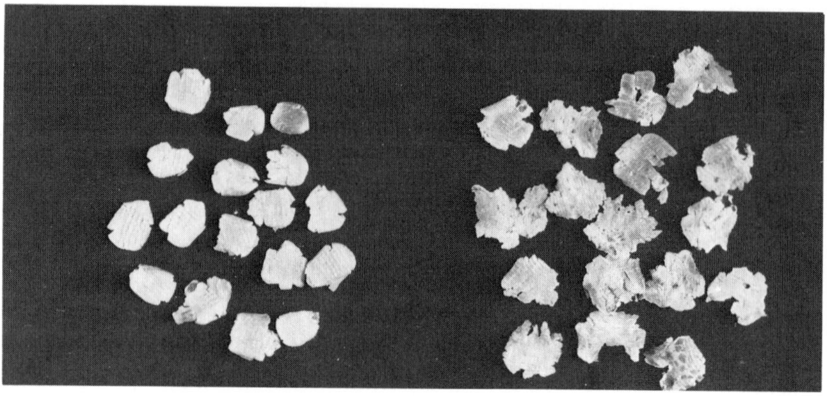

Fig. 6.2 Examples of undesirable (thick flake) and desirable (thin flake) steamed and flaked corn. (Photo courtesy of BEEF Magazine.)

flaking of barley also resulted in increased gain of cattle (3.10 versus 2.88 lb/day) but no improvement in the efficiency of feed conversion (7.3 versus 7.2 lb feed/lb gain).

II. PROCESSING OF ROUGHAGE

The author (Perry, 1972) has written a chapter on the effect of feed preparation and has included a section on the effect of processing on the nutritive value of roughages. It is pointed out that many types of processing can affect the nutritive value of roughages, but because of the low economic value of many roughages, it is often questionable as to whether the improvement brought about can justify the cost involved.

TABLE 6.8

Effect of Grinding, Cracking, Flaking, or Pelleting Corn for Beef Cattle[a]

	Corn treatment			
	Ground	Cracked	Flaked	Pelleted
Number of cattle	20	20	20	20
Initial weight (lb)	588	585	577	588
Daily gain, 126 days (lb)	3.30	3.40	3.60	3.50
Daily concentrates (lb)	18.1	19.5	18.3	18.5
Feed per pound gain (lb)	6.5	6.7	5.8	6.3

[a] Hentges *et al.* (1966).

A. Grinding

Grinding of roughages is restricted almost entirely to ground corn cobs. When properly balanced, ground corn cobs have an energy equivalent comparable to that of medium-quality hay. Unground corn cobs have practically no feeding value. Because of the abrasive nature of corn cobs, they are extremely hard on a hammer mill.

Grinding of hay has little value for beef cattle except as a prerequisite to pelleting or complete mixing of diets. However, the grinding of hay produces so much dust that this causes a feeding problem. Grinding of hay is not practical for dairy cattle because it results in lowered butterfat content.

B. Pelleting

Pelleting of roughages has received a great deal of interest. The Illinois Station indicated greatly increased nutritive value of hay as a result of pelleting it. Subsequent research indicated this benefit from pelleting could be realized only with poor-quality hay—there was no benefit from pelleting high-quality hay.

Pelleting of corn cobs will improve their nutritive value; similarly, dehydrating and pelleting whole plant corn silage will improve its value. However, the cost of processing such products for pelleting is almost prohibitive.

C. Wafering

Wafering of hay holds some promise. In this process the stems are broken somewhat and the resultant product is about 3 inches in diameter and from 1 to 3 inches in thickness. This process has been employed in preparing hay for shipment to other countries.

D. Predigesting

Predigestion using alkali, such as sodium hydroxide or potassium hydroxide, will increase the nutrient availability of low-quality roughages by as much as 33%. However, the cost is very high.

E. Dehydration

This process is restricted almost entirely to extremely high-quality roughages such as alfalfa. The energy cost is too high to permit its use under other conditions.

F. Ensiling and Hay Making

The use of these processes with roughages will be discussed separately in Chapter 8, Section II.

REFERENCES

Beeson, W. M., and Perry, T. W. (1958). The comparative feeding value of high-moisture corn and low-moisture corn with different feed additives for fattening beef cattle. *J. Anim. Sci.* **17**,368.

Burroughs, W., Self, H. L., and Geasler, M. R. (1971). "Iowa Cattle Feeding Trials with High Moisture Corn," Grain Feeders Seminar. Iowa State University, Ames.

Embry, L. B. (1971). Grain processing for feedlot cattle. *7th Annu. Okla. Cattle Feeders Semin.*

Hale, W. H., Cuitun, L., Saba, W. J., Taylor, B., and Theurer, B. (1966). Effect of steam processing and flaking milo and barley on performance and digestion by steers. *J. Anim. Sci.* **25**,392.

Hentges, J. F., Jr., Cabezos, M. T., Moore, J. E., Carpenter, J. W. and Palmer, A. Z. (1966). The effect of method of processing on nutritive value of corn for fattening cattle. *Fla., Agric. Exp. Stn., Mimeo Ser.* **AN67-4**.

McLaren, R. J., and Matsushima, J. K. (1971). Extruded corn compared with flaked corn and whole corn in finishing rations. *Colo. State Univ. Beef Nutr. Res. Bull.* **918**,14.

Matsushima, J. K. (1970). Feed processing for feedlot cattle. *Anim. Nutr. Health* May, p. 8.

Matsushima, J. K., and Montgomery, R. L. (1967). The thick and thin of flaked corn. *Colo. Farm Home Res.* **17**,4.

Matsushima, J. K., McLaren, R. J., McCann, C. P., and Kellog, G. F. (1969). Processing grains for feedlot cattle—extrusion, flaking, reconstituted and high moisture ensiled. *Colo. State Univ., Beef Nutr. Res.* **28**.

Merrill, W. G. (1971). The place of silage in production rations—feeding high-moisture grain silages. *Proc. Int. Silage Conf., 1971*, p. 156.

Mudd, C. A., and Perry, T. W. (1969). Raw cracked *vs.* expanded gelatinized corn for beef cattle. *J. Anim. Sci.* **28**,822.

National Academy of Sciences (1973). "Effect of Processing on the Nutritional Value of Feeds." Nat. Acad. Sci., Washington, D.C.

Perry, T. W. (1972). Feed preparation. *In* "The Feedlot" (I. A. Dyer and C. C. O'Mary, eds.), Chapter 9. Lea Febiger, Philadelphia, Pennsylvania.

Riggs, J. K., Sorenson, J. W., Jr., Adams, J. L., and Schake, L. M. (1970). Popped sorghum grain for finishing beef cattle. *J. Anim. Sci.* **30**,634.

Schake, L. M., Garnett, E. T., Riggs, J. K., and Butler, O. D. (1970). Micronized and steam flaked grain sorghum rations evaluated in a commercial feedlot. *Tex., Agric. Exp. Stn., Tech. Rep.* No. 23.

Totusek, R., Franks, L., Basler, W., and Renberger, R. (1967). Methods of processing milo for fattening cattle. *Okla., Agric. Exp. Stn., Misc. Publ.* **79**,79.

Wilson, B., and Woods, W. (1966). Influence of gelatinized corn on beef animal performance. *Neb., Agric. Exp. Stn., Prog. Rep.* p. 22.

II

Feedingstuffs

7

Pasture and Other Forages

The world's most abundant renewable source of available energy is that of roughages and forages. Fortunately, ruminant animals can utilize this most abundant energy; monogastrics such as swine, poultry, and people are unable to utilize forage materials. Historically good forage has supplied the most economical feed for beef cattle and now that feed grains have become so valuable, forage as a feed "cost-cutter" has become even more important.

More land area in the United States is in "pasture" of some sort than in all of the rest of the crops combined. Furthermore, beef cattle lead all classes of livestock in the consumption of pasture grasses, utilizing about one-third of the permanent pastures and over three-fourth's of the range areas. The ranges, pastures, and haycrop land in the United States have a combined area of more than one billion acres, which is equivalent in size to 18 states the size of Iowa. Only about one-tenth of such grazing land is suitable for cultivation which means at least 90% has only one possible use—that of the production of forage.

I. NUTRITIVE VALUE OF PASTURE AND FORAGES

It is extremely difficult to place a numerical value on the nutritive value of these feeds. A variety of factors affect their nutritive value and therefore it is important that a discussion of factors affecting the nutritive value of pastures and forages be included in this section.

A. Type of Plant

Legumes versus nonlegume plants for forage differ greatly in their potential nutritive value. Generally, the leguminous crops contain a much higher level of protein than nonlegumes, at comparable stages of maturity. Furthermore, legume crops tend to produce more tonnage of digestible dry matter per acre than many typical pasture grasses. However, there are several drawbacks to leguminous pasture. First, rather high quality and high fertility land are needed for legumes, which entails a high cost assessment per unit weight of dry matter produced from legumes.

Nearly all legume pasture crops have a tendency to cause bloat in beef cattle, unless such crops are diluted sufficiently with nonleguminous plants. One pasture legume which does cause bloat is birdsfoot trefoil. However, this plant is not commonly grown throughout the country.

B. Stage of Maturity of Plants

Stockmen utilizing pastures in their programs should take maximum advantage of the changing patterns of digestible nutrients in pasture crops due to stages of maturity. Young grasses growing rapidly may contain as much as 15–17% crude protein, on a dry matter basis; young growing legumes may contain 18–22% crude protein. However, as pasture plants mature, the level of protein declines drastically. Undoubtedly cattle feeders may fail to consider the high level of protein from their pasture. Although beef cattle can obtain more protein from their pasture than is actually needed, often cattle feeders feeding grain to finishing cattle on pasture supplement diets with protein. For example, a dry pregnant beef cow requires only 6% of her dietary dry matter as crude protein. Yet almost any grass at any stage of maturity contains that level, or more. A cow nursing a calf requires from 9 to 10% of her dry matter as crude protein; typical dry grass hay contains about 8.5% protein. On the other hand, many lush grass pastures contain nearly double that level.

Pasture, which is allowed to mature and head, contains very much lower levels of protein, so the matter of keeping pastures grazed and/or clipped sufficiently to prevent "heading out" is essential to maintain higher protein and digestible energy levels. This can often be controlled in rapidly growing spring pastures by cutting a crop for hay.

A second characteristic of young pasture, compared to older and more mature pasture, is the crude fiber and lignin content. As plants approach the seeding stage, the level of lignin in such plants increases rapidly—lignin is nearly 100% indigestible. Therefore, the available energy—or total digestible nutrient content—of growing plants is highest when they are young and growing rapidly, and lowest when they are mature.

Mineral content of plants remains fairly constant from early growth through maturity. For example, the phosphorus content of plants may remain quite high until such times as the plants have matured and have been subjected to extreme weathering, especially leaching rains.

Pasture has only one vitamin of interest to ruminant animals. Growing pasture contains carotene which can be converted, in at least limited quantities, to vitamin A by beef cattle. Mature and bleached forage has very little carotene.

The potential B-vitamin content of pastures and forages is of significance in beef cattle feeding only to young calves in which rumen development is not complete.

Growing plants contain no vitamin D as such. Once vegetation is cut from its stalk and allowed to dry in the sun, a fat-soluble material known as ergosterol is activated by the sun's ultraviolet rays to form a calciferol, or vitamin D_2. It should be borne in mind that cattle out of doors in the sunshine obtain vitamin D through the activity within their own body. Ultraviolet irradiation from the sun's rays acting upon oil droplets on their hair convert 7-dehydrocholesterol to activated 7-dehydrocholesterol, which is vitamin D_3. When cattle lick themselves, or each other, they consume the activated vitamin D_3.

C. Effect of Soil Fertility on Nutritive Value of Pasture and Forage

Naturally the yield of forage dry matter per acre is affected by the fertility of the soil. However, the composition may also be affected. Good examples of this point are cobalt and magnesium. Forage crops will grow on soils which are from borderline to markedly deficient in cobalt and magnesium. The crops produced under such soil conditions will be deficient in either cobalt or magnesium, and so cattle consuming such forages will experience a cobalt or magnesium deficiency in their diet. Certainly such dietary deficiencies can be corrected with supplemental cobalt or magnesium.

II. TYPES OF PASTURES AND FORAGES

A classification of pastures and forages which designates their general life expectancy and/or general location is in order.

A. Permanent Pastures

This type of pasture is usually found on land that cannot be used for cultivated crops, largely because of topography or moisture. With minimal care, such pastures last indefinitely. Most farms have at least some such land which is fit only for permanent pasture. Most of the rangeland and forest pasture lands come under this category.

B. Rotation Pastures

These pastures are a part of a program of crop rotation. Many years ago it was felt corn could not be grown continuously on land. Therefore several crop rotation programs were studied, such as a 4-year rotation of corn, corn, oats, and clover, respectively. Some such rotation programs covered 5 years; a sod crop, such as clover, was often included. After the first hay crop had been harvested, the clover field was pastured until such time as it was plowed for the next year's corn crop. Inclusion of one year's forage crop in corn cropping systems is the exception due to the low economic return from forages compared to that derived from corn or soybeans.

C. Temporary Pastures

Temporary pastures are seeded for use for very short periods of time. They are provided when regular permanent or rotational grazing is not available. Examples are Sudan grass or Sudex seeded in the spring for summer grazing (Fig. 7.1), or oats and rape seeded for spring and summer grazing. Rye may often be seeded following bean or corn harvest for fall and winter grazing (Figs. 7.2 and 7.3).

D. Winter Wheat Pasture

This type of pasture is restricted largely to Kansas, Oklahoma, and other states in the winter wheat belt where cattle are shipped in to graze the wheat which has

Fig. 7.1 Sudax, an annual, provides excellent summer grazing. (Photo by J. C. Allen and Son.)

Fig. 7.2 Good pasture plus shade provide an excellent environment for beef cows and calves. (Photo courtesy of BEEF Magazine.)

Fig. 7.3 Brood cows and calves on adequate pasture require only water and supplemental minerals. (Photo by J. C. Allen and Son.)

been seeded for next year's crop. This has a dual benefit in that it provides pasture for the cattle and also pasturing the wheat causes the plant to "stool" out and form a more dense mat of growth.

E. Harvested Crop Residue Material

This is typified by a corn stalk field after the picker–sheller combine has harvested the corn crop. Stalks, pods, and leaves remain after a soybean crop has been harvested. Straw and chaff remain following the combining of small grains. Unfortunately larger and larger farming equipment has resulted in the removal of more and more fences so that many corn stalk fields cannot be pastured because of the lack of fences.

Many attempts have been made to bunch or package corn plant residue for cattle feeding. This has not been too satisfactory except when the corn plant residue is ensiled.

III. PASTURE CROPS

One method of categorizing pasture crops is to divide them into legumes and nonlegumes.

A. Legumes

1. Alfalfa

Alfalfa, or lucerne, as it is called in many parts of Europe, is well adapted to a wide range of climatic and soil conditions. It responds to fertilization and water along with good cultural practices, including inoculation with nitrogen-fixing bacteria. It is one of the most important forage plants in the United States since it has the highest feeding value of commonly grown forage crops. It produces double the protein produced by clover and many times that produced by the nonlegumes.

It is not used commonly as the sole pasture crop for beef cattle because of its bloat-causing effect (Fig. 7.4). It is much more satisfactory as a pasture crop for beef cattle when it is grown in a mixture with nonlegumes such as bromegrass or orchardgrass or even bluegrass. Alfalfa has many natural enemies in the insect field as well as in the bacterial world, and thus it is often difficult to maintain a stand of alfalfa.

2. Bur Clover and Spotted Bur Clover

These are relatives of alfalfa. They are weak-stemmed plants resembling the clovers. California bur clover contributes to the range pastures of the California

Fig. 7.4 Alfalfa provides excellent pasture, but care should be exercised to prevent bloating losses. (Photo by J. C. Allen and Son.)

and Arizona foothills. In the southeastern states, spotted bur clover is the primary species. All bur clovers are unable to stand the rigors of winter and thus are restricted to the warmer climes indicated above. In the western states, bur clover comes up "volunteer" in range pastures whereas in the southern states spotted bur clover must be seeded. However, once it is seeded it stays in the sod indefinitely.

3. Red Clover

Red clover grows abundantly in the typical Corn Belt country as well as in the entire northeastern quarter of the United States. Furthermore, it grows well as a winter annual in the southeastern sector, and under irrigation in the six or seven far western states. Most varieties grown in the United States fall into the category of "medium" red clover, although Mammoth red clover is grown quite widely also. Red clover is seldom used as the sole pasture crop but rather it is normally grown in mixtures with grasses.

4. Alsike Clover

This is a perennial which contributes to cattle pasture mixtures. Because of its adaptation to wet soils, Alsike is good for establishing pasture sod on wet natural meadows or where irrigation is used. It may persist in bottomlands along creeks or rivers where alfalfa or red clover are unable to survive. It is well adapted to quite cool climates. It can tolerate a greater soil acidity than clover or alfalfa, but nevertheless responds to limestone applications.

5. *Ladino and Other White Clovers*

These are perennials and are grown widely throughout the world. This group is one of the most nutritious and palatable of all the legumes. The white clovers are usually grown in association with other legumes or with grass, or with complex mixes of both. In comparison to the white or "Dutch" clovers, Ladino produces from two to four times as much material. The white clovers usually appear as "volunteer"—especially in the cooler northern pastures whereas Ladino clover must be seeded. White clover in its varying forms is one of our most important pasture legumes. Cattle relish it. Interestingly, white clover pasture is not too desirable as a horse pasture since it causes excessive salivation or "slobbering" of horses.

6. *Lespedezas*

The lespedezas are especially well adapted to the southeastern one-fourth of the United States. It is an excellent summer pasture, wherever it can be grown, for all classes of livestock. If grazed sufficiently to prevent the plants from heading, it is very palatable and nutritious. Its maximum carrying capacity comes along at about the time many of the permanent pasture grasses begin to go dormant. It is an excellent fattening pasture; beef and dairy cattle will gain two pounds per day on lespedeza pasture. The high quality of the lespedeza pasture may be due at least in part to (a) low water content, (b) extreme palatability, (c) high nutritive value, and (d) the fact that it does not cause bloat. A lespedeza pasture which has matured and has seeds is a very high energy aftermath roughage for beef cattle.

7. *Birdsfoot Trefoil*

Essentially grown in the northeastern one-quarter of the United States and also along a narrow strip of the western coast, birdsfoot trefoil has not had a wide acceptance as a pasture legume because of the extremely slow development of the seedlings. Therefore in mixtures with clovers and grasses, birdsfoot is not able to compete. It perhaps has some resemblance to alfalfa. However, no cases of cattle bloat are known to have occurred on birdsfoot trefoil. It has a wide soil tolerance to fertility and acidity. It is a perennial which reseeds itself, even when grazed closely. It is especially compatible with Kentucky bluegrass and such stands may stay in existence and in balance for many years.

8. *Vetches*

The vetches are most common in the southeastern one-quarter of the United States. There are several varieties of vetch including hairy, madison, common, Hungarian, narrowleaf, purple, and bard. They are especially well adapted as cover crops for land exposed in highway construction because of their matting characteristic. They are usually considered as winter annuals, reseeding themselves each year. It is during this period of the year that land in the South is not growing cash crops such as cotton or peanuts, and thus the vetches are excellent crops at that time. All vetches are edible and palatable to cattle but only those with hard seed and good seeding habits are recommended for use in permanent pastures. Only hairy and smooth vetches fit these qualifications.

B. Nonlegumes

1. *Bluegrasses*

Characterized by Kentucky bluegrass, these nonlegumes are common in the northeastern one-quarter of the United States, where there are cooler conditions with adequate rainfall. It is adapted to better drained loams of limestone origin such as exists in Virginia and Kentucky. One of the drawbacks to this pasture grass is the marked periodicity in growth and development whereby greatly reduced production occurs from mid-July through August. It is well adapted to mixtures with legumes, especially the white clovers and birdsfoot trefoil, but it is too competitive with alfalfa and red clover in pasture mixtures. It is extremely "washy" in the first few weeks of the spring season, but this tendency disappears within a few weeks. Following its dormancy period in late summer, it responds to fall rains and produces well in September and October. Unpastured bluegrass that is allowed to grow and fall over is a most excellent forage crop for beef cattle in the late fall months.

2. *Bromegrasses*

These grasses are adapted to cool climates or to regions in which cool seasons prevail a portion of the season. The Greek derivative *Bromus* signifies oats, and thus the name oats grass often has been applied. There are at least 60 varieties of bromegrass but the most common is called smooth bromegrass. The main region of growth in the United States corresponds with the so-called Corn Belt, but extends quite a bit farther north. It is fairly resistant to drought but in severe drought becomes dormant much the same as other grasses. Smooth brome is used both alone and in mixtures of other grasses and legumes. In mixtures with legumes, the first cutting often goes for hay, followed then by grazing.

Bromegrass is one of the most palatable of the grasses and maintains its palatability and nutritive value at much later stages of growth than most grasses, except perhaps bluegrass. It produces more aftermath than timothy.

3. *Reed Canarygrass*

Adapted to the northern portion of the United States and southern Canada, reed canarygrass is especially resistant to flooding conditions and thus is adapted to seeding in gullies, along stream banks, and around farm ponds to prevent water erosion. Its natural habitat is in rather poorly drained pasture areas. It starts growth early in the spring, but it should not be allowed to grow too rank because it loses palatability rapidly. It does not seem to cohabit well with legumes because its shading effect discourages growth of the legumes. An exception here is that Ladino clover will exist with canarygrass if the grazing and clipping management do not permit too tall growth of the canarygrass.

Varying reports have been given relative to its palatability but most cattlemen report more than satisfactory consumption.

4. *Timothy*

Timothy is adapted to a cool and humid climate, thus it is restricted primarily to the northeastern portion of the United States. It is grown primarily for hay but is generally included in pasture seeding mixtures, especially with clover. Such mixtures are generally harvested in the first growth of the year for hay, at the time the timothy heads bloom. Subsequently it can be grazed as a pasture mixture. Young succulent timothy is often relished more than bluegrass by livestock. Timothy is gradually crowded out of permanent pasture by bluegrass.

5. *Orchardgrass*

Orchard grass is best adapted in the central states of the United States because of its tendency to winterkill in severe winters. It is able to persist in "thinner" soils than timothy or bluegrass, but responds to greatly increased yields of dry matter upon fertilization. Orchardgrass is a long-lived perennial if conditions are favorable. Its growth in the spring is earlier than that of other grasses and as much as three weeks earlier than timothy. It tends to grow more evenly throughout the drought conditions of summer months than is true for the other grasses.

It is especially used for pasture in the Piedmont and mountain areas of Georgia and North Carolina. In Missouri, it is adapted to a blend with lespedeza and clovers.

6. *Bermudagrass*

This grass is widely distributed throughout the tropical and subtropical countries of the world, and in the United States it is of prime importance in the southern one-third to one-half of the country. It grows best in areas where the mean temperature is at least 75°F. It is more drought-resistant than some of the other grasses typical of the south. Bermudagrass responds well to fertilization and moisture. It will give the greatest quantity of nutritious product under continuous grazing if it is not allowed to grow over 4 to 6 inches tall. In times of most luxuriant growth, it is possible to remove a cutting of Bermudagrass hay—and such hay has a feeding value comparable to that of timothy. As the age of the Bermudagrass plant increases, leafiness, crude protein, fat and ash decrease, but dry matter yield, crude fiber, lignin, and nitrogen-free extract increase.

7. Dallisgrass and Bahiagrass

These two grasses are adapted primarily to the southeastern United States. Dallisgrass is usually the first grass of the spring season and grows more or less continually through until fall. It cohabits well with a legume and thus provides more uniform grazing for a longer period of time than any other pasture combination. If it is not mowed or grazed judiciously, growth ceases with flowering and seeding.

Bahiagrass ranks intermediate between Bermudagrass and Dallisgrass in productivity and palatability. It is used primarily for pasture on the sandy soils of the Gulf Coast, and it furnishes little grazing during the winter months. Bahiagrass will stand more frequent harvest with good production than does Dallisgrass.

8. Fescue

Fescue is characterized by many species but the two most common varieties are Alta fescue and Kentucky 31 fescue. The fescue group has a wide distribution but the greatest concentration of growth is in the southeastern United States with another area along the coasts of Oregon and Washington. In the southeast, fescue is especially important as a winter pasture crop and does not persist as well in the hot summer environment. Fescue establishes one of the strongest sods which is thus resistant to breakthroughs from grazing traffic. Its winter growth habits make it important in a grazing program designed for year-round use.

There are some peculiarities about fescues that make it imperative that special management techniques are followed. At certain phases of their growth the fescues become rather unpalatable and cattle will not eat them unless forced to. There is also the matter of so-called "fescue poisoning" in which animals grazing on fescue tend to become lame. They will eventually develop "fescue-foot" which is a deformed hoof which may eventually slough off.

At the Southern Indiana Center, brood cows grazing fescue pasture appeared to breed, have normal gestation periods, and to produce healthy calves. However, there was little or no mammary development accompanying gestation so that at time of parturition there was no milk for the baby calves to nurse. Establishment of from 15 to 30% of legumes in a fescue pasture appears to counteract the periodic poisonous aspects of fescue pasture for beef cattle. Furthermore, the feeding of supplemental feed, such as limited hay or energy feeds, also seem to prevent any adverse effects from fescue pasture alone.

Fescue pasture perhaps will tolerate more grazing abuse than any other pasture sod. Heavily overgrazed pastures seem to come back with excellent growth after cattle are removed.

9. Wheatgrasses

Grown primarily in the central and northern Great Plains, the intermountain regions, and the Rocky Mountains wheatgrasses are perennials and cool season plants growing over several millions of acres of rangeland in the United States. They produce early season grazing in these areas, or as long as the climate remains cool. However, the hot dry season makes the wheatgrasses essentially dormant. In addition to their grazing potential, the wheatgrasses serve as a deterrent to wind and water erosion. Crested wheatgrass furnishes earlier spring and later fall pasture than native range plants and thus serves to supplement range grazing. Oftimes a hay crop can be cut from crested wheatgrass at about the time it heads out.

10. Sorghums

Sorghums for pasture are perhaps best typified by sudan grass and hybrids thereof. The sorghum plants are resistant to drought and thus they are grown primarily in the southern Great Plains area. Sudan grass grows most rapidly in hotter weather when most other pasture crops are dormant. Cattle relish sudan grass. Pasturing of the crop should be delayed until the plants have grown to 18 or 20 inches. Sudan grass is especially adapted to rotation grazing; while one plot is recovering from having been grazed, the other is being grazed into the ground. Since sudan grass stores energy in its roots, there is no danger to the plant in grazing it down to the ground since the plant will recover. There is always the potential danger of prussic acid poisoning for cows consuming sudan grass, but the most critical times are in the early stages (2- and 3-leaf stages) or immediately after the plants have been frosted or frozen. Mature sudan grass or that which has been cut for hay or silage has no danger for poisoning cattle.

C. Residue Material

The harvested crop residue materials (straw, chaff, and corn stover) may become somewhat of a problem to dispose of since environmental regulations do not permit massive burning of such products. All of these products may be picked up from the ground and formed into packages and transported to shelter. However, because all such products are actually byproducts of grain and seed production, normally they are left to decay in the field behind the grain harvesting machinery. All of these products have a low true feeding value for cattle although there may be potentially much greater feeding value. Many attempts have been made to make the basic cellulose constituent of straw, chaff, and stover more available to cattle. However, since these products have such a low economic value initially, its almost a waste of money to invest more than token quantities of time and money in attempting to upgrade their nutritional potential.

Since the basic nutrient contained in straw chaff and stover is cellulose (or energy potential) it is imperative that the other nutrients such as protein, vitamins and minerals be provided to balance the diet. It is almost impossible for a brood cow to maintain her weight on straw, chaff, or corn stover, alone. This may not be too critical for pregnant beef cows.

8

Hay and Haylage Making

One of the very basics for beef cattle production is optimum use of pasture—especially in cow and calf production. However, because of the diversity of climatic conditions where beef cattle are grown, during many months of the year it is not possible to provide pasture. Therefore, storage of herbage becomes an integral part of such an operation. Furthermore, there are periods when herbage does not grow due to lack of sufficient moisture and therefore harvested feed or ungrazed pasture must be provided for cattle.

Herbage is routinely stored either in the wet form as silage or else in the "air-dried" form as hay. This chapter will deal with the latter—hay—as well as the "half wet-half dry" product known as haylage.

I. HAY

A. Stage of Growth and Its Effect on Composition and Yield of Nutrients

Three basic measures of forage production are most desirable, namely, (1) maximum dry matter per acre, (2) maximum protein per acre, and (3) minimum crude fiber per unit dry matter. Unfortunately, it is almost impossible to combine all three of the above objectives at their respective maximums or minimums. Thus, a compromise among the three objectives is most typical. For example, generally, as a forage crop is managed for maximum dry matter yield per acre, the fiber content per unit dry matter increases. Thus as the plant reaches greater maturity—and thus more dry matter—fiber content increases to strengthen the more mature plant. The average crop tends to increase in carbohydrate content, and both digestible nitrogen-free extract (NFE) and less digestible crude fiber, with increasing maturity. The protein portion of the plant is synthesized early in the life of the plant; as the plant matures, the proportion of protein declines. This is more marked for the grasses than is true for the leguminous crops.

Alfalfa is a prime example of the effect of age on the value of the forage. Even in the pre-bloom stage, alfalfa contains about one-third of its dry weight as fiber whereas at the same time the leaves of alfalfa contain only about 12% fiber; by the flowering stage, nearly half of the alfalfa stem is fiber. This demonstrates how the potential nutritive value of alfalfa hay declines with advancing maturity.

B. Chemical Changes in Hay Drying

The more rapid the drying technique, the less the loss of nutrients. In contrast, prolonged drying can result in great losses of nutrients. One of the greatest causes for losses of nutrients is oxygen. Most hay crops are growing actively at the time they are cut. Therefore, respiration may continue for quite some time after the cutting process. The drying process slows down respiration. Rewetting the cut material, as by overnight dew or light rain, will renew the respiration process. The major nutrient loss is that of the soluble sugars, or the NFE portion. Almost all sucrose and dextrose may be lost, along with some losses of starch and dextrin if exposure to the weather is prolonged. An interesting reverse trend appears to operate in the early wilting stages. For example, alfalfa sugar content will increase from 4 to 7% when the dry matter is rising from 10 or 20% up to 35–50%, in the wilting process. Thus half-wilted hay contains sufficient sugars for the ensiling process. At about this point (50% dry matter) sugar content will decrease as the drying process continues.

The conversion of growing crops into hay does not necessarily result in loss of protein unless the curing and drying process is prolonged, or if there is intermittent raining and drying. However, inevitably there will be conversion of varying amounts of true protein into nonprotein forms. There is a loss of ether extract but there is no loss of crude fiber or ash.

The loss of vitamins, or of their precursors, is critical. Vitamin A does not exist as such in growing plants, although its precursor(s) in the various forms of carotene does. Animals can convert carotene to vitamin A, in varying degrees, based on the type of carotene and type of animal. The carotenes are especially susceptible to destruction in the curing and drying process. Furthermore, the rate of destruction of the carotenes is quite rapid at higher temperatures, higher moisture, and increased exposure to sunlight. In contrast, exposure of hay to sunlight results in the conversion of the vitamin D precursor, ergosterol, to active vitamin D^2, or (calciferol).

Although various B vitamins are contained in green plants and also in hay, excessive exposure of hay to rains and leaching tends to destroy much of this vitamin B activity.

C. Losses in Hay Due to Leaching

Rain has very little effect on newly mown herbage because the protective waxy coating on the plant is quite effective. However, when the plant dries, the waxy coating becomes ineffective and the cell walls lose their capability of selective flow of water and nutrients back and forth across the membrane wall. Therefore, rain water will dissolve out and wash away varying degrees of the soluble nutrients, according to the amount of exposure to rain. The less soluble and less digestible nutrients will, therefore, remain.

The loss of nutrients due to leaching may be slight to excessive depending on the nature of the weather. Records have shown losses as high as 25% or even more, of protein due to leaching alone; nitrogen-free extract loss may run as high as 35%. Leached material will be higher in crude fiber than nonleached material.

D. Hay Storage Methods

1. Sheltered Storage

Any forage crops dried to a moisture level not in excess of 12–15% will suffer little or no loss for several years if protected from rain and sunlight by a roof. The main changes which occur in properly stored dried hay are due to decreased intensity of color and associated losses of carotene. The requirements for optimal storage of dried (12–15% water) hay are, (1) protection from the weather, (2) packing to reduce the oxidation of carotene by air, and (3) protection from excessively high temperatures.

2. Big Package Haymaking

This represents a mechanized method of harvesting, handling, and feeding of hay. By this technique, cattle feeders can increase mechanization and reduce labor, not only for the haymaking operation but for the movement and feeding operation as well. One of the real banes of making, storing, and feeding hay has always been the tremendous amount of "back-breaking" manual labor involved. Furthermore, haymaking in late May and early June often is in direct conflict with critical row crop management, and so the haymaking process usually loses out in the competition for the farmer's time. Recently, the advent of machinery for this purpose have provided for a very high degree of mechanization in the haymaking program. Furthermore, because of the nature of several such hay package types resulting from these newer machines, outside storage of hay seems quite practical, although they are subject to wind and weather from early summer through the winter feeding period. Undoubtedly this means varying degrees of deterioration of the quality of the hay, especially around the outside of the packages and even all the way through, if not packaged properly. Also, certain obvious losses from big packages occur when a group of animals is given access to one or more of the big hay packages. This occurs from trampling, refusal, overconsumption, and urinating and defecating on the packages. Nevertheless, haymakers and livestock feeders are deserving of factual evidence of just how well such innovations get the job done and at what cost.

Research was conducted by Purdue University (Smith *et al.*, 1974) to study

TABLE 8.1

Weathering Losses in Grass and Legume Hay Stored Outside in Various Package Forms[a]

Type of package	Average package weight (15% moisture) (lb)	Portion of each package weathered (%)[c]	Total digested nutrients[d]		TDN loss due to outside storage (% of total)
			Unweathered core (%)	Weathered outside (%)	
Grass hay[b]—Year 1					
Hesston 10 stack	1195	12.6	53.4	35.1	8.83
Vermeer 605 bale	1089	14.6	54.0	39.4	8.20
Hawkbilt 480 bale	560	22.2	54.8	39.7	12.59
Small round bale	35	20.6	55.3	33.0	16.87
Average (big packages)		16.4	54.1	38.1	9.87
Grass hay[b]—Year 2					
Hesston 10 stack	1007	9.6	60.0	42.2	6.82
Vermeer 605 bale	1185	7.0	58.9	42.5	4.60
Hawkbilt 480 bale	683	16.8	57.9	40.4	11.29
Small round bale	37	20.2	60.0	42.6	13.45
Average (big packages)		11.1	58.9	41.7	7.57
Alfalfa hay[b]					
Hesston 10 stack	1377	8.1	57.0	33.9	7.36
Vermeer 605 bale	1097	10.7	56.5	34.2	9.14
Hawkbilt 480 bale	728	19.6	56.6	31.9	17.75
Average (big packages)		12.8	56.7	33.3	11.42
Average—two years					
Hesston 10 stack		10.1			7.67
Vermeer 605 bale		10.8			7.31
Hawkbilt 480 bale		19.5			13.87
Average (overall)		13.5			9.62

[a] Purdue University data.
[b] Grass hay stored from June to Nov. and from June to Feb.; alfalfa hay stored from Aug. to Feb.
[c] Separation of hay into weathered and unweathered parts was done by hand.
[d] TDN estimated on a basis of *in vitro* dry matter disappearance.

weather deterioration and feeding losses of hay packaged by three commercially available big package haymaking machines: Hesston Stakhand 10 stacker, Vermeer 605 baler, and Hawkbilt 480 baler. In addition, a much older Allis Chalmers small round baler was utilized. However, although this section presents research data, farmer experience conceivably might be somewhat different than that reported herein.

Based on the above Purdue University studies, it seems that large hay packages stored outside will keep quite well. However, variations in snow and rain undoubtedly will alter the results obtained under varying conditions.

a. Weathering Losses. The research data (Table 8.1) indicate weathering losses were not too severe and represented primarily the top four to six inches of the square (Hesston) stacks and the outside two to four inches of the round (Vermeer and Hawkbilt) bales. Percentagewise, the amount of weathering in the big package was comparable to that in the smaller round bales. Total digestible nutrient content of the weathered portion of the hay decreased about 17 points for grass hay and 23 points for alfalfa hay.

b. Moisture Content for Packaging. Along with the storage study, an experiment was conducted to determine the effect of hay moisture level on its keeping quality. Hay was baled and stacked at 28 and 34% moisture as well as at a "normal" range of 20–25%. The higher moisture hay heated and molded regardless of package type, whereas hay packaged at from 20 to 25% moisture kept quite well.

Based on this research, a loss of no more than approximately 10% of the total digested nutrients (TDN) due to weathering would justify outside storage of big packages as contrasted to building storage. However, note from Table 8.1 that much larger losses of TDN from outside storage of the smaller bound bale can be expected.

Large packages stored outside in a concentrated area should be so arranged as to allow good air movement around all sides. This means packages should be at least a foot apart. If the packages touch, water often gets trapped between and obvious increased quality deterioration will result.

Based on the evidence reported above, hay to be stored in big packages should contain less than 25% of moisture just as has been recommended for haymaking methods in use prior to the advent of big hay packages.

E. Hay Feeding Methods

1. Feeding Big Hay Packages

Two methods of pasture-feeding big hay packages to beef cows were investigated in the Purdue research. In both cases the Hesston, Vermeer, and Hawkbilt packages each were fed one at a time to groups of 20 cows. Under one system, the packages were placed on the ground and no attempt was made to restrict the animals' access to the hay (Fig. 8.1). With the other system, the hay was fed from three-sided racks on concrete (Fig. 8.2). Although neither method limited intake, additional hay was not made available until the cows had consumed all the previously fed hay they would eat.

Within each of the two types of feeding systems, hay requirement per cow per day (which reflects both hay consumed and hay wasted) was essentially the same for the three types of packages (Table 8.2). Between the two systems, however, the per-cow hay requirements were significantly different—being considerably higher when big packages were fed on the ground without racks than when fed on concrete with racks (average 31.3 versus 23.7 lb). The nearly one-third more hay required by the no-rack method can be viewed as the loss due to waste and trampling. The amount of hay refused or wasted with the rack-on-concrete method averaged only 3.7% of the total package by weight.

Just how much hay will be wasted when feeding large bales or stacks without racks may depend, to some extent, on environmental conditions. The ground was frozen only about one-half of the period during which the data were collected. Amount of waste should be less under dry or frozen conditions than when the soil is wet and muddy. It should also be borne in mind that the 1200 to 2000 pounds of hay were fed to the groups of 20 cows on pasture at one time. A larger number

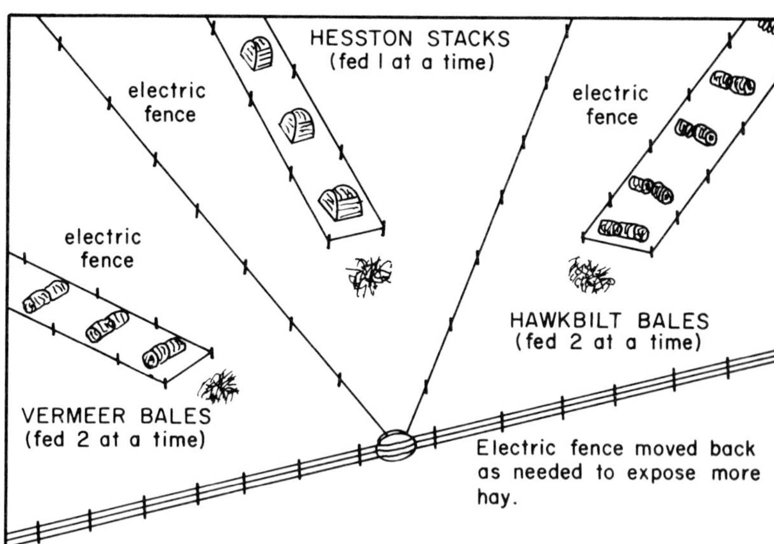

Fig. 8.1. Hay packages are placed on the ground for unlimited access by the animals to the hay.

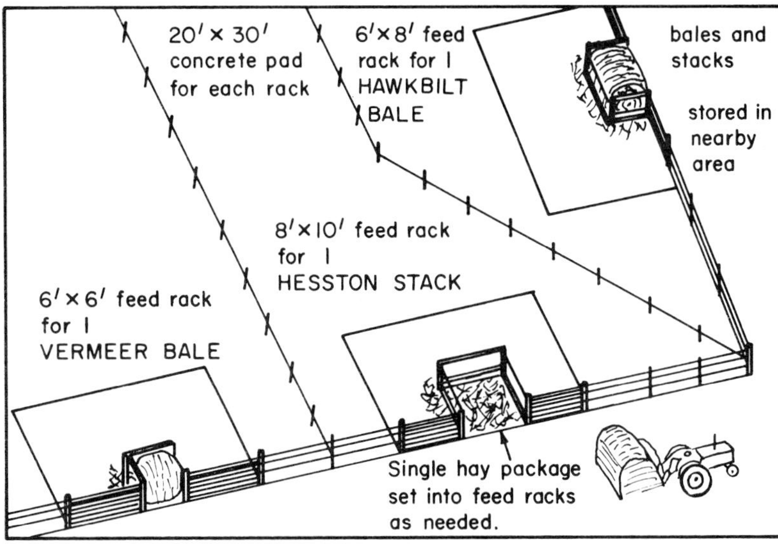

Fig. 8.2. Feeding of hay packages from racks on concrete.

of animals may have consumed this quantity much sooner and thus with less waste.

This experiment also showed that pasture-feeding big-package hay compares favorably with field grazing of conventional round bales. From previous research on the grazing of round bales left scattered in the field, it was found that hay requirements were 52 lb per cow-day with unrestricted field grazing, but only 33 lb when cows were restricted to a 2 or 3 weeks' supply at one time. These figures, compared to the 23.7- and 31.3-lb hay requirements with the rack and no-rack big-package feeding methods, respectively, indicate a slight advantage of large packages over conventional bales (in terms of hay requirement) if hay is to be pasture-fed.

One further observation from this study should be noted. When big packages are fed on pasture, considerable trampling occurs at each package site. This may destroy the grass and/or legume stand around the stack, especially if the ground is wet when the hay is fed. Reseeding in these areas may be necessary.

2. Limit Feeding Hay on Pasture

Under the no-rack pasture feeding system referred to above, each group of 20 cows was given access to from 1200 to 2000 lb of hay which represented one large package. This then represented more than one day's supply—probably 3 to 5 days (see Table 8.3).

Subsequent research was conducted to find out if limiting the amount of hay

TABLE 8.2

Hay Requirements and Losses for Pasture-Feeding Various Big Hay Packages with and without Racks[a]

Type of package	Number of packages fed	Average package weight (15% moisture) (lb)	Hay required per cow per day (%)	Hay required over rack feeding method (%)	Hay refused or wasted (%)[b]
On ground without hay racks					
Hesston 10 stack	12	1588	33.4	35.0	—
Vermeer 605 bale	16	1056	28.2	22.6	—
Hawkbilt 480 bale	25	615	32.3	38.6	—
Average			31.3	32.1	
On concrete with hay racks					
Hesston 10 stack	9	1579	24.7	—	2.34
Vermeer 605 bale	11	1315	23.0	—	4.40
Hawkbilt 480 bale	11	923	23.3	—	4.43
Average			23.7		3.72

[a] Purdue University data.

[b] Based on actual weigh-back of refused material in and around the hay racks on concrete; accurate weigh-back not obtained on the sod.

TABLE 8.3

Feeding Requirements and Losses for Four Levels of Hay Fed from Big Packages without Racks on Pasture, Compared to Rack Feeding on Pasture[a]

Type of feeding system	Approximate hay per cow per feeding (lb)	Hay required per cow per day (15% moisture) (lb)	Hay refused or wasted (%)[b]	Hay required over rack feeding method (%)
Rack feeding on pasture	—	19.2	4.7	—
No-rack feeding on pasture				
1 Day supply per feeding	20	21.7	10.9	12.2
2 Day supply per feeding	40	22.7	24.9	33.1
4 Day supply per feeding	80	26.5	31.0	44.9
8 Day supply per feeding	160	33.1	34.3	52.3

[a] Purdue University data.
[b] Based on actual weigh-back of refused material around the feeding site.

available to the animals would influence hay wastage, using a no-rack pasture system. Four groups of cows were allotted four different levels or days' supply of hay per feeding as follows:

Group 1—1-day supply, or 20 lb per cow
Group 2—2-day supply, or 40 lb per cow
Group 3—4-day supply, or 80 lb per cow
Group 4—8-day supply, or 160 lb per cow

A fifth, or control group, was also fed on pasture but from a three-sided rack.

The hay used in this experiment was packaged and fed as Vermeer 605 bales, averaging about 1200 lb per bale. Feedings were made to each group when the animals had consumed all the previously fed hay that they would eat.

With the rack feeding method (control group), hay refusal or wastage was found to be about 4.5%, typical of that reported in the previous research. Without racks, losses increased in proportion to the amount of hay available at each feeding—from 11% on a 1-day supply to more than one-third of the total for an 8-day supply.

3. Drylot Feeding of Hay

In addition to the pasture studies discussed above, research was conducted to compare several methods of feeding big-package hay in drylot. Again using Vermeer 605 bales, one method was to place the bales on a concrete pad without attempting to limit access to them. Another technique was to employ a movable hot wire which was moved back toward the stack a few inches each day. For purposes of comparison, limited hay feeding from a rack on concrete was studied, and also conventional square bales were fed on concrete without a rack at a predetermined rate of 17.7 lb of hay per cow per day.

TABLE 8.4

Hay Requirements and Losses for Unrestricted and Restricted Drylot Feeding of Vermeer Bales, Compared to Conventional Bales on Drylot and Vermeer Bales[a]

Type of hay package and feeding method	Winter season	Hay required per cow per day (15% moisture) (lb)	Hay refused or wasted (%)[b]	Hay required over ad lib rack feeding method (%)
Conventional square bales limit-fed on concrete without hay rack	Year 1	17.7	4.0	−23.2
Vermeer bales limit-fed on concrete and bare soil using a "hot" wire	Year 1	18.8	Nil	−18.3
	Year 2	19.8	Nil	− 3.4
Vermeer bales fed ad lib on concrete without hay rack	Year 1	28.0	—	+21.8
Vermeer bales fed ad lib on concrete with hay rack	Year 2	20.5	3.0	—
Vermeer bales limit-fed on concrete with hay rack	Year 2	18.7	2.7	− 9.1

[a] Purdue University data.
[b] Based on actual weigh-back of refused material on concrete in the feeding area; missing information indicates accurate weigh-back not possible.

Control of access to hay by either a rack of by "hot" wire proved effective in limiting animal access to hay and appeared to be workable approaches to limit feeding of big hay packages in drylot. The rack probably presents fewer management problems. Tall stacks do not lend themselves well to the "hot" wire because cattle tend to eat in under the stack and subsequently the top topples. However, round bales work quite well with the "hot" wire (Table 8.4).

Fig. 8.3. Cubing of hay facilitates its handling, but does not alter its nutritional value. (Photo courtesy of BEEF Magazine.)

Fig. 8.4. Baled hay can provide feed when snow covers winter pasture. (Photo by J. C. Allen and Son.)

Big hay packages should be restricted in some manner or else the cattle soon climb onto it and soil it (Figs. 8.3, 8.4, and 8.5).

II. HAYLAGE

Haylage is defined as any crop stored or ensiled at approximately 40–50% of moisture. The most common crops utilized in haylage making are the usually recognized forage crops such as alfalfa, and similar legumes, and forage grasses including timothy, sudan grass, and even immature small grains, such as oats or wheat. The most distinguishing characteristic of haylage to differentiate it from

Fig. 8.5. A portable hay feeding rack on wheels permits movement of winter feeding sites.

silage is its moisture content. Whereas most haylages are processed to contain from 40 to 50% of moisture, most silages contain from 60% up to 70 or 75% of moisture, for example, in the case of direct cut alfalfa.

A. Processing of Haylage

The processing of forage corps to attain a maximum of 40–50% of moisture is merely a wilting situation. In other words, such plants destined for haylage making are cut from their stems and allowed to lay on the ground until sufficient moisture has evaporated. Then the partially dried material is picked up, chopped to very small lengths, and ensiled. As in the case of hay making, the evaporation process will progress at a much more rapid rate in hot dry weather; crops lying in the swath will lose moisture more rapidly than those bunched into windrows; and mechanically crushing the stems at time of mowing will cause more rapid evaporation of water.

Unless the operator feels confident of his capability to estimate moisture of forage fairly accurately, it is recommended that a simple moisture tester be purchased. Such information as to sources of an inexpensive moisture tester, can be obtained from the local county agent. The principle of one such tester is that a sample of 100 gm of the product is placed in the machine and exposed to infrared heat. The machine is equipped with a small fan which blows the heated air through the product. In approximately 8 to 10 minutes the product should be sufficiently dry to weigh. Since the original weight was 100 gm, the final weight in grams will also represent the percentage of dry matter in the original product.

Subtracting dry matter percentage from 100 will give percentage moisture in the original product.

B. Ensiling Haylage

Ensiling haylage is critical because this much drier product (50% ± moisture and 50%± dry matter) will not pack and thus exclude air nearly as well as will more moist silage crops (60% moisture, or more). Although haylage can be stored successfully in pit and tower silos, it is perhaps a better policy to store it in a limited-oxygen sealed storage structure. It is important that the length of individual pieces of the chopped material for haylage making be quite small to aid in good packing and thus exclude as much air as possible.

1. Processing

The ensiling process of haylage fermentation is different than is that in the case of higher moisture forage silage making. Many assume the real advantage of haylage over the same crop made into silage is the difference in weight of water lost and thus a labor-saving effect. However, a most important nutritional difference between the two methods is the type of fermentation. In haylage making a much lower pH (4.5–4.8) is attained, compared to that for forage silage making (5.0 plus) because the carbohydrates from the wilted forage are more readily available for more favorable fermentation. Unwilted forage put into a silo will produce a much higher level of less palatable butyric acid.

2. Choice of Crops

Crops which are desirable for haylage making are many. However a few which are commonly used include alfalfa, clover, oats, wheat, barley, and several grasses. Haylage material removed from storage cannot be any better quality than that which was stored originally. Therefore, care should be taken to be sure a very high quality product at optimum time of harvest is made into haylage.

Alfalfa should be harvested at about the time the plants are approximately one-tenth in bloom. Fortunately, alfalfa and clover have a much longer period of optimal harvesting a high quality product. On the other hand, the cereal grains such as oats or wheat have a very short period for optimum harvest for haylage making—perhaps more in the range of from 2 to 3 days. Oats and wheat need to be cut at the "soft dough" stage of the grain, and this condition does not persist very long.

Considerable research has been conducted using oats as a hay or silage crop. In fact, the practice of harvesting the entire oats crop in the early dough stage and drying it in the swath for haymaking is a fairly well-established practice. In the 1950's, the University of Illinois published several years' research showing the increased feeding potential of harvesting the oats crop as full-moisture (nonwilted) silage. However, farmers have not followed the practice of ensiling full-moisture crops except corn and milo.

With the advent of more sophisticated limited oxygen storage, the system of wilting forages down to approximately 50% moisture and then ensiling them became plausible and farmers have begun to follow such a program. As indicated in the introduction of this section, alfalfa was the principal crop to be treated in this manner. Dairymen especially prefer alfalfa haylage for their cattle.

The oats crop is grown in many areas of the country as a grain. In fact, in South Dakota the acres planted to oats exceed those planted to all other feed grains except corn. Naturally the question should arise as how best to harvest, store, and feed the oat crop. South Dakota State University (Anonymous, 1977) has conducted research to identify the value of the oats crop stored as haylage. Stage of maturity of the grain varied from the milk stage in low areas of adequate moisture to late dough stage for drier areas. The forage was harvested with a windrower and subsequently two windrows were raked together prior to pickup. Alternate rows were baled for hay or chopped for forage. Moisture content at time of storage was 13% for hay and 52% for haylage; protein content was 16.6 and 16.3, respectively, on a dry matter basis. Growing steers (673 lb initially) were fed hay or haylage in amounts to ensure that feed would be available at all times. In addition, free-choice minerals were available. Note from Table 8.5 that daily dry matter intake was comparable on each oat hay and oat haylage. However, cattle fed the same amount of dry matter as haylage gained 28% faster (2.28 versus 1.78 lb per day) than cattle fed oat hay. Therefore, cattle fed oat haylage had a much more desirable efficiency of feed conversion than was the case for those fed oat hay.

TABLE 8.5

Oat Hay versus Haylage for Growing Cattle[a]

Item	Oat hay	Oat haylage
Number of animals	14	14
Days fed	134	129
Initial weight (lb)	673	681
Final weight (lb)	911	975
Daily gain (lb)	1.78	2.28
Difference (lb)		+.50
Daily forage (dry basis) (lb)	24.6	24.7
Feed per 100 lb gain (dry basis) (lb)	1386	1084
Difference (lb)		−302
Composition of forage		
Dry matter (%)	87.2	48.4
Protein (dry matter basis) (%)	16.6	16.3

[a] Anonymous (1977).

TABLE 8.6

Effects of Hay Quality and Supplementation on Gains of Stocker Calves (Average 2 Years)[a]

	Type of hay				
		Tall fescue supplementation			
Item	Orchardgrass–red clover	0	1.5 lb soy	4.0 lb corn	0.9 lb soy 3.1 lb corn
Number of animals	18	19	19	19	19
Total gain (lb)	188	42	106	116	162
Daily gain[b] (lb)	1.45	0.30	0.80	0.88	1.24
Hay consumption/day (lb)	14.9	12.6	12.5	10.1	10.9

[a] Hendrix *et al.* (1978).

[b] Average of 130 days each year. In second year, cattle on unsupplemented tall fescue failed to gain and, after 99 days, that treatment was discontinued.

III. INFLUENCE OF HAY QUALITY ON STOCKER CATTLE GAIN

Research conducted at Purdue University (Hendrix *et al.*, 1978) demonstrates the wide range of growth response by stocker cattle due to the quality of hay offered them. In two years of research, weanling steers (450–500 lb initial weight) were fed either second-cutting mixed orchardgrass and red clover hay or first-cutting tall fescue hay. The mixture of orchardgrass and red clover contained approximately 16% protein and was assessed to be 60–65% digestible; tall fescue hay contained just half as much protein (7–8%) and was estimated to be about 50% digestible. Since the tall fescue hay was estimated to be less nutritious, a series of supplementations were compared to determine what might be necessary to make it equal to the orchardgrass–red clover hay mixture (Table 8.6).

Based on the performance data, orchardgrass and red clover mixed hay was superior to tall fescue hay for growing stocker cattle. Supplementing the poorer quality fescue hay with either soybean meal or corn, or both, resulted in improved performance by cattle on the tall fescue hay treatments.

In beef operations where hay is the primary winter feed, especially for growing young cattle, major attention should be given to including nutritious plant mixtures as well as to proper harvest dates in order to obtain high-quality winter feed. In situations whereby good quality forage is not available, proper adjustment by supplementation should be made, especially in stocker calf feeding programs.

REFERENCES

Anonymous South Dakota State University, (1977). ''Cattle Feeders Day Report,'' p. 11. Brookings.

Hendrix, K. S., Lechtenberg, V. L., and Hawkins, K. G. (1978). ''Indiana Cattle Feeders Day,'' p. A-39. Purdue University, West Lafayette, Indiana.

Smith, W. H., Lechtenberg, V. L., Parsons, S. D., and Petritz, D. C. (1974). Suggestions for the Storage and Feeding of Big Package Hay. *Purdue Univ. Coop. Ext. Serv.* **ID-97.**

9

Silage and Crops for Silage

The production and use of silages is especially adaptable in the beef cattle feeding operation. Furthermore there are many economic advantages to the use of silage, such as: (a) nearly a doubling of the harvested nutrients per acre, in the case of the corn plant, (b) the maintenance of a high-quality product for long periods of time, (c) early harvest to clear the land for fall plowing or for second cropping, (d) high adaptability to automation, and (e) less wastage of feed, as compared to feeding dry stalks or hay.

I. SILAGE MAKING

The principle of silage making is to exclude as much oxygen as quickly as possible so that anaerobic fermentation may take place.

It is relatively easy to produce a highly acceptable feedstuff in the form of whole plant corn silage. Consequently, one is apt to run onto all sorts of recipes for producing the product. Undoubtedly, the ideal product has the following characteristics:

1. Appearance of plant—Silk is dead, leaves are green except for bottom four or five leaves, and husk is in the process of turning brown.
2. Kernel condition—Dented, glazed, fairly firm, and well past any liquid consistency internally when the kernel is crushed.
3. Moisture content of product—60–67% moisture, ideally.
4. Final product—pH, 3.8–4.1.
5. Length of cut (for beef cattle)—A fine cut makes for ideal fermentation and acceptability. A particle length of about ½-inch is excellent.
6. Color—A brown color is desirable. The addition of nitrogenous products (urea or ammonia products) tends to maintain a more intermediate greenish color.
7. Aroma—A general pleasant acidic aroma is desirable. A caramel-like

aroma is undesirable and indicates the material may have been too dry and that sufficient air seeped in to cause slightly elevated temperatures. Even though a caramel odor may seem desirable, it is not as good a feed for the beef animal. Also, any suggestion of a rancid butter smell suggests too much butyric acid was formed in the ensiling process; cattle will usually not accept this type of silage. The rancid odor is associated with harvesting the plant too early while it is quite immature.

Once material is ensiled and packed, the plant cells will continue to respire until all oxygen is utilized. If no more oxygen is made available, then the anaerobic activity takes over and acts on available carbohydrates to form several organic acids, particularly, lactic, and acetic (if the silage is too wet, butyric acid also will be formed. In addition, some ethyl alcohol is formed. Acid production proceeds until a pH of approximately 3.8–4.2 is obtained for the cereal silages (corn, oats); for legume silages a pH of 4.5–4.8 is the usual.

As was indicated above, the presence of available sugars is necessary for the desired fermentation to take place. Sometimes there may not be enough natural sugars in legumes for good silage making so it is necessary to either add a good carbohydrate source such as cracked shelled corn (200 lb per ton of silage) or else to wilt such crops down to about 50% moisture before ensiling (see Chapter 8 concerning haylage making).

A. Preservatives

Additions of preservatives are not necessary with whole plant corn or sorghum silage since such crops contain sufficient available carbohydrates from which the anaerobic bacteria can manufacture all the preserving acids that are needed. However, preservatives may be desirable for such silages as the unwilted legume crops. It was noted above that the addition of cracked shelled corn at the rate of 200 pounds per ton of chopped legume would provide sufficient carbohydrates for the desirable fermentation process. Furthermore, practically none of the corn energy supplied for this preservation process would be lost, but rather it would be available to the animals when they consumed it.

Other preservatives have been formulated for aiding in the preservation of the leguminous nonwilted crops made into silage. Although the use of any such preservative is not common, reference to them is worthwhile.

1. Acids

A Finnish scientist (A. I. Vertanen) developed the patented A.I.V. method which consists of a mixture of HCl and H_2SO_4. These caustic acids mixed together prevent the growth of the undesirable bacteria which in turn would cause mold. However, the expense of adding these products plus the danger of handling these acids make the A.I.V. method of preserving legume silages in this country impractical.

2. *Sodium Metabisulfite*

This is a dry powder which can be added to the green fresh chopped legume forage at the rate of 8 lb per ton of material. When this material is mixed with the moist green chopped material, SO_2 is released as a gas. Sulfur dioxide is a very efficient antibacterial preservative and is even used in small quantities to preserve human foods. The author has extensive experience in the use of sodium metabisulfite in preserving nonwilted legume silage, and several years' research has shown it to be an excellent preservative.

3. *Sulfur Dioxide Gas*

This product is supplied under pressure in steel cylinders and is administered by releasing such gas via a probe inserted into the packed green chopped material at two- to four-layer intervals as the silo is being filled. Furthermore, the operator performing this task must wear a gas mask each time he is in the silo treating the material. About 5 lb of SO_2 gas per ton of green chopped material is required. It was from the discovery of the protective effect from SO_2 gas that the more convenient dry sodium metabisulfite method was developed. However, as was indicated above, no form of special preservation of legume forage in the making of silage compares with the wilting to about 50% water and then ensiling as low-moisture silage or "haylage."

B. Special Enzymatic Substances

Several such products are on the market today. However, validated research results are still lacking.

C. Nutrient Additives

Nutrient additives for the ensiling process have attracted some interest and at least limited usage and therefore warrant discussion in this section.

Although high-quality corn silage is relatively easy to make and preserve, a great deal of research effort has been made to determine whether additives would improve its utilization by cattle.

1. *Urea*

Naturally, almost all silage feeding is restricted to ruminant animals which have the capability of utilizing limited amounts of urea as a source of crude protein. In other words, cattle through the aid of the bacteria inhabiting their rumen can synthesize high-quality amino acids—and subsequently true protein—from urea. This is discussed in detail elsewhere in the text.

The Ohio Agricultural Experiment Station first reported the effect of adding urea to corn silage at the time it was being ensiled. This apparently changes the fermentation pattern very little, if any. However, since it is present in the silage at the time of feeding, it perhaps might save a cattle feeder from having to provide a separate source of supplemental protein. Urea is generally added at 10 lb per ton of green chopped whole corn plant at the time of ensiling. The addition of 10 lb of urea is the equivalent of 28 lb of crude protein. Since whole corn plant chopped for ensiling contains about 38% dry matter (760 lb), which, in turn, is 7.5% crude protein (57 lb of protein) the addition of 28 lb of crude protein from urea gives a total of 85 lb of crude protein in the 760 lb of dry matter, in a ton of corn silage. This amounts to an 11% protein product, on a 100% dry matter basis; on an air-dry basis (10% water) the silage contains 10% protein; on an "as is" basis (38% dry matter) such treated corn silage would now contain 4.3% crude protein as compared to the untreated corn silage which would contain about 2.8% crude protein. In other words, fortification of one ton of 38% dry matter corn silage with 10 lb of urea will increase by about 50% the crude protein content, thus providing sufficient total protein for some cattle feeding situations.

2. *Ground Limestone*

It was pointed out previously that the principle of silage preservation was the formation of organic acids which prevented the proliferation of undesirable mold-forming bacteria. In the case of whole plant corn silage, acid formation proceeds until a status quo pH level of approximately 3.8–4.0 is reached. The Ohio Experiment Station researchers postulated that if some of the initial acids formed were neutralized with a basic product, such as ground limestone, the acid-forming process in the silo might be extended until ultimately the desired 3.8–4.0 pH was reached. This proved to be the case. Subsequent Ohio research with limestone in corn silage at the time of ensiling showed such silage was a slightly better feed than ordinary corn silage (Table 9.1). Wherever convenient, the addition of 10 lb of feed grade limestone per ton of green chopped whole corn plant at the time of ensiling, is a good practice to follow. In fact it is desirable to add 10 lb each of limestone and urea per ton.

3. *Ammonia*

Ammonia-containing materials which have been used to increase the nitrogen content of corn silage include aqueous ammonia solutions, aqueous ammonia–mineral solutions, aqueous ammonia–mineral–molasses solutions, anhydrous ammonia gas, and cold flow ammonia. Corn silage which has been treated with ammonia contains increased concentrations of lactic and acetic acids, higher pH values, and greater amounts of true protein. Such silage is more stable than untreated silage when exposed to air. The temperature of ammonia–minerals suspension-treated silage (Pro–Sil) remained unchanged 48 hours after aeration

TABLE 9.1

Effect of Limestone Additions to Whole Plant Corn Silage on Organic Acid Formation[a]

Silage treatment	Year	pH	Moisture (%)	Acetic acid[b] (%)	Lactic acid[b] (%)
Control	1958	3.8	69.3	1.24	6.38
1% limestone added		4.3	71.1	2.56	8.13
Control	1959	3.9	68.5	1.44	5.86
0.5% each of limestone and urea		4.5	66.6	1.83	10.42
Control	1960	—	65.5	1.41	7.58
0.5% each of limestone and urea		—	64.5	2.14	10.32

[a] Klosterman *et al.* (1963).
[b] Dry matter basis.

while untreated silages reached a high temperature within 24 hours. Data from trials to compare feedlot performance of cattle fed various nitrogen sources are compared in Table 9.2. These data indicate ammonia treatment of whole corn plant at time of ensiling is quite satisfactory. Silage made with Pro–Sil perhaps was the most consistent performer of the ammonia supplementation sources and came nearer duplicating the benefits of soybean mean supplementation. It should be observed from Table 9.2 that no supplementation of corn silage with nitrogen results in greatly depressed performance. Michigan State University researchers under the leadership of H. M. Henderson developed ammonia–mineral

TABLE 9.2

Corn Silage Supplemented with Various Nitrogen Sources (Summary of 27 Trials)

Comparison of N source	Daily gain (lb)	Feed per lb gain (lb)
1. No supplemental N	1.32	9.3
Soybean meal	2.95	5.3
Number of trials	4	4
2. Urea	2.16	7.4
Soybean meal	2.33	7.0
Number of trials	9	9
3. Pro–Sil silage	2.22	7.5
Soybean meal	2.38	7.7
Number of trials	11	11
4. Ammonia-treated silage	2.46	6.5
Pro–Sil silage	2.57	6.1
Soybean meal	2.55	6.3
Number of trials	3	3

suspension (A.M.S.) to be used as an additive to green chopped whole corn plant at the time it was being ensiled. The product is patented by the Michigan State University Research Foundation. Its composition is shown in Table 9.3. The product is normally added at a level sufficient to bring the protein content of whole plant corn silage up to 12.5–13% crude protein, on a dry matter equivalent basis. The author researched this product in two separate feeding trials with growing and finishing beef steers. One trial is summarized in Table 9.4 (Perry *et al.*, 1976). In this research, the regular corn silage tested 7.3% crude protein, on a 100% dry matter basis. The green chopped whole corn plant for the A.M.S comparison was treated with an average of 61.5 lb of A.M.S per ton of green chopped whole corn plant at time of ensiling. The treated corn silage analyzed 12.8% protein on a 100% dry matter basis, or 11.5% on an "air dry" basis (90% dry matter), which is more than sufficient protein for almost any age weanling calf and for which 57% of the protein is from a natural source. The results from the first trial shown in Table 9.4 speak for themselves and indicate the A.M.S-treated silage was an excellent total feed for growing and finishing beef cattle. The second year's work gave similar results, which verified the nutritional adequacy of this approach.

TABLE 9.3

Composition and Analysis of Ammonia–Mineral Suspension[a]

Ingredient/element	Percentage composition
Composition	
Anhydrous ammonia	14.540
Ammonium salts of phosphorus and sulfur	20.790
Calcium chloride brine	20.402
Sodium chloride brine	23.454
Minerals	13.314
Molasses	7.500
	100.000
Analysis	
Crude protein equivalent	84.991
Calcium	4.314
Phosphorus	1.500
Sulfur	1.338
Sodium	4.052
Chlorine	3.939
Zinc	0.057
Copper	0.013
Magnesium	0.049
Iodine	0.0015
Cobalt	0.0004

[a] Ammonia–mineral suspension (A.M.S.); patented product known as Pro-Sil.

TABLE 9.4

Comparative Utilization of Corn Silage with or without Ammonia-Mineral Suspension[a]

Item	Type of corn silage: Regular no additive	Treated with A.M.S.
Number of lots	I and III	II and IV
Number of cattle	48	48
Growth rate		
Initial weight (Oct. 31) (lb)	678	678
Weight Feb. 11 (lb)	897	912
Gain, 103 days (lb)	219	234
Daily gain (lb)	2.12	2.27[b]
Daily feed		
Corn silage (lb)	49	50
Supplement 64 (lb)	1.5	0
Total daily protein intake (lb)	2.32	2.34
Feed per 100 pound gain		
Corn silage (lb)	2297	2204
Supplement 64 (lb)	71	0
Dry feed per 100 pound gain (lb)	918	815
Feed cost per lb gain (cents)[c]	26.5	24.8

[a] From October 31, 1975 to February 11, 1976; 103 days.

[b] Cattle fed A.M.S. silage (Lots II and IV) gained significantly faster ($p < 0.05$) than cattle fed regular silage plus supplemental protein (Purdue Dry 64 Supplement).

[c] Based on the following feed costs: corn silage, $18.00 per ton; A.M.S.-treated corn silage, $22.50 per ton; shelled corn, $83.60 per ton ($2.34 per bu.); Purdue Dry 64 Supplement, $165.00 per ton.

4. *Microbial Inoculation*

The formation of desirable lactic acid from available carbohydrates, under anaerobic conditions, is dependent upon the presence of adequate lactobacilli to dominate the microbe species. It has been hypothesized that sometimes there are not adequate lactobacilli present to give a headstart to this type of fermentation. This theory is, therefore, the basis for many types of cultures to be added to whole corn plant at ensiling time. This area of research has become quite complex and various types of lactobacilli are being compared. Generally, treatment of corn silage with microbial additives has given variable responses ranging from no effect to significant improvements in nutrient preservation and animal performance (Owen, 1971). It appears the greatest advantage for a microbial additive may be with low-soluble carbohydrate forages such as legumes. The addition of microbes to whole plant corn silage has not been consistently beneficial.

5. *Formic Acid and Formaldehyde*

Lactic acid-producing bacteria are tolerant to acidic conditions whereas some of the less desirable bacteria are not as tolerant. Therefore, the addition of formic acid might permit the lactobacilli to operate but prevent competitive undesirable bacteria from getting into the act. Formaldehyde limits fermentation to near zero—particularly during the first few hours following ensiling—because of its bactericidal actions. Thus, it might limit the reduction of the loss of soluble sugars which occurs during routine fermentation. These products have been used in Germany for a number of years but have not attracted much interest in the United States. Furthermore, these products would appear to be limited in potential to silage crops other than corn or sorghum.

II. HARVESTING OF SILAGE

Harvesting of silages at the proper stage of maturity is desirable. The corn plant has a relatively long period at which high-quality whole plant silage may be made. However, the real peak season for this is related to the moisture level ranges of a maximum of 70% and a minimum of 55%. At moistures higher than 70%, there are a number of disadvantages. One is an excessive seeping of water containing dissolved nutrients from the silo—especially in a tower silo wherein a column fifty, sixty or even eighty feet tall is virtually squeezing the water from the silo contents. A second drawback is that increased quantities of butyric acid

Fig. 9.1. An earth bank silo that has been in use every year for the past 25 years.

are formed which give the silage a slimey appearance and an odor of rancid butter. Usually, when the whole corn plant contains 70% or more of water, the kernels are not very well developed and so the full nutrient potential of the corn plant has not yet been developed.

Harvesting whole corn plant for ensiling at moisture levels much lower than 55% percent is quite possible. However such a low-moisture product (cornlage) probably should be stored in a limited oxygen structure to prevent excessive losses due to molding and spoiling (Fig. 9.1). The author has demonstrated that whole corn plant harvested with less than 35% moisture and ensiled was equally as digestible as that harvested with 60–65% moisture, but several research papers have indicated apparent depressed palatability of very late cut silage as indicated by decreased intake by cattle.

III. SILAGE DIETS

A. Balancing a Corn Silage Diet

Corn silage alone is not a balanced diet. Therefore consideration must be given to what is essential to balance it.

Since it is a roughage, it is sufficiently low in digestible energy that it does not support maximum rate of growing and finishing beef cattle. Therefore, varying levels of grain may be added to it to meet the 10–90% of the digestible energy content of the diet. Increased levels of grain in a corn silage-type diet results in an increased rate of gain. However, increased grain feeding and decreased silage feeding may result in increased feed cost per pound gain under most price relationships.

1. Protein

Protein is basic. Since corn silage is deficient in protein, supplemental protein is needed. Some of the earlier research had indicated supplemental protein from plant sources, for example, soybean meal, linseed meal, or cottonseed meal, was superior to that from urea. However, more recent research data indicates that well-formulated high urea protein supplements may be used with corn silage quite satisfactorily. Much of the research indicates cattle fed high urea supplements with high silage diets gain less rapidly than when natural protein supplement is fed.

2. Vitamin A

Vitamin A should be included in a corn silage diet. Although corn contains a relatively high level of carotene, it has been demonstrated that beef cattle do not convert carotene to vitamin A very efficiently. Vitamin A may be provided either by feeding from 20,000 to 30,000 IU in the ration or by intramuscular injection of 1 million IU for each 100 days such cattle are on feed. The total injection for the entire feedlot life of the animal may be made in one initial injection. As long as finishing cattle have access to sunlight (from which vitamin D is activated within their bodies) there is no need for supplemental vitamins other than vitamin A.

3. *Minerals*

Supplemental minerals, such as calcium, phosphorus, salt, cobalt, and possibly zinc should be provided in the supplemental program. This may be provided in a portion of the diet fed, or minerals may be provided in a box on a free-choice basis. A good free-choice mineral consists of 2 parts dicalcium phosphate to 1 part trace mineralized salt.

Additional salt may be provided in a second compartment of the mineral box.

4. *Antibiotics and Hormones*

Beef cattle fed corn silage diets respond more to supplemental antibiotics, in terms of acceleration of growth, than do cattle fed all concentrate diets. When medium to high corn silage diets are fed, inclusion of chlortetracycline or oxytetracycline at a level of 75 mg per head, daily, has increased the rate of gain of cattle according to Purdue research.

The stimulating effect of effective hormones and hormonelike substances does not appear to be affected by type of diet, unless either energy or protein is lacking. Therefore the use of such hormones and hormonelike substances which are approved by the Food and Drug Administration of the United States government may be used to good advantage, as long as legal withdrawal periods and such other legal provisions are observed.

5. *Varying Energy Levels*

A number of universities have conducted research to determine optimum ratios of corn and corn silage in the finishing rations of beef cattle. In general, growing yearling steers gain nearly as rapidly on predominantly corn silage diets as they do on high concentrate diets. On the other hand, calves show more discrepancy in growth rate between high silage and high concentrate diets than do yearling cattle. Researchers at the University of Minnesota have summarized corn silage feeding trials from 17 university experiments in which various levels of corn and corn silage have been fed to steer calves (Goodrich *et al.*, 1974). From these data, feedlot performance at eight corn silage levels was predicted, using the regression equations developed from such feeding trial data. A modification follows (Table 9.5).

B. Costs of Corn Silage

It is difficult to arrive at a method of calculating the cost of corn silage which is universally acceptable. However, one method is to calculate the cost on the basis of amount of corn contained per ton of corn silage.

TABLE 9.5

Performance of Steer Calves Fed Various Levels of Corn Silage (Summary of 17 University Experiments)[a]

Item	Percentage of total ration dry matter as corn silage dry matter							
	10	20	30	40	50	60	70	80
Daily gain (lb)	2.52	2.49	2.43	2.36	2.28	2.17	2.05	1.91
Daily feed dry matter								
Corn grain (lb)	12.9	11.7	10.3	8.8	7.2	5.5	3.8	2.1
Corn silage (lb)	1.5	3.1	4.8	6.4	8.1	9.6	10.9	12.1
Supplement (lb)	0.9	0.9	0.9	0.9	0.9	0.9	0.9	0.9
Total (lb)	15.3	15.7	16.0	16.1	16.2	16.0	15.6	15.1
Daily corn silage at 35% dry matter (lb)	4.3	8.9	13.7	18.3	23.2	27.4	31.1	34.5
Feed/lb gain, dry matter								
Corn grain (lb)	5.12	4.70	4.24	3.73	3.16	2.53	1.85	1.10
Corn silage (lb)	0.61	1.26	1.97	2.74	3.55	4.42	5.33	6.31
Supplement (lb)	0.33	0.36	0.37	0.37	0.39	0.41	0.44	0.47
Total (lb)	6.06	6.32	6.58	6.84	7.10	7.36	7.62	7.88
Feed per 600-lb gain, air dry basis								
Corn grain (88% dry matter) (lb)	3490	3218	2890	2543	2154	1725	1261	750
Corn grain (88% dry matter) (bu)	62.3	57.4	51.6	45.4	38.5	30.8	22.5	13.4
Corn silage (35% dry matter) (lb)	1045	2160	3377	4697	6085	7577	9137	10817
Corn silage (35% dry matter) (tons)	0.5	1.1	1.7	2.3	3.0	3.8	4.6	5.4
Supplement (88% dry matter) (lb)	255	245	252	252	266	280	300	320
No. Days for 600-lb gain	238	241	247	254	263	276	293	314

[a] Ohio, South Dakota, Illinois, Michigan, Purdue.

The Statistical Reporting Service of the United States Department of Agriculture reports for the State of Indiana, for the five years (1969–1973) an average yield of 96.6 bushels of corn per acre and 15.1 tons of whole plant corn silage. This is equivalent to 6.4 bushels of corn per ton of corn silage. Bushels of corn per ton of silage (6.4) × value of corn in the field ($3.00 per bushel − 35¢ harvest, drying, and storage cost = $2.65) = 6.4 bushels × $2.65, or $16.95 for the corn contained in one ton of silage. Add the cost of harvesting and storing a ton of corn silage ($4.00) to come up with a total cost of $21.00 per ton for the corn silage.

Based on the above calculations, comparable prices for corn silage, at varying market prices for shelled corn would be as shown in Table 9.6. Table 9.7 presents the feed cost per 100 lb of gain for beef calves fed corn and corn silage at various prices. Table 9.8 indicates the feed budget requirement for feeding corn and corn silage to steer calves.

C. Feeding Corn Silage to Reduce Beef Costs

There are several beef feeding programs which may be based on corn silage. Some of these programs are described below.

TABLE 9.6

Cost per Ton of Corn Silage, Based on Varying Prices of Corn[a]

Market price of #2 corn per bushel ($)	Price of corn, per lb (¢)	Value of 1 ton of corn silage ($)	Value of 1 lb corn silage (¢)
1.00	1.78	8.16	0.41
1.25	2.23	9.76	0.49
1.50	2.68	11.36	0.57
1.75	3.12	12.96	0.65
2.00	3.57	14.56	0.73
2.25	4.01	16.16	0.81
2.50	4.46	17.76	0.89
2.75	4.91	19.36	0.97
3.00	5.35	20.96	1.04
3.25	5.80	22.56	1.13
3.50	6.25	24.16	1.21
3.75	6.69	25.76	1.29
4.00	7.14	27.36	1.37

[a] Based on market price of shelled corn minus 35¢ per bushel for harvesting, drying, and storing. Multiply this figure times bushels per ton of corn silage (6.4 in Indiana, 1969–1973) plus $4.00 per ton for harvesting and storage of silage.

TABLE 9.7

Feed Cost per 100 Pounds of Gain for Beef Calves Fed Various Proportions of Corn and Corn Silage with Corn at Various Prices

	Percentage of total ration dry matter as corn silage dry matter							
Price of corn	10	20	30	40	50	60	70	80
Daily gain (lb)	2.52	2.49	2.43	2.36	2.28	2.17	2.05	1.91
Days to gain 600 (lb)	238	241	247	254	263	276	293	314
Nonfeed cost/100-lb gain[a]	$8.00	8.05	8.14	8.25	8.39	8.65	8.87	9.20
$2.00 per bushel corn[b]								
Corn grain	20.77	19.14	17.21	15.14	12.82	10.28	7.50	4.46
Corn silage	1.27	2.63	4.11	5.72	7.40	9.22	11.12	13.16
Supplement	3.13	3.42	3.52	3.52	3.71	3.90	4.18	4.47
Total feed cost	25.18	25.19	24.83	24.38	23.93	23.40	22.80	22.09
Total cost[a]	33.18	33.24	32.98	32.63	32.31	32.05	31.67	31.28
$2.50 per bushel corn[b]								
Corn grain	25.96	23.91	21.50	18.91	16.01	12.84	9.37	5.58
Corn silage	1.55	3.20	5.01	6.97	9.02	11.24	13.55	16.05
Supplement	3.13	3.42	3.52	3.52	3.71	3.90	4.18	4.47
Total feed cost	30.64	30.53	30.03	29.40	23.93	27.98	27.10	26.10
Total cost[a]	38.64	38.58	38.17	37.65	32.32	36.60	35.97	35.30
$3.00 per bushel corn[a]								
Corn grain	31.13	28.68	25.79	22.68	19.21	15.41	11.24	6.69
Corn silage	1.81	3.74	5.86	8.14	10.54	13.13	15.84	18.75
Supplement	3.13	3.42	3.52	3.52	3.71	3.90	4.18	4.97
Total feed cost	36.07	35.84	35.17	34.34	33.46	32.44	31.26	29.91
Total cost[a]	44.07	43.89	43.31	42.59	41.85	41.09	40.13	39.11
$3.50 per bushel corn[b]								
Corn grain	36.37	33.50	30.12	26.50	22.44	18.00	13.13	7.81
Corn silage	2.11	4.36	6.81	9.47	12.27	15.28	18.43	21.82
Supplement	3.13	3.42	3.52	3.52	3.71	3.90	4.18	4.97
Total feed cost	41.60	41.28	40.45	39.49	38.42	37.18	35.74	34.10
Total cost[a]	49.60	49.33	48.59	47.74	46.81	45.83	44.61	43.30

[a] Nonfeed costs represent interest and overhead and is calculated on the basis of $4.23/cwt. gain plus 9.5¢ cost for yardage per head per day. [b] Costs are based on feed per pound gain from Table 9.5 times the price of corn or silage per pound from Table 9.6. Supplement cost is $190.00 per ton, or 9.5¢ per lb.

TABLE 9.8

Feed Budget Requirement for Feeding Various Levels of Corn and Corn Silage to Steer Calves (to Gain 600 lb)

	Percentage of total ration dry matter as corn silage dry matter							
	10	20	30	40	50	60	70	80
Daily gain (lb)	2.52	2.49	2.43	2.36	2.28	2.17	2.05	1.91
Days to gain 600 lb	238	241	247	254	263	276	293	314
Feed per 100-lb gain (lb)								
Corn grain	582	536	482	424	359	288	210	125
Corn silage	174	360	563	783	1014	1263	1523	1803
Supplement	37.5	40.8	42.0	42.0	44.3	46.7	50	52.3
Feed for 50 calves to gain 600 lb								
Corn grain (bushels)	3118	2871	2582	2271	1923	1542	1125	670
Corn silage (tons)	26	54	85	118	152	190	229	271
Supplement (tons)	5.6	6.1	6.3	6.3	6.6	7.0	7.5	7.8
Feed for 100 calves to gain 600 lb								
Corn grain (bushels)	6236	5742	5164	2542	3846	3084	2250	1340
Corn silage (tons)	52	108	170	236	304	380	458	542
Supplement (tons)	11.2	12.2	12.6	12.6	13.2	14.0	15.0	15.6

1. A Growing Winter Ration

This program is primarily for calves that will be turned to pasture the following spring, and can be built around a predominantly corn silage ration. Corn silage in itself is not a complete ration and must be supplemented in order to provide a balanced ration. Research at Purdue University has shown that a good growing ration for calves weighing approximately 500 lb at the start may consist of 30 lb of corn silage, 1.5 lb ground shelled corn, and 2 lb of a 32% protein supplement such as Purdue Supplement A or 1 lb of a 64% high urea supplement (Fig 9.2).

Calves will gain about 1.67 lb per day during the winter period and will not carry too much flesh to go on to pasture the next spring if a full feed of grain will be fed on grass. It will cost from 35 to 40¢ per pound of gain to grow cattle on such a ration, compared to a growing ration based on almost any other type of feed program, which will cost nearly 50% more.

Fig. 9.2. Wherever it is available, corn silage can help lower the cost of many cattle finishing programs. (Photo courtesy of BEEF Magazine.)

2. A Fattening Ration For Yearlings

This program can be built around an almost entirely corn silage ration. Let us suppose you bought calves one year ago and carried them through the winter in order to gain 1.67 lb per day and through the pasture season to gain 1.33 lb per day. Or, you may have bought yearling cattle directly from range country. They should have a good-sized frame and sufficient capacity to do well on a predominantly corn silage ration in the feedlot.

Research shows that yearling steers weighing from 600 to 750 lb will eat from 50 to 60 lb of corn silage per day plus a balancer of 1.5 lb of shelled corn and 2.0 lb of Purdue Supplement A or 1 lb of a 64% high urea supplement. If you started with reasonably sized cattle they will gain from 2.25 to 2.50 lb per day and at the end of the six months in the feedlot will be ready for market.

These cattle will produce choice carcasses if you started with choice feeder calves. The fat will be white and hard (the kind of beef cuts the consumer likes to buy in the supermarket).

2. Fattening Calves

The production of more economical gains on a ration of corn silage in order for calves to gain 2 lb or more and to get fat within 7–7½ months should include some corn. Experience has shown it is a good idea to almost divide your calf silage fattening program into two parts. During the first 3–4 months corn silage and corn in an average ratio of 2 lb corn silage to every 1 lb of shelled corn should be fed. You will also need to feed 2 lb of a balancing 32% supplement such as Purdue Supplement A, or 1 lb of a 64% high urea supplement.

A calf starting at a weight of 500 lb should eat an average of 20 lb of corn silage, 10 lb of shelled corn and 2 lb of Purdue Supplement A per day during the first part of the feeding period. This is the period when you really produce economical gains. During the latter 3–4 months of the calf fattening program built around corn silage, gradually change the ratio of corn silage to shelled corn to about 1:1. In other words, lower the corn silage to about 15 lb per day and raise the shelled corn to about 15 or 16 lb per day and hold the level of supplement to about 2 lb per day (if you are feeding a 32% supplement). There have been no damaging effects, such as overeating or going off feed, which may be encountered in a predominantly grain fattening program. Such finished cattle will hang a choice carcass, carrying hard white fat well distributed within the muscling.

D. Corn Silage versus Sorghum Silage

Because corn and sorghum may compete for crop acres, since only one may be grown at a time comparative evaluation of the two crops for ensiling in the whole plant form is desirable. However, there are basically two types of sorghum, namely the forage and grain types. An evaluation of the two types of sorghum

TABLE 9.9

Corn Silage versus Forage Sorghum Silage versus Grain Sorghum Silage in a Growing Ration for Beef Calves[a]

	Treatment		
	Corn silage	Forage sorghum silage	Grain sorghum silage
Number of lots	2	2	2
Number of cattle	12	12	12
Daily gain (164 days)	1.67	1.08	1.63
Daily feed			
Silage (lb)	31	37	34
Gr. corn (lb)	1.8	1.8	1.8
Protein suppl. (32%) (lb)	2.0	2.0	2.0
Air dry feed per pound gain (lb)	8.7	12.6	10.0

[a] Initial weight, 575 lb.

silage against corn silage was conducted at Purdue University, in growing beef cattle (Table 9.9).

Cattle gained equally well on corn silage and grain sorghum silage, but less rapidly on forage sorghum silage. However, exceptionally large tonnages of material may be harvested from forage sorghums as contrasted to corn silage or grain sorghum silage, and corn silage will yield more than double the tonnage one can harvest from grain sorghum silage.

The improvement of sorghum grain by feeding it as moist and ensiled is more dramatic than that for corn (Table 9.10). In the seven trials reported from Texas, a range in improvement of feed conversion of 11–26% was found, but no real effect on rate of gain was observed.

E. Cereal Silages

A number of cereal crops have been harvested in the late milk, early dough stage, and ensiled for cattle feed. Naturally the tonnage yield per acre is much less than for corn silage, but this method of harvest presents some advantageous alternatives. One of the real problems encountered in harvesting the small grain crops for ensiling is that the number of days in which optimum quality is maintained is very short—perhaps only two to four days. Too early harvest results in less than maximum starch deposit in the seeds and too late harvest results in excessive lignification, and thus greatly reduced digestibility.

Research at Purdue has shown that even though the cereal crops may be ensiled and fed to growing beef cattle quite satisfactorily, their feeding value is quite inferior to whole plant corn silage (Table 9.11).

TABLE 9.10

High-Moisture Sorghum Grain for Beef Cattle (Summary of Seven Experiments)[a]

Experiment number	Dry grain	Moist grain: Early harvested	Moist grain: Reconstituted	Advantage for moist milo
Daily gain (lb)				
1	2.28	2.42	—	0.13 lb
2	1.60	1.65	—	0.05 lb
3	1.98	1.96	—	−0.02 lb
4	2.13	2.09	—	−0.04 lb
5	2.28	—	2.09	−0.19 lb
6	2.24	—	2.42	0.18 lb
7	2.79	—	2.79	0.00 lb
Grain dry matter per pound gain (lb)				
1	6.57	5.38	—	18%
2	2.30	1.70	—	26%
3	7.32	5.50	—	25%
4	5.30	4.21	—	21%
5	6.57	—	5.86	15%
6	8.09	—	6.90	16%
7	6.52	—	5.80	11%

[a] Riggs and McGinty (1970).

TABLE 9.11

Small Grain Silages Compared with Corn Silage for Growing Beef Cattle (Purdue University Research)[a,b]

	Corn silage	Barley silage	Oat silage	Wheat silage
Number of cattle	9	10	10	10
Daily gain (154 days) (lb)	1.94	1.52	1.11	1.30
Daily feed				
Silage (lb)	32	35	30	30
Corn (lb)	2.6	2.6	2.6	2.6
Protein suppl. (32%) (lb)	1.0	1.0	1.0	1.0

[a] Initial weight, 500 lb.
[b] Perry *et al.* (1954).

TABLE 9.12

Performance of Growing Steers Fed Wheat, Barley, Oats, or Corn Silage[a]

Silage	Average daily gain (lb)	Daily feed intake (lb)[b]	Feed/lb gain (lb)[b]	Silage dry matter (%)	Silage CP (%)[b]	Forage harvest date
Trial 1 (122 days)						
Corn	1.92	15.3	8.1	33.4	8.7	—
Blue Boy wheat-head[c]	1.53	14.4	9.5	35.7	13.6	June 11
Parker wheat-head	1.43	13.4	9.5	36.6	13.2	June 9
Trial 2 (100 days)						
Corn	2.48	18.6	7.5	40.9	8.3	—
Barley	2.28	17.4	7.7	32.9	9.5	June 1
Arthur wheat	2.09	17.4	8.5	36.9	7.5	June 8
Parker wheat	1.54	14.7	9.6	34.6	7.4	June 6
Parker wheat-head	1.75	14.8	8.1	41.2	9.9	June 7
Corn + Parker wheat-head	2.22	18.0	8.1	—	—	—
Trial 3 (90 days)						
Corn	2.83	19.3	6.8	34.8	9.1	—
Barley	2.60	17.5	6.9	34.8	12.0	May 29
Arthur wheat	1.91	15.0	7.9	32.2	11.2	June 5
Blue Boy II wheat	1.56	14.3	9.5	36.9	11.2	June 5
Eagle wheat	1.91	16.3	8.5	34.3	9.6	June 5
Trial 4 (87 days)						
Corn	2.45	18.8	7.7	37.8	7.8	Aug. 29
Barley	2.70	18.9	7.0	35.3	11.5	June 4
Arthur wheat	2.32	18.3	7.9	36.9	10.8	June 13
Eagle wheat	2.11	16.9	8.1	37.8	8.4	June 14
Eagle wheat-milk stage	1.97	16.0	8.2	33.8	9.9	June 5
Trial 5 (89 days)						
Corn	2.52	19.1	7.6	37.2	8.3	Aug. 20
Barley	2.33	19.5	8.4	35.7	9.0	June 2
Arthur wheat	2.06	18.7	9.1	39.2	11.2	June 7
Sage wheat	1.96	19.2	9.9	41.2	8.3	June 9
Trio oats	1.09	14.6	13.5	30.1	12.6	June 17
Lodi oats	1.02	14.7	14.5	31.2	10.1	July 2

[a] Bolsen and Oltjen (1977).
[b] Calculated on 100% dry matter basis.
[c] Upper one-half of plant.

The cereal crops (wheat, barley, and oats) can, however, provide alternate silage sources for beef cattle and may replace at least some of the sorghum and corn silage portions of the diet. Kansas State University (Bolsen and Oltjen, 1977) reported the data from five growing trials with beef cattle in which various cereal silages were compared to corn silage. In the five comparisons, steers were fed according to appetite, twice daily, a diet of 86% of the respective silage and 14% balancing supplement calculated on a 100% dry matter basis. Comparative performances are summarized in Table 9.12. The authors assessed a comparative feeding value for barley, wheat, and oats silage. Using a base of 100% for corn silage, the following relative values were assessed: barley silage, dough stage, 98

± 6%; wheat silage, dough stage, 81 ± 15%; oats silage, dough stage, 48 ± 2%. These evaluations are based on four barley silages in four trials, twelve wheat silages in five trials, and two oats silages in one trial from rate and efficiency of gain data.

Based on the Kansas research, their recommendations concerning utilization of cereal crops for silage were:

1. Harvesting and feeding cereals as silage produces more beef per acre than the same crop harvested as grain.
2. For the best silage, ensile cereals at 60–65% moisture.
3. As cereals mature from boot to dough stages, silage yield increases but silage crude protein decreases.
4. Harvest cereals in the mid-dough stage of maturity for maximum TDN and beef production per acre.
5. Winter wheat, winter barley, and spring oats have similar dough-stage silage yields—6–9 tons per acre.
6. Cereal silages are usually about 2 percentage units higher in protein than corn or sorghum silages.

When fed to growing cattle in high silage rations:

7. Barley and corn silages are about equal in feeding value.
8. Wheat silages support about 80% the level of performance of corn silage.
9. The higher the grain content of wheat, barley, and oat silages, the higher the silage feeding value.

When fed to finishing cattle in high grain rations:

10. Wheat and corn silages support similar feedlot performance.

F. Corn Stover Silage

When corn is picker-shelled with a combine and hauled from the field, nearly one-half of the cattle feeding value remains in the field as husk, stalk, cob, and some shelled corn. Research has been conducted which has demonstrated that green corn stover silage is an excellent feed for beef cattle, and more nutritious than silage which has been made from the more mature stover (Table 9.13). In this comparison, the "green" stover silage was harvested on September 15, at a time when high moisture (32% water) ear corn was being harvested for ensiling. This was at the same stage of maturity one would harvest whole plant corn for ensiling. The "mature" stover was harvested for ensiling October 25, when the stalks were brown and quite dry and the ear corn contained 20% water. Such silage required added water.

G. Dehydrated and Pelleted Corn Silage

Although possibly because of the high price of energy there would be little interest in the potential of dehydrating corn silage, research has been conducted

TABLE 9.13

Mature and Green Corn Stover Silage versus Regular Whole Plant Corn Silage for Growing Beef Calves[a]

	Mature stover silage	Green stover silage	Green stover silage plus corn	Whole plant corn silage
Number of calves	20	20	20	20
Daily gain (128 days) (lb)	1.12	1.37	2.32	2.20
Daily feed				
Silage (lb)	32	35	21	37
Ground ear corn (lb)	—	—	11[b]	—
Protein suppl. (32%) (lb)	3.5	3.5	3.5	3.5

[a] Initial weight, 500 lb.
[b] Level of corn calculated to be present in whole plant corn silage consumed by the fourth treatment group.

to learn whether its nutritional value for beef cattle was altered by the dehydrating and pelleting process. The data in Table 9.14 demonstrates cattle will gain more rapidly on dehydrated and pelleted whole plant corn silage, but probably this is due only to the increased dry matter consumption capability of the dehydrated product. The efficiency of air dry feed conversion between regular whole plant corn silage and dehydrated and pelleted whole plant corn silage was nearly identical.

TABLE 9.14

Fresh Whole Plant Corn Silage versus Dehydrated and Pelleted Corn Silage for Growing Beef Calves[a,b]

	Fresh corn silage	Dehydrated pelleted corn silage
Number of lots	2	2
Number of cattle	12	12
Daily gain (140 days) (lb)	1.77	2.13
Daily feed		
Silage (lb)	29.4	
Pellets (lb)		13.7
Roughage converted to 11% moisture (lb)	10.6	13.7
Protein suppl. (32%) (lb)	3.0	3.0
Feed per lb gain (air dry)	7.7	7.8

[a] Purdue Agricultural Experiment Station, unpublished.
[b] Initial weight, 440 lb.

H. Opaque-2 Corn Silage

The improvement in the nutritional value of opaque-2 shelled corn for rats and swine due to the increase in percentage of lysine and tryptophan brought speculation as to its advantage in the ruminant animal. Furthermore, possibly such whole plant corn silage might have increased feeding value for cattle. Two metabolism trials and two feeding trials were conducted at this station (Thomas *et al.*, 1975) to evaluate this newer type corn silage. Metabolism studies indicated no improvement in utilization of opaque-2 whole plant corn silage by beef cattle. Furthermore, in feedlot comparisons, cattle fed regular whole plant corn silage gained significantly faster than comparable cattle fed opaque-2 whole plant corn silage.

I. Blighted and Drought-Damaged Silage

Infection of corn crops with southern corn leaf blight (*Helminthosporium maydis*) reduces yield and may affect quality of derived feedstuffs. The plant is killed before normal maturity, the ears are not filled, kernels are small, grain test weight is decreased, and ears may become infected with fungi. Silage making of the entire plant is probably the only hope for salvage of the crop—in fact it is recommended that silage making be instituted just as quickly as possible after the infection has been identified since the damaging effect of blight increases quite rapidly. The material should be chopped finer than normal and additional water may be added at time of ensiling since one of the characteristic damages of the blighting effect is the drying out of the plant.

Ensiling of the blight-damaged whole corn plant will prevent further deterioration and thus represents a satisfactory salvage procedure for such situations. Digestibility trials with 650-pound beef steers indicated no significant difference between blighted and normal corn silage for beef cattle feeding (Caldwell and Perry, 1972).

Drought-damaged corn plants can be salvaged in like manner for making maximum possible use of such products.

J. Brown Midrib Corn Silage

Corn plants containing the brown midrib mutant genes *bm3* contain about one-third less lignin (2.3 versus 3.5%, on a dry matter basis) than "normal" corn plants. Since the amount of lignin present in plants has a direct bearing on digestibility of such plants by ruminants, this lower lignin plant has spurred interest in its potential as silage. In research conducted at Purdue (Colenbrander *et al.*, 1977), it was demonstrated when corn silages were fed with only balancing protein supplement—and no additional shelled corn—steers fed brown midrib whole plant corn silage gained slightly faster (Table 9.15) and required about

TABLE 9.15

Performance of Steers Fed Normal or Brown Midrib Corn Silage[a,b]

	Silage treatment			
	No added corn		Full-feed corn	
	Normal silage	Brown midrib silage	Normal silage	Brown midrib silage
Number of cattle	12	12	12	12
Daily gain (112 days) (lb)	1.98	2.16	2.32	2.29
Daily feed				
Silage (lb)	26.7	26.6	13.6	13.6
Cracked corn (lb)	—	—	7.5	7.0
Supplement (lb)	1.9	1.9	1.9	1.9
Feed per 100-lb gain				
Silage (lb)	1410	1229	586	594
Cracked corn (lb)	—	—	322	304
Supplement (lb)	99	87	81	92
Dry feed per 100 pounds gain (lb)	614	536	570	558

[a] Keith *et al.* (1979).
[b] Initial weight, 533 lb.

13% less dry matter unit gain (536 versus 614 lb per 100 lb gain) indicating a superior feeding value for the brown midrib silage. However, when additional shelled corn was fed, there was no advantage for one silage over the other. In addition, yields of dry matter produced per acre have been less for the brown midrib corn plant.

K. Miscellaneous Products for Silage

Apple pomace, consisting of hull, core, peeling, and seeds, contains about 20% dry matter as contrasted to nearly double that (35–40%) in whole plant corn silage. However, the TDN content for beef cattle is listed as 15% (Anonymous, 1971), as contrasted to about 21% TDN for whole plant corn silage. When both are related to the same dry matter basis, the TDN content per unit dry matter is quite comparable. Generally it is advisable to mix an absorbent dry material with the apple pomace at time of ensiling in order to prevent excessive nutrient loss from apple pomace. Rice hulls have often been used at a ration of about 12 part pomace to 1 part dry rice hulls. It is important that the spraying history of the apples be known in order to avoid serious contamination of cattle carcasses from spray residues. Many such spray residues are fat soluble and will be deposited in the carcass fat.

Pea vines from pea canneries are often available in sufficient quantities that they may be ensiled for preservation as cattle feed. The dry matter content may run as high as 24–28% and on a dry matter basis the crude fiber is 31% and protein 13.1%. Because of its lower energy content, it is worth only about 75% the value of whole plant corn silage, but the addition of from three to four pounds of corn per day to the beef steers diet makes the combination similar in value to whole plant corn silage (Fig. 9.3).

Fig. 9.3. Cull potatoes (top) being stacked (bottom) in a pit silo in northern Wisconsin.

TABLE 9.16

Effect of Stage of Maturity of Harvest of Corn Silage on the Yield of Digestible Dry Matter

Date of silage harvest	Days after planting	Average weight of 4 steers (lb)	Daily consumption dry matter per 100 lb body weight (lb)	Moisture content (%)	Yield of dry matter per acre			
					Total plant (tons)	% Digested	Ear (%)	Digested dry matter (tons)
Experiment 1								
Aug. 24	101	664	1.1	80.9	5.1	71.1	47.0	3.6
Sept. 7	115	704	1.2	75.9	5.7	74.8	53.9	4.3
Sept. 21	129	717	1.2	71.7	6.4	72.4	64.6	4.6
Oct. 5	143	719	1.2	66.7	7.0	67.3[a]	57.0	4.7
Oct. 19	157	783	1.1	53.6	6.5	70.4	57.8	4.6
Nov. 2	171	814	1.3	47.5	7.2	72.4	53.7	5.2
Nov. 30	191	838	1.1	39.2	6.4	72.0	55.7	4.6
Dec. 28	227	840	1.1	38.9	3.9	71.3	81.9	2.8
Jan. 25	255	880	1.2	25.0	3.5	72.2	93.3	2.5
Feb. 22	283	904	1.2	19.7	3.5	68.4	92.5	2.4
Experiment 2								
Sept. 6	125	482	1.5(a)[b]	67.6	4.2	79.4	62.3	3.3
Sept. 20	139	495	1.5(a)	71.8	3.3	75.8	65.1	2.5
Oct. 4	153	512	1.7(a)	61.9	3.4	76.2	59.7	2.6
Oct. 18	167	528	1.2(b)	60.8	2.9	74.7	63.1	2.2
Nov. 1	181	601	1.1(c)	42.8	3.1	79.8	55.5	2.5
Nov. 15	195	673	1.1(c)	46.4	2.8	77.2	69.1	2.2
Dec. 27	237	664	0.9(d)	33.0	2.1	75.2	73.3	1.6
Feb. 14	286	689	1.2(b)	—	—	77.3	—	—
May 24	385	810	0.9(d)	15.1	1.3	78.7	—	1.0

[a] Significantly different ($p < 0.05$) than the other figures.

[b] Figures with letters in parentheses differ from those with the next lowest value, significantly ($p < 0.01$).

Sugar beet tops are often available in large quantities, as a by-product of the sugar beet industry. Although many forms of utilization are possible, one of the best methods of storage is by ensiling. The value will vary widely primarily due to the variance in moisture content. This product should not represent more than one-half of the total roughage since it has a laxative effect on cattle. They blend well with a dry roughage such as hay—preferably a legume hay.

IV. VALUE OF CORN SILAGE MADE FROM PLANTS AT VARIOUS STAGES OF MATURITY

Earlier in this chapter it was pointed out there is an "ideal" stage for the harvest of the corn plant for making whole plant corn silage. But what about corn which gets a bit more mature before you can harvest it? Also, could you go out with your field chopper later in the fall and refill with silage after you had fed out the first cutting from one or more structures? Two years research were conducted at Purdue University to study the nutritive value of whole plant corn silage which had been ensiled at various times during season. The ensiled product was then evaluated by digestion trial technique utilizing beef steers to demonstrate how well it was utilized by cattle. The silage was made from an Indiana corn field seeded at the rate of 24,000 seeds per acre during the first part of May. In one of the two years, a drought caused lowered yields of dry matter. However, for the two years, maximum yields per acre were: dry matter, 7.9 and 4.2 tons (equivalent to 22.5 and 12 tons of 65% moisture silage); ear corn, 4.6 and 2.7 tons (156 and 91 bushels No. 2). Harvests were made starting in the early dough stage (101 days postplanting) and extending through winter for as long as 385 days postplanting. Harvests were made at biweekly or monthly intervals. Since some of the later harvests contained quite low levels of moisture, it was necessary either to add water or to store the product in limited oxygen access storage to prevent excessive mold formation. The latter storage technique was utilized.

From Table 9.16 it is obvious the dry matter that can be harvested from an acre for corn silage declines with time, due primarily to wind losses. However, the digestibility of the dry matter harvested remained quite high for all of the harvest dates. Palatability—or acceptability—was not affected by date of harvest in Experiment 1, but acceptability declined with later harvested silage in the second experiment. Generally, it is concluded that the quite mature corn plant does not make as palatable silage as does more immature plants.

REFERENCES

Anonymous (1971). "Nutritional Data of United States and Canadian Feeds." Natl. Acad. Sci., Washington, D.C.

Bolsen, K., and Oltjen, J. (1977). "Cattle Day Report," pp. 43 and 48. Kansas State University, Manhattan.

Caldwell, D. M., and Perry, T. W. (1972. Nutritional value of blighted corn as silage or as grain. *J. Dairy Sci.* **55,** 1302.

Colenbrander, V. F., Beeson, W. M., Perry, T. W., and Keith, E. A. (1977). Performances of feedlot cattle fed brown midrib corn silage. *Purdue Annu. Cattle Feeders Day Proc.* p. 5.

Goodrich, B. D., Crawford, D. W., Thoney, M. L., and Meiske, J. C. (1974). "Influence of Corn Silage Level on the Performance and Economic Returns of Steer Calves," Res. Rep. B-195. Dept. Anim. Sci., University of Minnesota, Minneapolis.

Keith, E. A., Colenbrander, V. F., Perry, T. W., and Bauman, L. F. (1979). Effect of feeding brown midrib corn silage with three levels of corn to feed lot steers. *Purdue Annu. Cattle Feeders Day Proc.,* p. 73.

Klosterman, E. G. *et al.* (1963). Corn silage for beef cattle. *Ohio, Agric. Exp. Stn., Bull.* **934.**

Owen, F. N. (1971). Levels of silage for beef cattle. *Proc. Int. Silage Res. Conf., 1st, 1971.* p. 79.

Perry, T. W., Beeson, W. M., and Mottler, M. T. (1954). Cereal silages for beef cattle. *Ind. Agric. Exp. Sta. Bull.* **42.**

Perry, T. W., Peterson, R. C., and Beeson, W. M. (1976). Comparative utilization of corn silage stored with or without an ammonia-mineral-suspension. *Indiana, Agric. Exp. Stn., Bull.* **132.**

Riggs, J. K., and McGinty, D. D. (1970). Early harvested and reconstituted sorghum grain for cattle. *J. Anim. Sci.* **31,** 991.

Thomas, V. W., Beeson, W. M., and Perry, T. W. (1975). Effect of normal vs. opaque-2 roasted normal corn and normal vs. opaque-2 corn silage for finishing beef cattle. *J. Anim. Sci.* **41,** 641.

10

Concentrates for Beef Cattle

The beef animal was endowed with the capacity to utilize tremendous quantities of roughage through the help of the billions of microscopic plants and animals which inhabit their rumen. In fact, cattle subsisted, grew, and reproduced on nothing but grass and similar materials as their only source of organic nutrients for centuries. However, as man domesticated cattle he introduced concentrated energy and protein sources. The major source of energy concentrates is that of the grains—primarily, corn, milo, barley, oats, and wheat. The major sources of protein concentrates include the oil meals, primarily soybean meal, cottonseed meal, and linseed meal, with a small assist from corn gluten feed and some wheat by-products.

The United States produces about one-third of the world's feed grains and consumes only 22%, leaving about 11% for export (Anonymous, 1978). In the case of corn, nearly one-half (46%) of that produced in the world is produced in the United States with no other country producing even as much as 10%; China is next with 8% (Table 10.1).

I. THE CEREAL GRAINS

All the "cereal grains" are high in starch and low in fiber. They are rich in energy and generally are quite palatable. The highest energy grains are corn, milo, and wheat, followed by barley and then oats. Generally, the balance of amino acids is poor for the cereal grains, with deficiencies, particularly, in lysine and tryptophan. However, since amino acid balance is not critical in cattle feeds—except for the very young calf, protein quality of the grains is not given consideration in evaluating them. Corn is especially low in total protein with an average value of from 7.8 to 9% whereas barley and milo (Fig. 10.1) may contain as much as 12% protein. The grains are extremely low in calcium but almost adequate in phosphorus.

None of the grains contain vitamin D and only yellow corn contains

TABLE 10.1

World Production and Consumption of Feed Grains[a]

	1975–1976 (million metric tons)
Production	
Canada	18.1
Australia	6.0
Argentina	12.0
South Africa	9.5
Thailand	3.3
USSR	65.0
Western Europe	80.7
Eastern Europe	56.6
Other foreign	146.8
Total foreign	398
United States	184.1
World total	582.1
Consumption	
United States	133.0
USSR	81.0
People's Republic of China	43.8
All other foreign	325.2
World total	583.0

[a] Anonymous (1978).

β-carotene—the precursor of vitamin A. However, research conducted by the author demonstrated finishing beef cattle do not convert corn carotene to vitamin A very efficiently.

A. Corn

The hull and the germ are the more well-defined entities of the corn kernel (Fig. 10.2). Just under the hull is a shallow layer, predominantly composed of gluten, which contains most of the protein. Toward the center of the kernel is a mixture of gluten and starch. Percentagewise, the corn kernel contains 55% starch, 16% water and 29% gluten, hull, and germ. Corn oil amounts to about 3% of the kernel weight.

Corn is rich in nitrogen-free extract which is nearly all starch. It is higher in ether extract than any cereal grain except oats. It is low in fiber and is highly digestible. Corn is the most palatable of all grains for beef cattle and more cattle are finished out on corn than all the rest of the cereal grains combined.

Part of the yellow of yellow corn is due to the carotenes but a part of it is due to xanthophyll, which is critical for imparting yellow color to the body fat of chickens.

Fig. 10.1. Milo, properly processed, is a most excellent cattle feed. (Photo courtesy of BEEF Magazine.)

Fig. 10.2. Corn is the No. 1 grain fed to livestock in the United States. (Photo courtesy of BEEF Magazine.)

There is no nutritional justification for grinding air-dry corn for beef cattle, but most nutritionists and cattle feeders recommend at least rolling of ensiled high-moisture corn. Almost any kind of heat processing or ensiling of high-moisture (25% water) corn will improve its nutritive value by about 10%. The reason for this is not clear but cattle feeders should consider roasting, steam flaking, micronizing, or moisturizing corn for cattle feeding in order to achieve the average 10% bonus.

Many types of yellow corn such as high lysine, opaque-2, and waxy maize have been studied, but none are quite equivalent to regular high yielding hybrid field corn for beef cattle.

The stage of maturity at which corn is harvested affects the total nutritive value of the kernel. Note from Table 10.2 that the starch and nitrogen-free extract (NFE) do not approach their maximum level until the corn grain is at least in the mid-dent stage. Therefore, unless an early frost, drought, or blight dictates a relatively early harvest, its best to not harvest corn too early. It is interesting to note also that protein is laid down quite early in the corn kernel.

The by-products of corn grain processing utilized as feedstuffs include corn gluten meal, corn gluten feed, corn germ meal, corn steep liquor, and distillers dried grains. Few, if any, of these products with the exception of corn steep liquor are used in beef cattle diets. Rather, almost all of the above by-products are especially well adapted to dairy cattle formulations. Corn steep liquor, a by-product of the corn starch industry, is well adapted to liquid cattle supplement manufacture wherein it can replace a portion of the cane molasses and supply some "natural protein" in such formulations.

Corn Types

Several new types of corn have been introduced in the past few years. One of these is waxy corn which is chemically different from regular corn in that nearly 100% of its starch is amylopectin (a many branched molecule) while that of normal corn is 25% amylose (an unbranched chain of glucose units) and 75% amylopectin. Another corn type is opaque-2, or high lysine corn, which makes it more nutritious for monogastric animals. A fourth type is opaque-2 plus floury-2 containing a combination of genes that result in increased lysine and tryptophan levels. Several universities have compared corn grain types for beef cattle. The University of Minnesota (Goodrich and Meiske, 1974) compared the four types of corn described below (Table 10.3). Although cattle fed waxy corn appeared to gain faster than those fed regular corn, increased consumption of waxy corn resulted in no difference in efficiency of feed conversion. The Minnesota researchers assayed the various types of corn for amino acid content, and these results are presented in Table 10.4. Typically, the opaque-2 corn types contained nearly one-third more lysine than regular corn.

The University of Nebraska researchers (Brady and Farlin, 1978) compared only waxy corn with regular corn in two trials. There was a slight trend for more

TABLE 10.2

Starch Cell Wall Constituents and Gross Energy Compositions of Corn Grain at Four Stages of Maturity[a]

	Concentration in dry matter			
Nutrient	Early milk	Early dough	Mid-dent	Mature
Dry matter (%)	21	36	55	77
Crude protein (%)	17	13	11	11
Ether extract (%)	3	4	5	5
Crude fiber (%)	5	3	3	2
Ash (%)	3	2	2	1
Nitrogen-free extract (%)	72	78	80	81
Starch (%)	47	55	59	63
Cell wall constituent (%)	28	25	16	64
Gross energy (kcal/kg)	4560	4540	4590	4580

[a] From Goodrich and Meiske (1969).

TABLE 10.3

Four Types of Corn Compared for Beef Cattle[a]

Item	Type of corn			
	Dent	Waxy	Opaque-2	Opaque-2 plus floury-2
Number of steers	16	16	8	8
Initial weight (lb)	680.0	679.0	683.2	671.5
Final weight (lb)	1077.0	1103.3	1083.4	1043.0
Average daily gain (lb)	2.63	2.81	2.65	2.46
Daily feed				
Corn (lb)	15.87	17.62	15.74	14.18
Supplement (lb)	0.78	0.88	0.79	0.71
Dry feed per pound gain (lb)	5.6	5.6	5.4	5.4

[a] Goodrich and Meiske (1974).

TABLE 10.4

Amino Acid Contents of Four Types of Corn[a]

Amino acid	Dent	Waxy	Opaque-2	Opaque-2 plus floury-2
Lysine	0.33	0.27	0.42	0.44
Histidine	0.29	0.24	0.28	0.29
Arginine	0.46	0.34	0.53	0.55
Aspartic acid	0.71	0.60	0.90	0.84
Threonine	0.33	0.29	0.32	0.32
Serine	0.44	0.39	0.40	0.39
Glutamic acid	1.94	1.65	1.61	1.61
Proline	0.89	0.81	0.80	0.80
Glycine	0.42	0.36	0.49	0.48
Alanine	0.77	0.68	0.63	0.62
Half cystine	0.02	0.03	0.02	0.03
Valine	0.66	0.50	0.60	0.59
Methionine	0.14	0.11	0.11	0.13
Isoleucine	0.40	0.34	0.37	0.36
Leucine	1.36	1.13	0.94	0.96
Tyrosine	0.19	0.18	0.20	0.17
Phenylalanine	0.49	0.54	0.43	0.40

[a] Goodrich and Meiske (1974). All values in percentage.

TABLE 10.5

Waxy Corn Compared to Regular Corn for Beef Cattle[a]

Number of animals		Daily gain (lb)	Gain: feed ratio	Feed: gain ratio
Steers				
60	Normal	3.28	0.162	6.17
	Waxy	3.65	0.177	5.65
48	Normal	3.39	0.183	5.46
	Waxy	3.60	0.188	5.32
180	Normal	2.22	0.112	8.93
	Waxy	2.33	0.113	8.85
Heifers				
120	Normal	2.11	0.144	6.94
	Waxy	2.09	0.143	6.99

[a] From Wilson *et al.* (1975).

efficient gains with the waxy corn. University of Illinois (Hatfield *et al.*, 1975) conducted four feeding trials with beef cattle in which waxy corn was compared against regular corn. In two of the trials, cattle fed waxy corn gained more rapidly whereas in the other two trials, cattle fed waxy corn gained comparably (Table 10.5). In addition, the Illinois researchers also conducted nitrogen balance studies. They concluded "based on growth, feed efficiency and nitrogen balance studies, waxy corn is as nutritious as regular corn" for beef cattle.

Purdue University (Thomas *et al.*, 1975) compared the four main types of corn in cattle digestion and nitrogen balance studies. Their conclusions were: (1) disregarding different protein levels, steers fed regular, waxy, and opaque-2 corn gained similarly, (2) steers fed opaque-2 corn ate least and thus required less feed per pound of gain, and (3) waxy corn had a lower digestion coefficient than the other three types. Thus, it seems cattle utilize the various types of corn similarly, and probably the various types of corn should be included in the diet on a least cost basis, irrespective of genetic makeup.

B. Concentrates as Other Food Sources*

In the Corn Belt, naturally corn is by far the number one feed grain for all classes of livestock fed. In fact, it is the number one feed grain for each and every class of livestock in the United States. Nevertheless, there are many areas of the United States in which transportation costs make other feed grains more attractive for beef cattle feeding and at certain times, even in areas where corn is

*Adapted from a talk presented by K. S. Hendrix, Purdue University, West Lafayette, Indiana, before the 1978 Purdue Feed Industry Conference, February 22.

grown, other feed grains may be competitive. We are now in an age of computers by which least-cost formulations can tell the livestock feeder quite rapidly whether he should use corn or whether other feed grains should be substituted as a part—or even all—of the feed grain portion.

Corn is fed primarily for its energy source since it contains the lowest protein content of all grains. Corn fits this role well because the energy source represents about 85 to 88% of the total dietary requirement. Therefore, in discussing corn replacements, one of the major considerations is energy replacement values of the likely candidates. This section will attempt to relate other products to corn as replacements in beef cattle feeding.

1. Grain Sorghums

These have become extremely important sources of energy grain in areas where corn cannot be grown. In other words, grain sorghum does not compete with corn, but rather provides an optional product in semi-arid areas of the country which are too dry for corn production. Thousands of cattle have been finished out to market in the Great Plains area of Kansas, Oklahoma, and the Texas Panhandle, on sorghum grain (also called milo).

The very large cattle feeding area that just a few years ago mushroomed around Hereford, Texas, owes its existence to the introduction of irrigation followed by the growing of milo. However, the development of much higher yielding adaptable corn hybrids probably may force milo growing out of that area.

The grain sorghums have a general nutrient makeup similar to that for corn, but require more rigorous processing. They are rich in starch and NFE, but low in protein and calcium; they are low in crude fiber and comparable in total digestible nutrients to corn. The seeds of grain sorghum are quite hard and therefore it is most important that the seeds of grain sorghum be crushed mechanically, before they are fed to cattle, either by coarse rolling or fine grinding. Milo responds well to high-moisture storage either from high-moisture harvest or reconstitution. Feed efficiency of high-moisture milo is 10–18% higher than that of dry milo when fed to beef cattle. High-moisture milo must be rolled or ground prior to feeding. If harvested in the high-moisture state, it can be stored either in the whole or processed form with equal feeding value. However, if dry milo is reconstituted, it must be stored in the whole form for at least 10 days and preferably 20 days prior to feeding to obtain an improvement in feeding efficiency.

In evaluating comparative methods for treating sorghum grains, usually no treatment for unprocessed sorghum is included as a "control" since the results from such a treatment would be so poor that it would not be a realistic comparison. Texas researchers (Shaake *et al.*, 1972) conducted two feeding trials in

West Texas to study the comparative value of four sorghum grain processing methods upon animal performance and carcass characteristics. Cattle fed steam-flaked, whole-reconstituted and rolled or rolled-reconstituted sorghum grain diets gained similarly and produced carcasses with similar characteristics. Similar results were obtained for cattle fed either steam-flaked or micronized grain sorghum. Cattle fed reconstituted and rolled grain sorghum diets had most efficient feed conversions. The greatest improvement in efficiency of feed conversion from ensiled high moisture grain is not an uncommon observation in such comparative research.

2. *Barley*

Barley is basically a cool weather relatively drought-resistant crop. Since there are two areas of the United States where these conditions exist, there are two general geographic cattle feeding areas related to barley grown in those respective regions. The northern part of the north central regional United States grows spring barley. This includes Minnesota and the Dakotas. California and Arizona grow the winter varieties. In fact, barley is sown on more acres in California than any other cereal crop.

Barley is usually listed in most feed ingredient tables under two categories—Pacific coast barley which has an average protein value of 8.7% and nonPacific coast barley which has an average protein value of 11 to 12% (Crampton and Harris, 1971). The source where barley was grown, therefore, can have a profound effect upon its feeding value for beef cattle. Its TDN content is listed as from 75 to 78% (shelled corn has 80% TDN).

Many experiments have been conducted which show that it is critical that barley be processed before it is fed to beef cattle. For many years this consisted primarily of grinding or dry rolling. However, more sophisticated methods of treating barley include steam flaking and reconstituting. The University of

TABLE 10.6

Dry-Rolled and Steam-Flaked Milo or Barley for Finishing Beef Steers[a]

	Milo		Barley	
	Dry-rolled	Steam-flaked	Dry-rolled	Steam-flaked
Number of steers	30	32	32	32
Average initial weight (lb)	572	572	579	576
Average daily gain (lb)	2.81	3.12	2.88	3.10
Average daily feed (lb)	22.4	24.8	20.8	22.7
Feed per 100 lb gain (lb)	830	792	722	732

[a] Hale *et al.* (1966). Average of two trials.

Arizona researchers (Hale *et al.*, 1966) compared dry-rolled and steam-flaked milo or barley diets (Table 10.6) for finishing beef cattle in two trials.

Ensiling of "high-moisture" (25–27% water) barley is also an excellent method of processing barley to make it a more valuable feed for beef cattle. The Northwest Experiment Station at Crookston, Minnesota published a two-year summary of research on the comparative feeding value of dry and ensiled high-moisture barley. Over 200 head of cattle were involved in this research. Cattle fed high-moisture barley gained an average of 8.6% faster (2.52 versus 2.32 lb per day) and required 9.3% less feed (805 versus 888 lb) per pound gain than cattle fed dry-rolled barley.

When barley is processed properly to obtain its maximum feeding potential, it is just about equal to corn in value. There may be some additional warnings about barley feeding that the inexperienced feeder should watch for. For instance, cattle tend to bloat more on barley than on some other grains. This problem can be somewhat reduced by mixing with other grains. Also, cattle tend to tire more quickly on long feedlot periods of barley feeding than is true for corn or for milo.

3. Wheat

Normally wheat has been too expensive to be considered as a livestock feed. However, it appears that utilization of more wheat can be justified in cattle diets on the basis of price comparison.

An analysis of wheat finds an average of 80% total digestible nutrients (TDN) and from 10 to 12% protein. So, on paper at least, wheat contains the same energy as corn (corn has about 80–81% TDN) and just a point or two more protein than corn (swine raisers have recognized for years that wheat is actually worth about 5% more than corn for growing–finishing hogs). However, wheat for cattle feeding takes a bit better management than does corn.

Just how does wheat differ from corn for beef cattle? The starch content of wheat is available in the cow's stomach so much more quickly than that from corn, and this available starch is converted to lactic acid. Cattle which develop "acidosis" or "enterotoxemia" are usually manifesting an accumulation of lactic acid in the rumen. The most severe manifestation is sudden death. Kansas State University has conducted a great deal of research on the subject, and they have shown a production of from 2 to 4 times as much lactic acid in the stomachs of cows fed certain types of wheat as from those fed corn (Table 10.7).USDA research (Oltjen *et al.*, 1966) compared various ratios of corn to wheat (90:0, 60:30, 30:60, and 0:90) in a diet containing 10% of molasses, urea, vitamins, and minerals. Cattle were fed either 60 or 90% wheat diets. Performance on all diets was comparable for the first 70 days. After the first 70 days cattle on the higher wheat diets decreased their feed consumption indicating they were tiring of their diets. Chemical analysis of the rumen fluids showed cattle on high wheat

TABLE 10.7

High Wheat Diets for Beef Cattle (Six Experiments)[a]

Item	Grain treatment		
	Rolled milo	Rolled wheat	Rolled wheat plus bicarb[b]
Feed intake (lb)	26.4	22.1	23.0
Average daily gain (lb)	2.68	2.66	2.87
Feed per pound gain (lb)	9.8	8.3	8.0

[a] Brethour (1977).
[b] Sodium bicarbonate, 100 grams, per head daily.

diets had 16% less propionic acid, which is the high energy VFA. Cattle on the high wheat diets had greater total VFA and ammonia nitrogen in their rumens.

However, wheat is an excellent feed for fattening cattle when handled properly. It is a good idea not to replace more than from one-third to one-half of the corn with wheat for fattening cattle. Although it is possible to feed higher levels, the 50% maximum is a pretty safe figure. Some who feed wheat will include a quarter pound of sodium bicarbonate (baking soda) per head daily which helps to neutralize the extra lactic acid produced from wheat-based diets.

In introducing cattle to this type of feed ground—or preferably rolled wheat should be used. Cattle should be introduced gradually to a full-feed with about one-third wheat to two-thirds corn, much the same as you would do with corn alone. If you know your wheat tests 12% protein, you can actually feed about one-third less protein supplement than you would feed with straight corn. Thus, if wheat is a good buy, compared to other feed grains, it has excellent feeding potential (up to about one-third to one-half the grain) for finishing beef cattle.

Wheat lends itself well to ensiling as a high-moisture grain. It should be rolled as it comes out of the silo, just before feeding.

Certain management practices have been learned in the feeding of what, which include the following:

1. Wheat-fed cattle should have feed in front of them at all times. Possibly hand-fed cattle which had intervals when no feed was in front of them suffered when they were on an all-wheat program.

2. In high grain finishing type rations where only 1–3 lb of dry roughage or 10–15 lb of silage are fed, limit wheat to no more than 30–35% of the total or a maximum of 50% of the grain portion. With high roughage growing type rations such as corn silage, haylage or hay, all of the supplemental grain in the ration can be wheat if limited to no more than 1% of body weight.

3. Avoid fine dusty particles. The finer the particle size, the more rapidly the wheat starch is digested, thus generating large amounts of lactic acid. A coarse-rolled flake is probably the best form since whole wheat is not desirable for beef cattle. Whole wheat tends to get rubbery as it is masticated. Pelleting or adding liquids is not of great help with finely ground wheat since neither would aid in slowing down the rapid starch digestion of wheat.

4. All-wheat diets appear to be better in warmer seasons when feed consumption appears to be more regular than it is in times of erratic cold weather and environmental pressure.

5. Exceptional care should be exercised in bringing cattle up to a full feed of wheat. The shift should be gradual rather than abrupt. In other words, take a few days longer before cattle are allowed a full feed of wheat.

6. Inclusion of 100 gm of sodium bicarbonate in an all-wheat diet has improved performance under Kansas experimental conditions. Sodium bicarbonate is a buffer and thus neutralizes any surges in lactic acid formation preventing acidosis. Other buffers might be substituted.

7. Watch total dietary protein content when feeding an all-wheat diet because wheat may contain 12% protein, and thus additional protein would be without benefit and conceivably could place an ammonia overload on the kidneys and liver. However, the other nutritional contributions by the protein supplement such as vitamins, minerals, and trace minerals, should not be forgotten if supplemental protein is omitted.

Kansas researchers also demonstrated (Table 10.8) that varying levels of wheat from low to high could replace rolled milo in cattle finishing diets, but once again demonstrated the benefit of 100 gm supplemental sodium bicarbonate per head daily.

4. Oats

Because of their high fiber (12%) and lower TDN (69%) oats have a much lower energy value than the four high-energy grains discussed earlier. However, oats are a most excellent source of supplemental energy and protein in growing diets and for lactating cows when supplemental energy is needed. Limited levels of oats may be included in finishing cattle diets, especially when such formulations do not contain very much fiber, or roughage.

Oats are considered to have about 80–85% the feeding value of corn. However, when they are in growing diets or when they constitute less than 30% of a finishing diet, oats may have nearly 100% the feeding value of corn, milo, or barley.

In growing diets or in brood cow diets where the level of grain feeding is quite limited, it is not necessary to process oats. In finishing diets it is important to grind or crush the oat kernel, since processing will improve utilization.

TABLE 10.8

Feeding Varying Levels of Wheat to Beef Cattle[a]

Item	Rolled milo	Low wheat	Medium wheat	High wheat	High wheat plus bicarb[b]
Daily feed					
Forage sorghum silage (lb)	11.3	10.3	10.2	9.5	9.5
Rolled milo (lb)	19.6	14.8	9.2	0.4	0.4
Rolled wheat (lb)	—	4.1	8.9	15.6	16.2
40% protein supplement (lb)	1.5	1.3	1.1	0.7	0.7
Dehydrated alfalfa meal (lb)	0.5	0.5	0.5	0.5	0.5
Ground limestone (lb)	0.2	0.2	0.2	0.2	0.2
Air-dry total (lb)	25.6	24.4	23.3	20.6	21.2
Initial weight (lb)	885	890	889	887	885
Final weight (lb)	1219	1198	1217	1194	1222
Daily gain (lb)	2.73	2.52	2.69	2.52	2.77
Air-dry feed per pound gain (lb)	9.4	9.7	8.6	8.2	7.6

[a] Brethour (1977).
[b] Sodium bicarbonate, 100 gm, per head daily.

5. *Rye*

Rye is used mainly for pasture or cover crop and little grain is fed to livestock. Contamination with ergot greatly hinders its use as a feed grain. It is considered to be one of the least palatable of all feed grains for livestock.

6. *Grain Screenings*

Screenings are composed largely of small and broken kernels, hulls, kernel tips, small pieces of cob, dirt, and weed seeds. Nutrient composition can be quite variable depending upon the level of cob and foreign material. Screenings relatively free of dirt and weed seeds will contain nearly as much energy as corn grain and may even be somewhat higher in crude protein.

Feeding management is probably the most important factor to consider when using screenings as a replacement for corn grain. The screenings are usually extremely fine and dusty, which may reduce their intake by cattle. In addition, if cattle are not fed some type of a roughage source (silage or hay), digestive disorders in the rumen (paunch) could occur when high levels of screenings are fed for long periods of time (over 100–120 days). Care should also be taken to ensure that screenings are relatively free of weed seeds. Certain weed seeds are highly toxic when fed to cattle.

To obtain maximum intake and feeding value from corn screenings, these suggestions are offered as guidelines:

1. Feed 25% of the concentrate as regular corn, preferably as whole shelled corn.
2. Feed at least 3–4 lb of coarsely chopped or long hay or 10–20 lb of silage daily. Using silage or haylage would be better than a dry roughage as this would reduce the dustiness of the ration.
3. Dustiness could also be reduced by using 3–5 lb of molasses with each 100 lb of screenings.
4. Buy and feed screenings on the basis of weight rather than volume. Screenings have a lower density than grain, thus weight per unit volume is lower.
5. Have the screenings analyzed for crude protein. It may be possible to reduce the amount of supplemental protein required in certain situations.
6. If supplemental protein is reduced, be sure to provide supplemental vitamin A and minerals (especially calcium). It may be practical for some feeders to inject cattle with vitamin A and provide a mineral mix containing equal parts of limestone, dicalcium phosphate, and trace mineral salt-free choice.

II. MOLASSES

Molasses is generally a by-product of the sugar industry—both beet and cane. However, other products resembling cane and beet molasses, both in appearance and in feeding value, have become available and competitive. Since the family of molasses products are worthwhile constituents in beef cattle formulations, they will be considered here.

A. Cane Molasses

"Blackstrap" or cane molasses is a by-product from the manufacture of sugar from sugar cane. The plant is grown world wide, so the by-product, cane molasses may be shipped for very great distances, primarily by ocean-going barges. Cane molasses is the residual remaining after as much sugar as possible has been crystallized from the sugar cane. It contains 55% sugar, 3% protein, and has a TDN value of about 53%. Cane molasses is extremely palatable to beef cattle, and is often included also for its dust-settling effect and for the pleasant aroma it imparts to feeds. Cane molasses can be offered on a free-choice basis in a tank or it may be incorporated into a portion of the ration, as in the protein supplement, or into the total ration. Fattening cattle offered cane molasses ad lib in a tank often will consume two to three pounds per head daily, in addition to their diet, but at the expense of their other dietary requirements. When fed in limited quantities, cane molasses has a high corn replacement value. In fact, in quantities up to a maximum of four or five pounds per head daily, cane molasses has a 100%, pound-for-pound replacement value for corn in heavy cattle fattening diets. Higher levels, however, are not recommended because the value of molasses above this level of feeding drops off very rapidly.

Liquid cattle supplements (Table 10.9) based on urea nitrogen have become very popular in cattle feeding because of (1) economics and (2) because they are really the only cattle supplements that can be fed safely ad lib.

Note from the table that two liquid supplement beef cattle formulas developed at the Purdue Station include 38.5 and 69.3% liquid molasses, respectively. No other vehicle for liquid cattle supplements has been discovered which can completely replace molasses in the formulation.

The author, in cooperation with Mr. E. G. Brown of Millbrook, New York, developed the "Brown-Perry Slurry System of Cattle Feeding" (Perry, 1972), in which all the ingredients are ground very finely by a hammermill and then incorporated into a liquid slurry system of molasses and water. The justification of this system is that it permits utilization of many and dusty ingredients which would not normally fit into a cattle fattening program (Table 10.10). This type of formulation permits the use of many less sophisticated materials such as for example, "tailings," which consist of dust and fine broken grain particles that are vacuumed off hundreds of carloads of corn, barley, and oats which come into New York City each day. Because of their dusty characteristics tailings are not very suitable for cattle feed, but they often contain 14% protein and have a potential of 80% plus TDN. Therefore, when they are mixed into a molasses-containing slurry, they become a very useful cattle energy feed.

TABLE 10.9

Formulas for Liquid Cattle Supplements Containing Molasses

	Level of protein (%)	
Ingredient	64[a]	32[b]
Molasses (%)	38.5	69.3
Liquid urea (32% N) (%)	29	14.5
Ammoniated polyphos (10-34-0) (%)	9	4.5
Distillers solubles (27% dry matter) (%)	9.3	4.6
Salt solution (28% salt) (%)	9.0	4.5
Calcium chloride (%)	1.2	0.6
Sodium sulfate (%)	1	0.5
Premix (%)[c]	3	1.5

[a] Purdue Liquid 64.

[b] Purdue Liquid 32.

[c] Premix, 2 million IU vitamin A; 435 gm zinc sulfate; 1 gm cobalt sulfate; 1.9 lb water.

TABLE 10.10

Typical "Brown–Perry" Slurry Cattle Feed Formulation[a]

Ingredient	Percentage in mix
Ground ear corn	27.8
Ground shelled corn	15.8
Starch	19.8
Molasses	7.9
TM salt	0.4
Peanut butter	5.0
Diammonium phosphate	0.4
Dolomitic limestone	0.4
Urea	0.8
Dehydrated alfalfa meal	5.9
Cookie meal	15.8
Total	100.0

[a] The above is fortified with vitamin A at a level of 1000 IU per pound. A 2500 pound batch of the above formulation is mixed with 2000 pounds of water from which it achieves its slurry consistency.

B. Beet Molasses

A by-product of the beet sugar industry, beet molasses actually is listed in most feed charts as containing more TDN than cane molasses (61 versus 53% for ruminants); also it contains more crude protein (6.7 versus 3%) but partly due to certain additives in the processing. Beet molasses and cane molasses contain comparable levels of sugar.

In beet processing, cleaned beets are stripped and sliced and then the soluble sugars are extracted with warm water, leaving wet beet pulp. After the liquid is concentrated by heating and evaporation, sugar crystallizes out and is separated mechanically, but considerable sugar remains dissolved. Thus the liquid is treated further with chemicals to remove additional sugar. The final liquid has a sugar level similar to that for cane, or "blackstrap," molasses. Beet molasses has essentially the same feeding value as cane molasses for beef cattle, but its tendency to have a laxative effect on cattle should be taken into consideration. The addition of certain alkaline salts and other laxative materials in the manufacturing process must also be considered. Since beet molasses might represent only one or two pounds per head, daily, this should not cause any major problem, providing it is introduced cautiously into the diet. In fact, with proper management, cattle weighing 700 to 1000 pounds may consume three to four pounds of beet molasses per head daily without adverse effects.

C. Condensed Corn Steepwater Solubles

This is produced by evaporation of the water in which corn kernels are soaked or steeped in the early stages of the corn manufacturing process. The steeping operation is comparable to cooking sweet corn in the home in which some soluble material is extracted from the corn and clouding the cooking or steeping water. This represents small fractions of soluble proteins and carbohydrates. However, when the steeping water is concentrated by evaporation, the resulting product has a consistency similar to other molasses products and contains about 10% protein—all from a natural source. Quite often this product is less expensive than cane molasses and thus may be incorporated into liquid supplements as a relative natural protein carrier. It does not have the palatability of cane molasses, but has a comparable energy value and a superior protein value.

D. Lignin Sulfonates

These and comparable products are made from wood as by-products of the pulping industry. At least fifteen different paper manufacturing companies are in operation in the United States and are producing the lignin sulfonate products. Chemically, lignin sulfonates are one or a combination of, ammonium, calcium, magnesium, or sodium salts of the spent sulfite liquor derived from the sulfite digestion of wood. Wood is reacted under conditions of heat and pressure with sulfur dioxide and the indicated salts. Part of the solids content of lignin sulfonates exist as undigestible lignin and part as digestible hemicelluloses, sugars, and as other fractions. On a strict TDN basis, lignin sulfonates are considered to have about one-half the energy of cane molasses. However, since such lignin sulfonates are used primarily as carriers in liquid cattle supplements, it is difficult to demonstrate any performance difference with cattle on liquid supplements formulated with cane molasses or lignin sulfonates. Lignin sulfonates generally are used at a maximum molasses replacement value of from one-fourth to one-third in liquid cattle supplements.

E. Other Types of Molasses-Like Materials

This category includes corn molasses, or hydrol, and citrus molasses. Wherever economics permit, either may be used to replace one-fourth to one-third of the cane molasses in beef cattle liquid supplement formulations without affecting cattle performance.

III. FAT

Extensive research on fat as a cattle feed has been conducted because of its very high energy content. Fat contains 2.25 times as much energy as carbohydrates or proteins. Theoretically, if feed-grade fat is less than 2.25 times as expensive per pound as carbohydrate feeds, it would be worthy of consideration as a feed ingredient. Historically, however, feed fats sell for more than 2.25 times the cost of energy grains. Much research effort has been expended in learning how fat might fit into beef cattle diets.

Fats have also been an important component of several species formulations, but perhaps the most critical formulation has been that obtained for broilers wherein high energy diets are very critical and quite difficult to achieve. Another group of formulations which utilize added fat are the various milk replacers. Of course, in at least some formulations it is sometimes possible to substitute a combination of high-energy fat and lower energy feedstuffs for the typical energy grains, but this does not occur too often.

Hale (1966) has summarized twelve feeding trials in which the addition of 4% fat and 1% dicalcium phosphate was researched (Table 10.11). In eleven of the twelve trials, a growth stimulation was achieved with an overall average of about 8% in favor of diets containing the added fat; since total feed consumption was only slightly less for cattle fed the fat and dicalcium phosphate supplemented diets, efficiency of feed conversion was 8% better for such diets.

Although the maximum level at which fats may be added to rather dry type diets has not been established, probably a level of 4 or 5% is the most practical level of supplementation. In high roughage diets containing dry hay and/or straw, it is possible to incorporate fat at levels as high as 20% (Hale, 1963).

Steam-rolled barley will produce about 7% faster gains with 11% less feed per pound of gain than will dry or steam-rolled milo. Feeding trials at the University of Arizona have demonstrated that the addition of fat to milo diets will enhance their quality to about that of barley. Generally added fat in the diet has resulted in a depression in appetite greater than would be anticipated from the increased energy content of the added fat. The exception to this has been research findings in the southwestern United States where the addition of up to 5% of fat to finishing beef cattle diet was practical. Such diets generally are fairly dry from having been steam flaked and then exposed to the very arid climate. In Corn Belt country a great deal of higher moisture feeds such as corn silage and high moisture grains are fed.

Research at Purdue (Hatch, 1971) would indicate moisture level of the diet may greatly affect the response one may obtain from adding fat to cattle diets. In contrast to the benefits obtained on dry diets (increased gain, improved feed efficiency), the benefits from adding fat to corn silage–high-moisture corn diets have been minimal at best (Table 10.12). The addition of 3, 6, or 9% tallow depressed rate of gain ($P < 0.05$) but also feed consumption so that efficiency of feed conversion actually favored all lots fed supplemental tallow. These results tend to agree with research at other experimental centers in which the feeding value of added fat is questionable for diets built around high-moisture feeds such as corn silage and high-moisture grains.

TABLE 10.11

Added Fat to Arizona Cattle Diets (Summary of Twelve Trials)[a]

Experiment no.	Number of days	Grain fed	Number of cattle	No added fat			4% Tallow added		
				Daily gain (lb)	Daily feed (lb)	Feed per lb gain (lb)	Daily gain (lb)	Daily feed (lb)	Feed per lb gain (lb)
1	126	Milo	32	2.51	25.2	10.0	2.74	23.8	8.7
2	126	Milo	32	2.74	25.6	9.4	2.97	27.0	9.1
3	140	B:M[b]	28	2.56	20.2	8.3	2.57	18.5	7.6
4	140	B:M	28	2.29	19.0	8.7	2.41	18.6	8.1
5	123	Milo	42	2.47	21.5	8.7	2.75	21.3	7.7
6	140	Barley	32	3.00	22.6	7.5	3.27	22.8	7.0
7	140	Milo	32	2.51	21.2	8.4	2.86	21.2	7.4
8	143	Milo	32	2.65	21.3	8.0	2.88	21.8	7.6
9	143	B:M	32	2.78	21.4	7.7	3.02	21.4	7.1
10	168	B:M	32	2.41	20.6	8.6	2.61	20.7	7.9
11	140	Milo	32	3.20	24.7	7.2	3.44	24.5	7.1
12	140	Barley	32	3.11	22.0	7.1	3.28	21.6	6.6
Average				2.69	22.1	8.4	2.90	21.9	7.7
Difference							+8%	−1%	−8%

[a] Hale (1966).

[b] B:M = barley and milo mixed.

TABLE 10.12

Feedlot Performance of Beef Steers Fed Levels of Added Fat on a High-Moisture Diet (233 Days)[a]

Item	Level of added fat (%)			
	0	3	6	9
Daily feed (lb)	15.9	14.5	14.2	14.2
Daily gain (lb)	2.22	2.00	1.98	1.96
Feed per pound gain (lb)	7.2	6.6	6.5	6.5

[a] Hatch (1971).

IV. MISCELLANEOUS ENERGY CONCENTRATES

These are manyfold and will be mentioned only briefly. For example, earlier in this chapter, the author discussed "tailings," a dusty by-product of grain cleaning, which has made an excellent cattle feed when its dusty characteristics were masked in a slurry feeding system. Wheat flour which had become damp in railroad cars, and thus were unfit for human food, made excellent cattle feed, when mixed into a slurry. Peanut butter, stale cake mixes, stale bread, broken cookies, stale candy bars, and almost any material which is (1) reasonably digestible by cattle, (2) does not contain toxic materials, (3) is not extremely unpalatable, (4) does not contain too much crude fiber, (5) and is economical, probably can be formulated and balanced into cattle feeding programs. Any "different" or unique material which meets the five qualifications indicated above probably should be investigated and evaluated as a potential cattle feed.

V. PROTEIN CONCENTRATES

Protein concentrates for beef cattle have been substituted for by nonprotein nitrogenous materials, in large part, by such products as urea, biuret and ammonia. However, natural or true proteins are used a great deal in spite of the innovative utilization of the urealike products. In the first place, nutritionists generally agree that urea products should not supply more than one-third of the total protein of the diet. Furthermore, in times of stress, probably natural protein should supply nearer 100% of the total protein requirement. For example, newly weaned calves, shipped for many miles, and unloaded at their feedlot destination should probably receive all their protein as natural protein. Many producers of purebred or show cattle—or any type of fancy cattle—limit the level of supplemental protein from nonprotein nitrogen to one-third or less, which is even less than the "one-third total" listed above.

Very young cattle whose rumen has developed very little do not have the microbial population to handle nonprotein nitrogen, and so, of course, all their supplemental protein should be of a natural and "true" protein source.

A. Cottonseed

Cottonseed is a by-product of the cotton crop. After removal of the fibrous cotton lint, the cotton seeds are broken open, allowing the seed kernel to drop out; subsequent vibration techniques separate out the seed kernel. Crushing of the seed kernel follows and oil is removed either by hydraulic pressure or extraction. After as much oil as possible has been removed, the remaining product—cottonseed meal or cake—is ready for processing as livestock feed. The use of the hydraulic method of "squeezing" out the oil results in rather large residual pieces called "cake," which could then be ground into "meal." The cake has always been popular with cattle feeders who feed out-of-doors, since it is not apt to blow away.

Since cottonseed meal is a by-product material, it would be possible to have widely varying qualities of meals between batches. In other words, it would be possible to dilute the true cottonseed meal with varying quantities of cottonseed hulls. It is important that formulators and feeders study all feed tags carefully to evaluate potential feeding values of all feedstuff ingredients. Knowledge of the protein assay alone is helpful because as the fiber content increases in a feedstuff, contents of the other ingredients will decrease. Nevertheless, it is best to know not only protein content, but also crude fiber content and as much data as possible, to aid in evaluation of the ingredient.

Cottonseed meal on today's markets usually contains 41% crude protein and from 12 to 13% crude fiber. It contains nearly 1% phosphorus, but is quite low in calcium (0.15%). Unless specially processed, cottonseed meal contains levels of gossypol which can be harmful to monogastric animals. However, this is no problem to beef cattle in which rumen development is nearly complete (350 to 400 pounds liveweight, or more) since the rumen can detoxify gossypol.

Cottonseed meal has a tendency to firm up the excreta, so it is an excellent protein supplement for diets which tend to be laxative in nature, as in the case of high corn diets—especially high-moisture, high corn diets.

Sometimes, though not often, cottonseed meal may be more economical per pound than corn, especially at points where it is manufactured. When it is fed at a level in excess of what is needed to meet the protein needs of the diet, it is worth slightly less per pound than corn. Thus, as a general rule, it is not good husbandry to feed more than is actually needed to meet protein requirements.

B. Soybeans and Soybean Meal

In the United States, soybeans have become an outstanding cash crop. The harvested bean is seldom used intact, but rather the oil and residual meal are separated for economic reasons. There is no nutritional reason why unextracted full fat soybeans should not be fed to beef cattle—it is just not sound economically. If it were economical to feed full fat-unextracted soybeans as a source of supplemental protein it would not be necessary to cook them since cattle can utilize raw soybeans. They should, however, be rolled, cracked, or ground for maximum utilization.

Solvent-extracted soybean meal containing 45% crude protein and less than 0.5% oil is often used as a source of supplemental protein for beef cattle when the economics permit. It is an outstanding source of natural supplemental protein and is quite palatable for cattle. Therefore, it is commonly the protein of choice when natural protein is desired. There are no special precautions about its use in cattle rations. It is the leading ingredient in protein supplement formulation, when natural protein is needed—just as corn is the main energy source in livestock feeding in the United States. Because of its excellent balance of essential amino acids, when mixed with corn, monogastric feed formulations utilize soybean meal extensively for a source of supplemental protein.

C. Linseed Meal

A by-product of the linen and linseed oil industry, linseed oil is derived from flaxseed and was the number one choice of supplemental protein for beef cattle formulations for many decades. Haircoats of cattle fed supplemental linseed meal always had a superior gloss or sheen, compared to cattle fed other sources of supplemental protein. The cause of the desirable haircoat effect with linseed meal-fed cattle has never been explained, but cattle herdsmen were well aware of its effect. Linseed meal does have a laxative effect on cattle.

The author demonstrated there was a peculiar and desirable synergism between linseed meal and urea such that a 32% protein cattle supplement containing linseed meal and 5% urea was superior to almost any other similar cattle supplement.

Linseed meal usually contains from 32 to 34% crude protein; therefore, more of it is required than is true for cottonseed meal or soybean meal to supply a given amount of protein. Linseed meal is produced in the cooler northern part of the United States and thus transportation costs plus scarcity often makes it more expensive per unit protein than other sources.

D. Miscellaneous Protein Sources

The rumen of the cow is so well adapted at utilizing foods and feeds that a whole host of miscellaneous protein sources have been used.

1. Animal Wastes as Cattle Feedstuffs*

a. Dehydrated Cage Layer Manure (DPW). Chemical composition of this product varies greatly, based largely upon how long it was stored wet, before processing. The greatest variation is the crude protein because of ammonia losses in wet manure stored for long periods. However, the "average" values for DPW are 10% moisture, 30% crude protein, and 12% crude fiber, which then categorize DPW as a bulky protein concentrate.

It has a digestible energy value of 2000 kcal per kg for cattle, making it equal "energywise" to good quality hay. The TDN value assigned to DPW is 53.2%. However, DPW is not normally fed as an energy source, but rather is used as a source of crude protein, practically all of which is NPN. The crude protein digestibility of DPW for ruminants is quite low, probably in the range of 50%.

b. Broiler Litter. Many factors contribute to a wide range of nutrient values given to broiler litter, particularly the type and quantity of litter used. Broiler litter also is valuable mainly for its nitrogen content; the "average" crude protein assigned to it is 30%, on a dry basis. Usually it contains 15% crude fiber which is primarily lignified. Litter also is quite high in ash (15%). About one-half the protein fraction is made up of true protein that is high in glycine but low in arginine, lysine, methionine, and cystine. Uric acid constitutes about one-half the total NPN. The digestible energy value is about 2440 kcal/kg, making it comparable to alfalfa hay, for ruminants.

Broiler litter is rather unpalatable to beef steers and, if they have the opportunity to choose, will practically ignore diets containing this product (Fontenot, 1971). Steers were fed diets containing 0, 25, or 50% dried broiler waste (Table 10.13). Dispite their apparent dislike for diets containing broiler material, statistical analysis failed to show significant depression in gain of cattle fed diets containing 25% of the material.

c. Dry Cattle Feedlot Waste. Three trials were conducted with Texas (Albin and Sherrod, 1975) cattle feedlot waste which dries out quickly due to the low humidity. In one trial, waste which had been scraped and stockpiled for a month was ground and included at levels of 0, 20, 40, or 60% of a high-energy, adequate protein diet. In trial II, the manure was composed for five days by bringing the moisture content up to 40%. The third trial consisted of adding unaltered feedlot waste to a low-energy, low-protein diet resembling a high roughage diet. In all three comparisons, as the percentage of feedlot waste was increased, apparent digestibility of all nutrients decreased. However, apparent digestibility coefficients for feedlot waste were higher when it was included in low-energy, low-protein diets, indicating its use is limited primarily to such diets

*Some of the material in this section was adapted from Bhattacharya and Taylor (1975).

TABLE 10.13

Levels of Broiler Litter in Beef Steer Diets[a]

Item	Litter in diet (%)		
	0	25	50
Initial weight (lb)	816	831	822
Daily gain			
Days 1–50 (lb)	1.84[b]	1.54[b]	0.50[c]
Days 51–121 (lb)	1.43	1.40	1.03
Total 121 days (lb)	1.60	1.47	0.81
Daily feed			
Days 1–50 (lb)	21.9[b]	19.2[b]	13.8[c]
Days 51–121 (lb)	20.3	20.4	17.2
Total 121 days (lb)	21.0	19.8	15.8

[a] Fontenot *et al.* (1971).

[b,c] Data in the same row with differing superscripts differ by $p < 0.05$.

TABLE 10.14

Wastelage as a Cattle Feed (102-Day Test)[a]

Item	Treatment number			
	1	2	3	4
Treatment				
Basal diet[b] (%)	100	60		
Cattle wastelage (%)			40	40
Corn (%)			60	60
Cottonseed meal (lb/day)				2
Manure (%)		40		
Performance				
Initial weight (lb)	627	623	627	625
Daily gain (lb)	2.68	2.62	2.70	2.90

[a] Anthony (1971).

[b] Corn, 75; CSM, 8; dehydrated alfalfa, 5; molasses, 10; salt, 1; defluor. phos., 1; vit. A was added also.

and has little replacement value in high energy-type diets. Composting feedlot waste lowered the digestibility of diets for organic matter and protein; cell wall digestibility was increased by composting.

d. Cattle Manure Wastelage. Perhaps the most common sense approach to feeding feedlot wste is to ensile it with a lower grade roughage such as stalks, cobs, straw or nonlegume hay on approximately a 50:50 basis, w/w. In other words, 50 lb of wet cattle feedlot waste to 50 lb of a drier material such as hay or straw is combined and then ensiled. Anthony (1971) used such a product (57 parts wet manure to 43 parts grass hay, ensiled) in cattle feeding research (Table 10.14). Manure feeding reduced the amount of basal feed required per unit gain

TABLE 10.15

Relative Feeding Value of Various Ingredients as Energy Sources in Beef Growing–Finishing Rations[a]

Feedstuff (air dry)	Value compared to corn grain (%)	Recommended usage level as replacement for corn grain (% of ration)
Alfalfa meal (17%)	60–65	20–50
Animal fat	160–180	0–5
Bakery product	110–115	0–30
Barley	88–90	0–100
Beet pulp, dried	88–95	0–50
Brewers grains	84–87	0–35
Corn, No. 2 yellow	100	0–100
Corn and cob meal	85–92	0–100
Corncobs	50	20–40
Corn gluten feed	90–95	0–50
Distilled dried grains	80–85	0–35
Hominy feed	95–98	0–75
Millet	90–100	0–50
Milo, grain sorghum	85–95	0–100
Molasses, cane or blackstrap	75–95	0–5
Oats	88–94 (75–85)	0–35 (35–100)
Rice grain	85–95	0–50
Rye	80–85	0–20
Screenings, No. 1	90–95	0–75
Screenings, No. 2	75–82	0–50
Screenings, commercial grade	45–65	0–35
Soybean millfeed	65–85	0–20
Wheat	100–105	0–50
Wheat bran	65–80	0–35
Wheat mids.	70–85	0–35

[a] Adapted from 1975–1976 Ingredient Analysis and Estimated Feed Value Tables for Beef and Sheep Growing–Finishing Rations (Morrison, 1975).

without reducing cattle gain. Substitution of straight shelled corn (treatment 3) or of corn plus cottonseed meal (treatment 4) for the basal diet, supplemented with wastelage further improved efficiency of gain.

e. Additional Sources. These include peanut meal, sunflower meal, coconut meal, safflower meal, and many others. The author in cooperation with the USDA has established the safety and nutritive value of crambe meal to replace up to one-half or more of the supplemental protein for finishing beef cattle.

Table 10.15 lists the feeding value of various feedstuffs in relation to corn grain and a suggested level of feeding each ingredient as a replacement for corn. Recommendations as to usage level are those that should result in the corresponding feeding value listed when fed in a balanced diet. In many cases, feedstuffs could be fed at levels greater than those listed. This, however, will in most cases result in somehwat lower feeding value and thus lower animal performance.

TABLE 10.16

Estimated Feed Value and Corn Replacement Value of Various Beef Cattle Dietary Ingredients[a]

Ingredient	Level advised as grain or grain replacement in diet	Value for cattle as replacement for No. 2 corn on equal poundage
A. Grains and low protein materials		
Alfalfa meal, 12% sun-cured	pelleted diets, 20–50	45–50
Almond hulls, no shells	0–30	70–75
Almond hulls plus shell meal	0–20	30–35
Bakery product, blended, dried	0–30	110–115
Barley, Midwest or East grown	0–100	88–90
Barley, as above, heavy weight	0–100	88–94
Barley, Pacific Coast	0–100	80–88
Beet pulp, dried	0–50	88–95
Beet pulp w. molasses, dried	0–50	88–95
Buckwheat, ord.	0–30	75–85
Citrus pulp, dried	0–50	80–88
Corn, waxy	0–100	100–105
Corn, high moisture (equal moisture basis)	0–100	100–105
Corn and cob meal	0–100	85–92
Corn and cob meal, high moisture	0–100	90–95
Corn, snapped (grain, cob, husk)	0–100	75
Corn gluten feed	0–50	90–95

[a] Adapted with permission from 1977 Cattle Feeders Planner, Feedlot Magazine, Miller Publishing Co., Minneapolis, Minn. Section author was Dr. S. H. Morrison.

VI. SUMMARY AND CONCLUSION

There are numerous grains and by-product feeds that can be fed to beef cattle economically if proper attention is given to processing and feeding management. In many cases some products (i.e., screenings) are quite variable in nutrient content and feeding value from batch to batch. To minimize the risk of reducing animal performance due to variation in nutrients and quality, these materials should be used as only part of the ration, since they are likely to have greater feeding value and their variability in quality will have less effect on animal performance. Care should be taken to properly supplement substitute feeds. In some cases, it may be advisable to provide higher than normal levels of supplemental minerals, vitamins, and protein. This provides a wide margin of safety when it is recognized that a particular feedstuff is quite variable in quality from batch to batch (Table 10.16).

TABLE 10.16 *(continued)*

Ingredient	Level advised as grain or grain replacement in diet	Value for cattle as replacement for No. 2 corn on equal poundage
Hegari grain	0–100	85–95
Hominy feed, yellow	0–75	95–98
Kafir grain, irrigated	0–100	85–95
Kafir head chops	0–100	70–75
Milo or sorghum grains	0–100	80–95
Milo head chop	0–100	75
Molasses, beet	0–25	75–95 (limited)
	25–45	40–50 (overfed)
Molasses, cane	0–45	(same as beet molasses)
Molasses, citrus	0–35	70–85 (limited)
Molasses substitute (wood by-product)	0–20	75–95
Oat millfeed, usual grade	0–25	25–35
Oat millfeed, high quality	0–35	45–50
Oats, good grade	0–35	88–94
	35–100	75–85
Oats, lower grade	0–35	82–92
	35–100	72–82
Potatoes, white, dried	0–35	95–98
	35–50	90–95

(continued)

TABLE 10.16 (*continued*)

Ingredient	Level advised as grain or grain replacement in diet	Value for cattle as replacement for No. 2 corn on equal poundage
Potatoes, sweet, dried	0–50	95–98
	50–100	85–92
Rice bran, solvent	0–35	67–75
Rice polishings	0–30	88–95
Rye grain	0–50	80–85
Screenings, grain	0–75	90–95
Soybean mill feed	0–20	65–85
Spelt and emmer	0–30	85–90
Triticale	0–50	95–100
Wheat, hard spring	0–100	100–103
Wheat, hard winter	0–100	95–97
Wheat, soft winter	0–100	96–98
Wheat, soft, Pacific	0–100	95–97
B. Animal and vegetable fats	0–5	160–180
C. Wet materials sometimes used		
Beet pulp, wet	0–70	10–20
Brewer's grains, wet	0–35	13–15
Cull carrots	0–20	10–15
Potatoes, white, fresh	0–35	20–25
Potatoes, sweet, fresh	0–50	25–33
D. Medium protein materials		
Alfalfa meal, 17 and 20%, Dehy.	20–50 (pelleted)	60–65
Alfalfa meal, 15	20–25	55–60
Avocado oil meal	0–20	60–65
Beans, field, navy	0–17	70–85
Brewer's grains, 22% protein	0–35	82–85
Brewer's grains, 26% protein	0–35	85–88
Coconut meal, hydraulic	0–50	87–90
Coconut meal, solv.	0–50	85–88
Corn germ meal, wet milled	0–25	90–95
Corn germ meal, wet mill, solv.	0–25	80–88
Corn gluten feed (21–22%)	0–25	80–88
Cottonseed, whole, unext.	0–30	110–120
Dist. dried grains (corn)	0–25	80–85
Dist. dried grains, w. solubles	0–35	80–85
Dist. dried rye grains	0–35	73–75
Dist. dried sorghum grains	0–35	78–80
Peas, cull field	0–40	100
Tomato pulp, dried	0–30	75
Wheat bran	0–35	65–80
Wheat middlings	0–35	70–85
Wheat shorts	0–40	72–86

REFERENCES

Albin, R. C., and Sherrod, L. B. (1975). Nutritional value of cattle feedlot waste for growing-finishing beef cattle. *Managing Livest. Wastes, Proc. Int. Symp., 3rd, 1975* p. 211.

Anonymous (1978). Feed marketing. *Feedstuffs Ref. Issue* **48,** No. 38, 25.

Anthony, W. B. (1971). Cattle manure as feed for cattle. *Proc. Int. Symp. Livest. Waste, 2nd, 1971* p. 293.

Bhattacharya, A. N., and Taylor, J. C. (1975). Recycling animal waste as a feedstuffs: A review. *J. Anim. Sci.* **41,** 1438.

Brady, S. E., and Farlin, S. D. (1978). "Waxy Corn for Beef Rations," Beef Cattle Day Rep., p. 14. University of Nebraska, Lincoln.

Brethour, J. R. (1977). Cattle feeding roundup. *Beef Mag.* **14,** No. 10, 32.

Crampton, E. W., and Harris, L. E. (1971). "Atlas of Nutritional Data on United States and Canadian Feeds." Natl. Acad. Sci., Washington, D.C.

Fontenot, J. P., Webb, K. E., Jr., Harmon, B. W., Tucker, R. E., and Moore, W. E. C. (1971). Studies of processing, nutritional value and palatability of broiler litter for ruminants. *Proc. Int. Symp. Livestock Waste, 2nd, 1971* p. 301.

Goodrich, R. D., and Meiske, J. C. (1974). "Corn Silage for Beef Cattle," Cattle Day Rep. B-200, p. 56. University of Minnesota, St. Paul.

Hale, W. H. (1963). "Grain Processing," Cattle Feeders Day Rep., p. 22. University of Arizona, Tucson.

Hale, W. H. (1966). The role of fat in cattle fattening rations. *Beef Cattle Sci. Handb.* **3,** 210.

Hale, W. H., Ciutin, L., Saba, W. J., Taylor, B., and Theurer, B. (1966). Effect of steam processing and flaking milo and barley on performance and digestion by steers. *J. Anim. Sci.* **25,** 392.

Hatch, C. F. (1971). Factors affecting the synthesis of protein from urea in beef cattle. Ph.D. Thesis, Purdue University, West Lafayette, Indiana.

Hatfield, E. E., Braman, W. L., and Robinson, G. N. (1975). "Chemical Properties that Affect the Nutritive Value of Corn," Cattle Day Rep., p. 20. University of Illinois, Champaign.

Morrison, S. H. (1975). Ingredient analysis and estimated feed value tables for beef and sheep growing-finishing rations. *Feedstuffs* **47,** 32.

Oltjen, R. R., Putnam, P. A., Williams, E. E., Jr., and Davis, R. E. (1966). Wheat vs. corn in all-concentrate cattle rations. *J. Anim. Sci.* **25,** 1000.

Perry, T. W. (1972). Adaptability is the key to a liquid feed system. *Feedlot Mag.* **14,** No. 11, 32.

Shaake, L. M., Riggs, J. K., and Butler, O. D. (1972). Commercial feedlot evaluation of four methods of sorghum grain processing. *J. Anim. Sci.* **34,** 926.

Thomas, V. M., Beeson, W. M., Perry, T. W., and Mohler, M. T. (1975). "Nutritional Value of Normal, Opaque-2, Waxy Corn and Other Types for Finishing Beef Cattle," Cattle Feeders Day Rep., p. 67. Purdue University, West Lafayette, Indiana.

Wilson, W. M. D., Hatfield, E. E., and Hixon, D. L. (1975). "Phosphorus Sources and Levels for Feedlot Steers," Cattle Feeders Day Rep., p. 24. University of Illinois, Champaign.

III

The Breeding Herd

11

Breeding Herd Nutrition and Management

It is not difficult, really, to meet the nutritional needs of the mature nonlactating beef cow, whether she is pregnant or not. Naturally, she does have definite nutrient requirements which must be met or impaired performance may be expected. Principal among her nutrient requirements which may require special attention are limited energy, protein, vitamin A, calcium and phosphorus; almost all other of her nutrient requirements will be met adequately. On rare occasion a magnesium deficiency or a copper or cobalt deficiency may be encountered. The National Research Council bulletin on Beef Cattle Nutrient Requirements (Perry *et al.*, 1976) has established (see Appendix) requirements for protein, energy, calcium, phosphorus, vitamins A and D, and a few trace minerals.

Lactating dietary requirements differ from nonlactating ones in requiring slightly increased levels of energy sources, nearly doubled levels of protein. calcium, and phosphorus, but no change in vitamin A. Reproduction in mature beef cattle generally is a most natural phenomenon requiring little extra quantities of nutrients. Often in the more pampered cow herds, it is a matter of overfeeding energy—and possibly protein and other nutrients—rather than underfeeding.

I. NUTRITIONAL NEEDS OF REPLACEMENT HEIFERS

Wiltbank *et al.* (1969) conducted research to determine the effect of energy levels fed to weanling heifers on age at puberty (or first appearance of estrus). Two levels of feeding were compared. Heifers on the "high" nutritional level received a full feed of concentrates plus hay whereas heifers on the "low" nutritional level were fed only protein supplement and hay. Heifers were fed the respective high- and low-energy diets from weaning until time of first ovulatory estrus (which was defined as puberty). Estrus was detected with the aid of sterilized bulls which had a marking grease pigment on their briskets.

The effect of energy level on age at first estrus—or puberty—is quite obviously shown in Table 11.1. The average age at puberty was 381 days for heifers on a high nutritional level of energy for both straightbred and crossbred heifers; but on the low level of nutrition, puberty did not occur until over six months (191 days) later for straightbred heifers and a month and a half later (43 days) for crossbred heifers. Average live weight at puberty was 658 and 726 lb for straightbred and crossbred heifers, respectivley, on high nutrition and 590 and 599 lb, respectively, on low nutrition.

The data in Table 11.1 bear out the generally accepted idea that there exists differences among breeds as to age for reaching puberty, with Herefords taking a bit longer. However, the data presented demonstrate nutrition probably has a much greater effect on age at puberty than breed.

Once puberty, ovulation, estrus, and fertilization have been achieved in the young heifer, then consideration must be given to her nutrition and management for the next 283 ± days of gestation.

The USDA (Wiltbank *et al.*, 1965) studied the effect of energy and protein levels in the diet of pregnant beef heifers on reproductive performance of heifers carrying their first calves. In the energy level comparison, heifers on high energy were fed their diet ad lib (25% timothy hay, 10% molasses, plus corn and cob meal and cottonseed meal to meet the desired protein level) whereas heifers on medium energy were fed 66% as much as that consumed by high-energy heifers. Those on the low-energy regimen were fed only enough to maintain body weight. Then within each of the three energy levels, protein was fed at high, medium, and low levels. At the high protein level, 0.23 lb digestible protein per 100 weight (cwt) was fed; at medium protein, 0.15 lb digestible protein per cwt was fed; and at low protein, 0.06 lb digestible protein per cwt was fed.

The heifers were kept on their respective diets through gestation except that the

TABLE 11.1

Age and Weight of Beef Heifers at Puberty as Affected by Energy Level Fed[a]

	H bull H cow	A bull A cow	A bull H cow	H bull A cow	Straight-bred	Cross-bred
Age at puberty (days)						
High nutrition	387	374	384	378	381	381
Low nutrition	660	483	416	402	572	424
Difference	273	109	62	24	191	43
Weight at puberty (lb)						
High nutrition	645	671	728	724	658	726
Low nutrition	614	565	594	524	590	559
Difference	31	106	134	200	68	167

[a] Wiltbank *et al.* (1969). H, Hereford; A, Angus.

TABLE 11.2

Gestation Energy and Protein Level in First Calf Heifer Diets, and Reproductive Performance[a]

Dietary level	Heifer weight one week before calving (lb)	Live calves (%) At birth	At 2 weeks	At weaning	Average birth weight (lb)	Calf's birth weight as % of dam's weight	Average gestation period (days)
High energy							
High protein	1151	80	20	20	54	5.1	277
Medium protein	1089	80	40	40	62	6.3	275
Low protein	873	67	50	50	46	4.9	277
Medium energy							
High protein	926	100	100	100	60	7.2	272
Medium protein	891	80	80	80	56	6.8	274
Low protein	760	80	80	67	54	8.1	282
Low energy							
High protein	625	75[b]	75	75	43	8.3	280
Medium protein	569	100	100	100	40	8.3	274
Low protein	533	67[c]	67	67	41	8.8	273

[a] Wiltbank *et al.* (1965).
[b] One premature calf at 7.5 months.
[c] Two premature calves at 6.5 months.

heifers on low energy were fed slightly more feed to compensate for the increase in weight due to pregnancy. This program continued for 180 days or until they were 90 days pregnant with their second calf. Beginning at this time, heifers were fed ad lib on a diet of 94% timothy hay and 6% cottonseed meal, plus vitamin A and minerals, through their second calving. This diet then tended to improve the condition of the low-energy heifers and to reduce the obesity of the high energy group (Table 11.2).

A large number of the calves born to heifers on the high-energy–low-energy protein regimen or on the high-energy–medium-energy protein regimen died either at birth or shortly thereafter. Many of these calves were presented for birth backward or in some other abnormal position. Birth weights of calves born to heifers on high-energy diets were not excessive and gestation period was comparable. Calving difficulty of heifers appeared to be related more closely to their own conditions rather than to size of the calf. Birth weights of the calves born to heifers on the low-energy diets were considerably lower than those of calves born to heifers on high- or medium-energy diets which contained high or medium protein levels.

Heifers fed high-energy or medium-energy diets with either high or medium protein levels produced the most milk and thus heavier calves at 60 days of age.

Low protein in the dam's diet caused decreased milk production and thus lowered 60-day calf weights. It is interesting to note from Table 11.3 that on low-energy diets, a high level of protein seemed to depress milk production of the dam and growth of the calf. Thus, if the energy level is low, it is best not to feed high levels of protein to the dam. Also, high energy alone to the lactating dam is not sufficient—she needs at least a medium level of protein to transform her energy into milk and thus into optimum calf growth.

Perhaps the most sensible way then to feed lactating first calf heifers is on medium energy and medium protein.

Differences in average body weights of the cows in the various groups used in the first calf studies tended to disappear after the cows were put on the hay and cottonseed meal before their second calves. At the time they went on the hay diets, some heifers differed in weight by 700 lb. Heifers on medium or low levels of energy and low levels of protein gained weight steadily from the time they were on hay until their second calf was born. After second calving, the weight losses of cows formerly on high energy-high protein or on high energy-medium protein diets tended to be greater than those of heifers on the other dietary regimens. The weight gains of the cows before second calving and weight losses after calving tended to minimize the large differences which had been built up due to the first calving treatments. Occurrence of estrus following second calving did not appear to be affected by previous programs, and there appeared to be little

TABLE 11.3

Growth of First Calves and Milk Production of Heifers as Affected by Energy and Protein Levels[a]

Dietary level	Weight gain of calves, first 60 days (lb)	12-hr milk production of dams at 60 days (lb)
High energy		
High protein	104	8.2
Medium protein	92	7.3
Low protein	57	6.2
Medium energy		
High protein	89	7.5
Medium protein	98	8.4
Low protein	48	6.5
Low energy		
High protein	38	3.4
Medium protein	42	4.3
Low protein	39	4.6

[a] Wiltbank *et al.* (1965).

Fig. 11.1. Yearling replacement heifers are fed a high silage ration during winter drylot conditions.

or no carryover effect in the second calves. Body size of the heifers was not affected permanently by the dietary regimens applied over their first calving.

From the preceding discussion, it is obvious the nutrient requirements of the replacement heifer from weaning through her first pregnancy and lactation are quite specific and must be adhered to if optimum performance is to be obtained (Fig. 11.1). It is fortunate, however, that even though her nutrient requirements are not met—or if they are oversupplied, as in the case of energy—probably no permanent damage has been done. Therefore, she will be capable of performing quite normally in her second and subsequent pregnancies and lactations if she is given the opportunity with reasonably sound feeding practices during her second and subsequent pregnancies. Possibly no other farm animal is as capable of living a normal healthy reproductive life as the beef female.

A replacement heifer should gain approximately one pound per day, or slightly more, from weaning at 400–450 lb through her first pregnancy. This means that if she calves at two years of age, she should be weighing near 900 lb for most of the English breeds; this should be a bit more for the heavier breeds. It is not too difficult to supply enough feed to cause a pound a day gain, but attention must be given to some of the details. The National Research Council (see Appendix) suggests a 440-lb heifer (weanling) gaining 1.1 lb per day should receive 1.25 lb of crude protein, 7.7 lb TDN, 14 gm calcium, 13 gm phosphorus, and 13 thousand IU of vitamin A per head daily. Furthermore, she has a feed capacity for about 13 lb of dry matter (15 lb of 87% dry matter equivalent). Thus, 15 lb of a diet containing 13% water, 8.5% crude protein, 51% TDN, plus calcium, phosphorus, and vitamin A would meet these requirements. Typically, good pasture or reasonably good-quality hay would meet all these nutrient requirements, with the possible exception of vitamin A. However, vitamin A may be purchased, and an intramuscular injection of 5 million IU of vitamin A per heifer would suffice for from 6 to 9 months.

Some sample feeding programs for the weanling heifer from weaning (450 pounds and 6 months of age) to breeding time (740 pounds and 15–16 months of age) follows below.

1. Good-quality pasture plus limited protein supplement (1 lb per head daily of a 32% protein supplement, or of soybean meal, or of dehydrated alfalfa meal pellets) if the pasture is fairly mature. If the pasture herbage is in short supply, then 2–3 lb of grain plus the protein supplement will be needed. A free-choice mineral mixture of two parts dicalcium phosphate to one part of trace mineralized salt should be available in a device which will prevent rain from getting to the mineral mixture.
2. In the season when pasture is not available, a bright clean hay containing a small amount of legume will be satisfactory. However, if the hay is not too good, then from 1 to 2 lb of grain should be supplied. The free-choice mineral mixture listed above should be included.
3. Corn silage fed on a limited basis (25 lb per head daily) plus 1.5 lb of a 32% protein supplement will cause a gain in excess of 1 lb per day—possibly 1.5 lb per day. If this program is followed, the free-choice mineral mixture should be included.
4. Alfalfa haylage, or small grain haylage, are excellent growing feeds for weanling heifers through the time of breeding. Good quality haylage is almost a complete diet in itself and needs little else except free-choice minerals.
5. Such by-products as corn stover, cornstalk fields, or straw are not very good feeds for young heifers at this age. Such feeding programs are deficient in energy, protein, calcium, phosphorus, and vitamin A. These feeding programs should be saved for the mature beef cow who is better able to utilize such feeds and whose nutrient requirements are not nearly as exacting as those of the developing heifer.

II. BEEF COW FEEDING PROGRAMS

The nutrient requirements for brood cows can be classified under four general headings, namely, (1) energy, (2) protein, (3) minerals, and (4) vitamins. These requirements are set forth in Table 11.4.

Two major considerations help determine the nutritional requirements of a dry mature pregnant beef cow or of a lactating beef cow. These are size of the cow and amount of milk produced. The National Research Council (Table 11.4), establishes the requirements for most given situations. For example, an 1100-pound mature pregnant cow (not in lactation) during the middle one-third of pregnancy should eat a minimum of 17.7 lb (air-dry) of feed which contains 0.92 lb protein, 8.6 lb TDN, 13 gm each of calcium and phosphorus, and 20,000 IU of vitamin A daily. What kind of diet, then, are we talking about? It is about 49% TDN and 5% protein plus calcium, phosphorus, and vitamin A. Almost any type of hay would meet these requirements. For example, a grass hay fed at the level

TABLE 11.4

Nutrient Requirements of Mature Beef Cows[a]

Body weight (lb)	Feed, 90% dry matter, per animal (lb)	Total protein (lb)	TDN (lb)	Ca (gm)	P (gm)	Vitamin A (thousand IU)
Dry pregnant mature cows, middle third of pregnancy						
880	14.8	0.80	7.3	11	11	17
990	16.4	0.86	7.9	12	12	19
1100	17.7	0.92	8.6	13	13	20
1210	18.9	1.00	9.2	14	14	22
1320	20.3	1.08	9.8	15	15	23
Dry pregnant mature cows, last third of pregnancy						
880	17.1	0.97	8.7	14	14	21
990	18.3	1.06	9.4	15	15	23
1100	19.9	1.12	10.0	15	15	24
1210	21.1	1.18	10.7	16	16	26
1320	22.6	1.25	11.2	17	17	27
Cows nursing calves, average milking ability, first 3–4 months						
880	19.4	1.78	10.4	25	25	21
990	20.5	1.89	11.0	26	26	23
1100	21.6	1.98	11.7	27	27	24
1210	23.1	2.13	12.3	28	28	26
1320	24.2	2.22	13.0	28	28	27

[a] A modification of Table 1B, National Academy of Sciences Publication, Nutrient Requirements of Beef Cattle (Perry *et al.*, 1976).

of 17–18 lb per day, along with a free-choice mineral mixture, certainly would meet all the requirements except possibly vitamin A.

A. A Year-Round Feeding Program for the Cow Herd

The best adapted feeding program for a cow–calf herd depends on (1) time of year for calving, and (2) feeds available. The two most common calving periods are fall and spring, with both systems built around the availability of a pasture program for as long as possible. Then when pasture is no longer available, harvested feeds are introduced. Naturally, the length of the pasture season will vary from one area of the country to another. In Indiana the typical pasture season extends from May 1 to November 1, with possible modifications on either or both ends of the season. In Southern climes the pasture season possibly may extend nearly year-round. However, for illustrative purposes, an examination of a Corn Belt cow herd program for feeding beef cows is provided.

1. Spring Calving (February 15 to April 15).

We will start with May 1 when grass pasture is ready. The calves are approaching six weeks of age by May 1 and, therefore,will start to benefit from the tender young grass. A real benefit from this program is that lush grass will provide the total needs—except for minerals—of the lactating beef cow. A free-choice mineral mixture is needed at this point and can be as simple as two parts dicalcium phosphate to one part trace mineralized salt. In areas where grass tetany is suspected, a free-choice mineral mix consisting of equal parts dicalcium phosphate, trace mineralized salt, magnesium oxide, and ground shelled corn should be provided. This phase could extend then from May 1 to November 1; by that time the calves have been weaned and the cows are no longer producing milk.

Starting around November 1 when the calves have been weaned, the cow feeding program then can revert to a maintenance program until about February 15 when the next crop of calves will be born. Any one of the following roughage programs, when a free-choice mineral mixture is fed, will be adequate for this nonlactating—but pregnancy—program (Table 11.5).

During lactation and before young lush pasture is available (February 15 to May 1), you can continue on the roughage program outlined for the dry period. However, at this time the cow will need added energy. This may be provided by adding 4 or 5 lb of grain per cow daily (Fig. 11.2). This may be difficult to provide because the cows would like to have access to four times that quantity. Therefore, distribution of such a small quantity becomes a management problem. However, this is the period of the very highest energy demand by the cow. If her requirements are not met, her milk production will fall off and calf performance will suffer. In addition, she needs added protein during the lactation period. Therefore, in addition to 4 or 5 lb of corn, approximately one additional pound of

TABLE 11.5

Roughage Feeding Programs for Nonlactating Pregnant Cows when Pasture Is Not Available

Feedstuff	Dry matter (%)	Intake (as fed) (lb/day)	Protein supplement (32–40%) (lb/day)
A. Silages			
1. Corn silage	35	38	1
2. Oat silage	30	57	—
3. Wheat silage	30	50	—
4. Haylage	50	32	—
5. Sorghum silage	35	40	1
6. Grass silage	30	60	1
7. Stalkage	40	60	1.5
8. Legume haylage	50	40	—
B. Hay			
1. Grass[a]			
a. Orchard	90	16	1
b. Timothy	90	16	1
c. Brome	90	16	1
d. Fescue	90	16	1
e. Bluegrass	90	16	1
2. Legume–grass, mixed			
a. Alfalfa	90	15	—
b. Red clover	90	16	—
C. Straws and others			
1. Oat or wheat	90	20	1.5
2. Corn cobs	90	20	1.5
D. Grazing			
1. Cornstalks	75–90	Ad lib	1.5
2. Dead grass aftermath	90	Ad lib	1.5

[a] If the grass hay is green and of high quality, no supplemental protein will be needed.

a 32% protein supplement should be provided. Often the protein requirement is provided by a 32% liquid supplement in a lick tank. Cows normally will not overeat liquid supplements provided in this manner.

When the cows and their calves are turned to young pasture (May 1), about all the supplement they will need is a mineral mixture, probably containing magnesium oxide. Later, if the pasture tapers off in quality, a lick tank may be needed to provide supplemental protein.

2. *Fall Calving (October 1 to December 15)*

In a fall calving system, the cow will be dry nearly all of the pasture season, and fair to good pasture will more than meet all her nutrient needs except possibly minerals. If the pasture is good, she may even gain a little extra weight.

Fig. 11.2. Cornstalks, properly supplemented, make excellent winter cow pasture. (Photo by J. C. Allen and Son.)

With the onset of lactation, additional energy and protein will be needed. Probably by this time pasture will not be available, and so a nonpasture lactation program such as that listed above for the spring calving cow should be implemented.

a. Meeting the Protein Requirements. Good pasture and mixed (legume and nonlegume) hays are excellent sources of protein. Normally, on good pasture supplemental protein is not required. Similarly, when high-quality legume hay is fed, probably no supplemental protein is needed. However, lactating cows not on pasture and dry pregnant cows fed poorer quality roughages probably need supplemental protein. The following are some suggested methods for supplying from one-third to one-half pound of supplemental protein per cow daily:

1. A minimum of 6 lb clover or alfalfa hay.
2. One pound of soybean meal, cottonseed meal or linseed meal.
3. A 32% protein supplement fed at a level of 1.5 lb.
4. A 32% protein liquid supplement fed from a lick tank.
5. Protein blocks fed ad lib.

The method selected should be based on economy and management required. One of the most economical and easiest to manage is a liquid supplement offered from a lick tank on a free-choice basis.

b. Vitamins. Probably the only vitamin requiring attention in diet formulations for mature beef cattle is vitamin A. A dry cow requires about 20,000 IU per day whereas a lactating cow needs approximately 25,000 IU daily. Any one of the following procedures will ensure adequate vitamin A for the brood cow:

1. At least a portion of the roughage which is (a) green, (b) high quality and (c) less than a year old will provide adquate carotene for the cow to synthesize vitamin A.
2. Add vitamin A via a dry or liquid protein supplement.
3. An intramuscular injection of 2.5 to 3 million IU of vitamin A will provide adequate protection for 6 months.
4. Add vitamin A to the free-choice mineral mix at a level to supply 20 to 25 thousand IU in one ounce of mineral. Although this is the least satisfactory method, still this may be the only one available under some circumstances.

c. Minerals. The following minerals should be given consideration for beef cows: calcium, phosphorus, salt, iodine, magnesium, zinc, and cobalt. Any one of the following may be employed to supply minerals:

1. If a commercial protein supplement is provided, it should contain most of the supplemental minerals required. As a rule of thumb, the complex protein supplement should supply at least one-half the daily mineral requirement.
2. Provide a free-choice mineral mixture even if a complex protein supplement containing minerals is provided. This gives cattle the opportunity to consume additional minerals, if they desire. For this, you may purchase a commercial mineral or you may prepare a homemade mineral mix consisting of equal parts dicalcium phosphate and trace mineralized salt. It is difficult to improve on this mineral mixture unless there is a potential grass tetany problem.
3. Grass tetany periods require that magnesium be provided. Magnesium oxide is a good source, but no source of magnesium is very palatable for cattle. Therefore, the source should be flavored or masked. A good free-choice mineral mix for grass tetany-suspect situations consists of equal parts dicalcium phosphate, trace mineralized salt, magnesium oxide, and ground shelled corn.

B. Feeding Systems That Meet the Cow's Winter Nutrient Needs

Various combinations of feedstuffs will supply the nutrients for a dry mature pregnant cow from the end of the pasture season (November 1) until calving

(February 15) and a lactating cow from calving until pasture (May 1) (Fig. 11.3). The following tabulation lists some feeding programs for an 1100-pound cow. (Free-choice minerals should be provided in all situations listed below.)

1. Dead grass aftermath, grazed	Nov. 1 to May 1
4 lb grain per day	Calving to May 1
Protein supplement (provide 0.5 lb protein)	Feb. 1 to May 1
Minerals, free choice	Nov. 1 to May 1
2. Grass hay (16 lb)	Nov. 1 to May 1
4 lb grain per day	Calving to May 1
Protein supplement (provide 0.5 lb protein)	Feb. 1 to May 1
3. Cornstalks (grazed)	Nov. 1 to Calving
Hay (all the cow will eat)	Calving to May 1
5–7 lb cereal grain/head/day	Calving to May 1
Protein supplement (provide 0.5 lb protein)	Feb. 1 to May 1
4. Cornstalks (grazed)	Nov. 1 to Calving
Corn silage (58 lb/head/day)	Calving to May 1
Protein supplement (provide 0.5 lb protein)	Jan. 1 to May 1
5. Grass hay (16 lb/head/day)	Nov. 1 to Calving
Corn silage (58 lb/head/day)	Calving to May 1
Protein supplement (provide 0.5 lb protein)	Jan. 1 to May 1
6. Corn silage (38 lb/head/day)	Nov. 1 to Calving
Corn silage (58 lb/head/day)	Calving to May 1
Protein supplement (provide 0.5 lb protein)	Jan. 1 to May 1
7. Grass hay (10 lb/head/day)	Nov. 1 to May 1
Alfalfa hay (6 lb/head/day)	Nov. 1 to May 1
5–7 lb cereal grain/head/day	Calving to May 1
8. Mixed clover and grass hay (16 lb/head/day)	Nov. 1 to May 1
Corn silage (20 lb/head/day)	Calving to May 1
9. Mixed clover and grass hay (16 lb/head/day)	Nov. 1 to May 1
5–7 lb cereal grain/head/day	Calving to May 1
10. Haylage (32 lb/head/day)	Nov. 1 to May 1
5–7 lb cereal grain/head/day	Calving to May 1
11. Stalklage (all the cow will eat)	Nov. 1 to May 1
Corn silage (20 lb/head/day)	Calving to May 1
Protein supplement (provide 0.5 lb protein)	Nov. 1 to May 1
12. Stalklage (all the cow will eat)	Nov. 1 to May 1
5–7 lb cereal grain/head/day	Calving to May 1
Protein supplement (provide 0.5 lb protein)	Nov. 1 to May 1

III. CROSSBREEDING AND COW PRODUCTIVITY

Crossbreeding has been recognized as a method of reaping the benefits of so-called "hybrid vigor" in many aspects of livestock production. Several experiment stations and farm records have demonstrated that crossbred calves will grow from 5 to 10% more rapidly than their herdmates. Purdue University (Mar-

Fig. 11.3. Large bales provided in racks hold roughage wastage to a minimum. (Photo by J. C. Allen and Son.)

tin *et al.,* 1975) published research date on the effectiveness of crossbreeding a dual-purposed breed (Milking Shorthorn) and a traditional beef breed (Aberdeen Angus). Matings were designed to produce contemporary animals of four breed combinations: (1) 72 purebred Angus (A), (2) 50 purebred Milking Shorthorn (MS), (3) 64 Angus sire times Milking Shorthorn dam (A × MS), and (4) 57 Milking Shorthorn sire times Angus dam (MS × A). The above cows were all bred to Hereford bulls for their first calf and to Charolais bulls for their second calf. The amount of hybrid vigor was determined as the difference between the two crossbred groups and the two purebred groups. Since no significant hybrid vigor effects on carcass traits were observed, the results shown in Table 11.6, will not make reference to them.

From the last two lines of the table, it may be observed that crossbred cows produced 73 lb additional calf weight per cow exposed to the bull which represents an increase of 27% over the average of purebred cows. This means four crossbred cows produced as much weaning weight as five purebred cows. The crossbred cows were heavier (however, see last line of Table 11.6). Therefore, based on the difference in cow weight, crossbred cows produced only 20% more weaned calf per hundredweight of cow weight than purebred cows. It, therefore, requires the production of five crossbred cows to equal that of six purebred cows.

TABLE 11.6

Summary of Two-Year Performance of Four Groups of Cows[a]

Item	Angus (A)	Milking Shorthorn (MS)	A × MS	MS × A
First calf				
Weaning wt (lb)	394	420	426	430
Percentage weaned	0.67	0.60	0.81	0.75
Pounds calf/cow exposed (lb)	263	252	346	324
Cow wt (lb)	703	701	771	742
Pounds calf/cwt cow (lb)	37.4	35.9	44.9	43.7
Second calf				
Weaning wt (lb)	427	451	472	463
Percentage weaned	0.65	0.55	0.73	0.70
Pounds calf/cow exposed (lb)	278	250	342	325
Cow wt (lb)	749	714	804	756
Pounds calf/cwt cow (lb)	37.1	35.0	42.6	43.0
Avg calf wt/cow exposed (lb)	271	251	344	325
Avg lb calf/cwt cow (lb)	37.3	35.5	43.8	43.4

[a] Purdue University data (Martin *et al.*, 1975).

The total effects of crossbreeding on efficiency seem to be an accumulation of a number of small effects. The crossbred cow usually breeds more promptly; she then mothers the calf better than the purebred cow. The increased milking ability of the Milking Shorthorn might be considered as vital here and possibly should be weighed heavily in making decisions concerning crossbreeding programs.

IV. EFFECT OF DIET ON ESTRUS

In contrast to the mature beef cow, the nutritional needs of the growing replacement heifer are much more rigid, and should be met if promising results are to be achieved. It is most desirable that beef heifers calve as two-year-olds if maximum return on investment is to be realized. However, this program requires adequate nutrition plus two other management considerations. Such heifers should be bred to bulls which are known to sire smaller calves and skilled assistance should be available at calving time (see Table 11.7).

The Oklahoma data show that cows calving as 2-year-olds live productively just as long and produce one more calf during a 12-year lifetime than cows which do not calve until they are 3 years of age. However, the heifers which calve as 2-year-olds will need additional assistance at calving time.

During the first winter of the developing heifer's life, it is important she gain about one pound per day if she is expected to reach puberty by 15 months of age.

From the Nebraska report it is obvious that breed, weight, daily gain, and heterosis, as well as age, affect the appearance of puberty—and first estrus. Therefore, in order to obtain as early estrus as possible, (1) feed replacement heifers to gain one pound per day following weaning; (2) although breed differences affect minimum desirable weight, it appears Angus and Shorthorn heifers should weigh 550 lb or more at 15 months whereas Herefords may need to weigh almost 100 lb more; and (3) hybrid vigor resulting from crossbreeding may assist in the appearance of 15-month estrus wherein the developing heifers have received less energy. At this time such age and weight effects on the appearance of first estrus in developing heifers of breeds other than those listed in Table 11.8 are not readily available.

The apparent heavier weight requirement for the appearance of puberty in Hereford heifers was emphasized by Wiltbank (1968) (Table 11.9) who studied puberty in such heifers at 50-pound weight increments from 400 pounds through 650 pounds live weight.

Turman *et al.* (1964) further demonstrated the effect of level of winter feeding of heifers on the onset of puberty and thus first appearance of estrus (Table 11.10). The Oklahoma data thus illustrate the importance of moderate to high energy levels for wintering weanling heifers if a high percentage of them are expected to reach puberty by 15 months of age. Breeding yearling heifers after 15 months of age will extend the next year's calving period beyond the desirable 60-day limit. Thus, such heifers will be expected to be late in calving for many years to come.

The data available do not suggest that creep feeding of suckling heifers has any advantage in causing the early appearance of puberty. In fact, there is some

TABLE 11.7

Age at Calving and Lifetime Performance[a]

Item	Age at calving: Two years	Age at calving: Three years
Number of cows	60	60
Number remaining at 12 years of age	41	42
Years in production	10	9
Calves weaned per cow	9.1	7.9
Percentage calf crop weaned[b]	86.7	85.2
Adjusted calf weaning weight (lb)	476	487
Heifers assisted at first calving	28	1
Cow mature weight (lb)	1148	1178

[a] Turman *et al.* (1964).
[b] Based on number of cows bred to calve each year.

TABLE 11.8

Age and Gain Effect on Beef Heifer Puberty[a]

	0.5 lb Daily winter gain[b]				1.0 lb Daily winter gain[c]			
	Puberty		% Estrus		Puberty		% Estrus	
Breed	Age (mo)	Wt (lb)	13 mo (%)	15 mo (%)	Age (mo)	Wt (lb)	13 mo (%)	15 mo (%)
Angus	13.1	518	33	77	11.2	572	76	92
Hereford	15.5	594	4	41	13.6	665	38	77
Shorthorn	13.7	500	25	91	10.9	544	73	92
A × H	13.2	550	41	91	11.8	628	74	97
A × S	12.1	508	71	94	10.2	556	87	100
H × S	12.4	522	64	94	9.7	542	97	100

[a] Wiltbank *et al.* (1965).
[b] Wintered on range plus 1 lb of 40% protein supplement per day.
[c] Wintered on 9 lb alfalfa hay plus 5 lb concentrate per day.

indication that heavy creep feeding of suckling heifers may actually be detrimental to their overall productive life. Supplemental feeding during the first and second winters of replacement heifers is more sensible than creep feeding. Furthermore, too rapid gains postweaning—such as heifers in a fattening lot gaining two plus pounds per day—can be detrimental to replacement heifers since one of the first sites for deposition in the developing heifer is around the reproductive tract.

Growthy heifers weighing 600 lb at time of conception will gain from 1 to 1.5 lb per day on pasture alone through the pasture season. During the second winter as they approach their first calving, they should be fed moderately but not

TABLE 11.9

Weight at which Hereford Heifers Reach Sexual Maturity[a]

Weight (lb)	Percentage showing heat
400	1
450	4
500	17
550	48
600	80
650	94

[a] Wiltbank (1968).

TABLE 11.10

Effect of Winter Feed Level on Age at Puberty[a]

	Winter feed level (% reaching puberty)		
Age in months	Low	Moderate	High
9	3	3	0
10	17	13	10
11	30	33	47
12	47	57	70
13	60	64	90
14	63	73	93
15	70	90	100

[a] Turman *et al.* (1964).

luxuriously. From the end of pasture season to time of parturition, a daily gain of 1–1.5 lb per day is much more desirable than either less or more.

Why are many cow herd managers reluctant to calve 2-year-old heifers? Problems of better feeding and management plus greater attention at calving time have been mentioned. However, another factor is the generally accepted opinion that 2-year-old first calf heifers are considerably more difficult to get bred back promptly than older cows. At least limited data would indicate the later onset of estrus following her first calf by 2-year-old heifers is that she did not receive adequate energy during the last 3 to 4 months of her first pregnancy. It is recommended, therefore, that one should be fairly certain the pregnant yearling heifers get from 8 to 10 lb of total digestible nutrients (TDN) or 12 to 14 Mcal digestible energy per day during the last stages of their first pregnancy. Since the nutrient requirements for the first calf heifer are so much more critical than for mature cows, it is an excellent management practice to feed yearling bred heifers separately from the mature cows.

Data presented by Wiltbank (1964) (Table 11.11) would indicate that it takes longer for younger heifers to ovulate—and thus show signs of estrus—for the first time after calving than is true for older cows. Therefore, in order to be assured the first calf 2-year-old heifer does ovulate promptly following calving, excellent nutrition and management are required.

Energy is without doubt the most important single factor limiting a high level of reproduction in the beef breeding herd, and it is of extreme importance for the replacement heifer through her second calving. In an experiment at the Nebraska

TABLE 11.11

Percentage of Different Age Cows Showing Estrus after Calving[a]

	Days after calving (%)							
Age of cow	40	50	60	70	80	90	100	110
5 years or older	56	70	82	88	92	95	100	100
4 years	26	47	63	72	88	100	100	100
3 years (on crested wheat)	14	32	43	64	77	79	86	92
3 years (on native range)	6	16	28	41	57	70	85	95

[a] Wiltbank (1964).

Station, yearling Angus and Hereford heifers were fed different levels of energy before and after calving as indicated in Table 11.12. The feeding programs indicated in this table began 140 days before the first anticipated calving.

After calving sterile bulls were used to detect heat, and cows in heat were bred by artificial insemination between 60 and 120 days postcalving. The average weight of the pregnant heifers at the start of the experiment was 780 lb. Heifers on moderate energy levels gained 133 lb prior to calving whereas those fed low energy level lost 36 lb during the same period of time. From the experimental data presented in Table 11.12, it is obvious that if energy must be limited

TABLE 11.12

Reproductive Performance of Two-Year-Old Cows on Different Levels of Energy[a]

Level of TDN					
Before calving	After calving	Number of heifers	Cows pregnant at end of 60-day breeding season (%)	Cows not showing estrus first 40 days of breeding (%)	Cows conceived by first service (%)
Moderate (8 lb)	High (23 lb)	42	81	7	62
Moderate (8 lb)	Moderate (13 lb)	37	70	5	65
Moderate (8 lb)	Low (7 lb)	42	64	19	53
Low (4 lb)	High (23 lb)	41	90	5	73
Low (4 lb)	Moderate (13 lb)	41	73	17	53

[a] University of Nebraska data (Wiltbank *et al.*, 1965).

drastically, it is better to limit energy prior to parturition. Heifers fed high levels of energy following calving exhibited estrus and were fertilized most promptly.

V. FREE-CHOICE LIQUID SUPPLEMENTS

A. Wintering Beef Cows

Liquid supplements in tanks equipped with lick wheels have been used extensively for providing supplemental nutrients for cows offered low quality roughages. A winter study was conducted by the University of Illinois at their Dixon Springs Agricultural Center (Saenger *et al.*, 1978) to learn more about consumption patterns and potential benefits to brood cows on such a program.

Four lots consisting of 32 crossbred cows each were fed grass hay in big round bales on winter pasture from late fall to late April. Each lot consisted of several breed crosses such as Angus × Hereford, Jersey × Hereford, Red Holstein × Hereford, and Simmental × Hereford. The calving season started in early March. Two of the four lots were given access to lick tanks in late January; the other two lots were not given protein supplement. The liquid supplement in the tanks contained 32% protein, all from nonprotein nitrogen, 1.5% phosphorus, and 30,000 IU of vitamin A per pound. All lots received minerals free choice.

Bale consumption was recorded throughout the time the cows received the liquid supplement. The hay was also sampled for crude protein and dry matter digestibility. The cows in each lot were weighed and assigned a condition score of 1 to 9 in January and again in April. Cows in one of the lots given the liquid supplement were observed for 2 days approximately every 2 weeks to determine the pattern of liquid consumption within the herd. Consumption was calculated one or more times per week by measuring the depth of the liquid in the lick tanks to determine liquid disappearance.

Consumption patterns of the cows varied greatly. Some of the cows visited the tank as often as seven times per 8-hour observation period; others made no visits. The time spent at the tank by individual cows ranged from 0 to 25 minutes per 8-hour observation period; average time per visit was 2.58 minutes.

From the start of the study, average daily consumption of liquid decreased from 4.51 to 0.05 lb for one lot and 3.69 to 0 lb for the other (Table 11.13).

Bale consumption also differed between lots. The two lots not receiving liquid supplement consumed 46 and 57 big round bales. The two lots receiving the supplement consumed 58 and 61 big round bales. The hay offered to the cows ranged from 5.25 to 8.58% in crude protein content and from 32.54 to 47.21% in digestible dry matter. During the last two weeks of the trial, some higher quality hay was fed, ranging from 14.88 to 16.45% in crude protein and from 44.81 to 52.69% in digestible dry matter.

TABLE 11.13

Average Daily Consumption of Liquid Supplement[a]

Date	Lot 1	Lot 2	Date	Lot 1	Lot 2
January 30	4.51	3.69	March 3	2.54	2.27
February 3	3.59	2.95	6	2.74	2.39
10	3.89	3.10	10	1.71	1.59
11	2.09	—	13	1.85	1.34
12	4.03	—	20	1.59	0.57
13	3.69	5.12	27	0.67	0.55
17	3.37	2.81	April 4	0.32	0.09
20	3.92	3.07	7	0.18	0.11
21	4.38	1.91	17	0.08	0.07
24	3.79	2.84	21	0.05	0.00
27	0.89	1.14			

[a] Consumption figures are the average daily consumption of liquid from previous measuring date. Values are in lb/day.

Scores for weight loss and change in condition are shown in Table 11.14. Lots 1 and 2 received the liquid supplement. Lots 3 and 4 did not receive the supplement and lost about 58 lb more than the supplemented groups. Condition scores of the unsupplemented cows decreased about 1 unit more than supplemented cows in a 1 to 9 scoring system.

B. Wintering Pregnant Heifers

University of Wyoming beef cattle researchers compared a commercially prepared 15% protein liquid supplement with a 15% protein range cube (Waggoner *et al.*, 1977). There were 82 heifers in each of the two comparative groups. Heifers in one group were hand-fed 2 lb of the dry supplement daily, along with

TABLE 11.14

Weight Loss and Condition in Wintering Beef Cows through Calving

	Free-choice liquid supplement			No liquid supplement		
	Lot 1	Lot 2	Average	Lot 3	Lot 4	Average
Number	30	31	61	31	31	62
Weight loss through calving (lb)	−98	−107	−103	−188	−135	−161
Condition change[a]	−1.5	−1.3	−1.4	−2.6	−2.2	−2.4

[a] Condition scores were 1 to 9, with 9 being the fattest.

TABLE 11.15

Liquid Supplement from Lick Tank versus Dry Supplement Hand-Fed for Wintering Pregnant Heifers[a]

	Dry supplement	Liquid supplement
Number of heifers	82	82
Initial wt. (lb)	738	746
Final wt. (lb)	802	827
Average daily gain, 84-day period	0.77	0.97
Days pregnant[b]	82.0	79.0

[a] Data from Waggoner *et al.* (1977).

[b] Number of days females were pregnant with their second calf, determined by rectal palpation on Oct. 3.

native hay and a salt–mineral mix; heifers in the other group had free-choice access to a lick tank with the 15% protein supplement, in addition to native hay and the salt–mineral mixture.

Performance during the 84-day period (Dec. 13 to March 4) is presented in Table 11.15. Heifers in group II consumed 2.20 lb of liquid feed per day and gained 0.97 lb per day. During this same period, group I received 2 lb of the 15% range cake and gained an average of 0.77 lb per day. The use of the lick tank to provide fortification of low-quality roughages appeared to be a sensible labor saving method as compared to hand-feeding the supplement every day.

VI. UREA FEEDING EFFECT ON CATTLE REPRODUCTION

Actually there are no scientific data to support any claims that urea feeding adversely affects reproductive capability in cattle. However, there are data which demonstrate urea is not harmful for reproducing cows by any method known to measure its effect. Michigan State dairy researchers (Ryder *et al.,* 1972) surveyed Dairy Herd Improvement Association (DHIA) cooperators for the State of Michigan and their conclusion stated, "Neither urea fed nor number of years urea was used in herds affected their reproductive efficiency" (Table 11.16). From 1400 DHIA herds in Michigan, over 900 completed questionnaires were obtained representing 85,281 individual lactations and 3,157 year-herd observations. Number of cows per herd averaged 48.6 and milk production per cow 12,707 lb per 305-day period. Average adjusted calving interval for all herds was 315.5 days. Urea was fed in 1709 of the 3157 herd years with an average intake per cow (in those herds where urea was fed) of 81 gm per day (0.18 lb). Of the total urea, 39.9 gm was derived from corn silage, 0.6 gm from high-moisture

TABLE 11.16

Relationship of Urea Intake to Calving Interval, Percentage Cows Sold as Sterile, and Milk Yields[a]

	Urea intake (gm/day)					
Average:	None	79.8	36.1	90.5	146.7	219.5
Item Range:	None	1–370	1–60	61–120	121–180	181+
Herd-year observations	1442	1715	760	653	219	83
Calving interval (adjusted) (Days)	314	316	313	318	316	314
Cows sold as sterile (%)	2.1	2.4	2.4	2.4	2.6	1.7
Milk per cow/year [lb (× 1000)]	13.0	12.9	12.5	13.0	13.1	13.1

[a] 600 DHIA herds from Michigan over a 5-year period. Data adapted from Ryder *et al.* (1972).

corn, 7.5 gm from dry grain, and 32.6 gm from commercial supplements. The maximum urea fed to any single herd was 370 gm per head daily (0.81 lb).

Urea feeding had no significant effect on calving interval length at any level compared. Similar to results with calving intervals, urea feeding had no significant effect on milk production although there was a trend for greater milk production from urea-fed cows at the higher levels of urea feeding.

The author has had the opportunity of observing several beef cow herds in which an unthrifty condition—and even small death losses—had been diagnosed as "urea toxicity" for cattle in which a portion of the supplemental protein was being supplied by urea. Upon questioning the manager, it appeared the cattle were obtaining from 1 to 2 lb of a 32% protein (mostly from NPN) supplement per head daily. However, the real problem was a marked energy deficiency. A brood cow needs about 85–90% of her digested nutrients for energy. Therefore, a diet grossly deficient in energy—but supplemented adequately with protein, vitamins, and minerals—can do very little for a cow. Under these conditions, she will literally lose weight, become unthrifty, and starve to death. Brood cows will labor to survive, but if there is not adequate forage to meet their energy requirements, they will "go down hill" quite rapidly.

Cattlemen's fears for the use of NPN as a source of crude protein in cattle diets are generally based on problems arising from misuse of NPN, improper diagnosis, or on stories which have no basis whatsoever.

Consumption of high levels of urea (100 gm, or nearly one-quarter pound, by a 500-pound calf) in a short period of time by an animal which was not adapted to urea can be fatal. However, this is inexcusable because adapted cattle can safely consume over twice this much. Mistakes can occur, for example, like cattle breaking into the feed room or spilling a bag of urea fertilizer when the planter boxes are being filled and cattle licking up the residue. If such a mistake is detected, drench each surviving 500-pound calf with about a gallon of vinegar—about two gallons for mature cattle. The dosage may have to be given once more a couple of hours later.

Actually, urea toxicity is a manifestation of ammonia toxicity. Urea breaks down in the rumen—like much of the true protein consumed—to ammonia. When this occurs at a normal rate, the rumen bacteria are capable of resynthesizing the ammonia into protein. When there is too rapid production of rumen ammonia, it is absorbed into the blood and carried to the liver. In great excesses, the system is overwhelmed with an ammonia burden and the animal may succumb.

REFERENCES

Martin, T. G., Spelbring, M. C., and Hodges, J. P. (1975). Effect of crossbreeding on cow productivity. *Proc. Indian Beef-Forage Res. Day, 1975* p. 55.

Perry, T. W., Garrett, W. N., Klosterman, E. W., Matsushima, J. K., Nelson, A. B., and Woods, W. R. (1976). "Nutrient Requirements of Beef Cattle," 5th rev. ed. Natl. Acad. Sci., Washington, D.C.

Ryder, W. L., Hillman, D., and Huber, J. T. (1972). Effect of feeding urea on reproductive efficiency in Michigan dairy herd improvement association herds. *J. Dairy Sci.* **55,** 1290.

Saenger, P. F., Weichenthal, B. A., and Cmarils, G. F. (1978). "Free Choice Liquid Supplement for Wintering Beef Cows," Cattle Feeders Day, Rep., p. 11. University of Illinois, Champaign.

Turman, E. J., Smithson, L., Pope, L. S., Renbarger, R. E., and Stephens, D. F. (1964). Factors affecting cow and calf performance. *Okla., Agric. Exp. Stn., Misc. Publ.* **57.**

Waggoner, J. W., Jr., Kaltenbach, C. C., Humphrey, W. D., Dunn, T. G., and Radloff, H. D. (1977). Liquid vs. dry supplements for wintering pregnant beef heifers. *Proc. Univ. Wyo. Stockmen's Homecoming Roundup, 1977.* p. 49.

Wiltbank, J. N. (1964). Proportion of different age cows showing heat after calving. *Fort Robinson Beef Cattle Res. Stn. Field Day* p. 47.

Wiltbank, J. N. (1968). Effects of heterosis on age and weight at puberty in beef heifers. *J. Anim. Sci.* **29,** 602.

Wiltbank, J. N., Bond, J., Warwick, E. J., Davis, R. E., Cook, A. C., Reynolds, W. L., and Hazen, W. M. (1965). Influence of total feed and protein intake on reproductive performance in the beef female through second calving. *U.S. Dep. Agric., Tech. Bull.* **1314.**

Wiltbank, J. N., Kasson, C. W., and Ingalls, J. E. (1969). Puberty in crossbred and straightbred beef heifers on two levels of feed. *J. Anim. Sci.* **29,** 602.

12

Forages and Environment Effect on Brood Cows

I. QUALITY OF PASTURE EFFECT ON COW AND CALF PERFORMANCE

Research was conducted at the Southern Indiana Forage Research Center (Lechtenberg *et al.*, 1975) to determine the effect of forage species on average daily gain and weaning weight of suckling calves and on cow performance. Orchardgrass, tall fescue, and mixed stands of tall fescue and legumes were compared. Ladino and red clover were established in tall fescue pastures by heavy grazing of the fall and winter pasture to remove most of the fescue vegetation and to retard spring growth of the fescue. The pasture to contain the legume mixture was seeded with one-half pound Ladino and five pounds of red clover per acre in March. Excessive growth of tall fescue was controlled subsequently by heavy grazing.

In the pasture comparison studies, grazing was begun in late April and the calves were weaned on October 6. Stocking rates were adjusted on each pasture by restricting the cattle to only part of the pasture during the first 60 days of grazing with the remaining portion being harvested for hay. Note from Table 12.1 the early growth was primarily fescue, but by mid-summer the legumes had become more prominent.

Performances of the suckling calves and their mothers grazing the three types of pastures are presented in Table 12.2. Daily gains of the calves were 1.76, 1.28, and 1.82 lb per day on the three respective pastures which were reflected in depressed weaning weights for calves grazing with their dams on tall fescue pasture. Note the similarity of response over the 3 year study. Cow gains, too, mirrored the poorer quality of the tall fescue pasture alone. It should be borne in mind there was no shortage of forage material in these studies since controlled grazing pressure caused cows to be removed at such time as when the grazing quantity became low.

TABLE 12.1

Percentage Legume in Tall Fescue–Legume Pastures during Grazing Season (Dry Weight Basis)[a]

Item	May 23	June 20	July 12	Average
Pasture 1	14.4	37.2	44.6	32.1
Pasture 2	10.3	42.4	27.0	26.6
Average	12.4	39.8	35.8	29.4

[a] Lechtenberg *et al.* (1975).

TABLE 12.2

Performance of Cows and Calves Grazing Orchardgrass, Tall Fescue, and Tall Fescue–Legume Pastures during a 3-Year Study[a]

	Orchardgrass	Tall fescue	Tall fescue–legume
Calf performance			
Average daily gain (lb)			
1st Year	2.02	1.48	2.05
2nd Year	1.62	1.17	1.68
3rd Year	1.66	1.21	1.73
Average	1.76	1.28	1.82
Weaning weight (lb)[b]			
1st Year	493	404	492
2nd Year	394	315	405
3rd Year	401	336	412
Average	429	351	426
Cow performance			
Average daily gain (lb)			
1st Year	0.63	0.17	0.73
2nd Year	0.36	−0.21	0.25
3rd Year	0.73	0.08	0.76
Average	0.58	0.02	0.58
Conception (%)			
1st Year	95	83	98
2nd Year	86	77	94
3rd Year	89	54	84
Average	90	72	92

[a] Lechtenberg *et al.* (1975).
[b] 205-day adjusted weight.

TABLE 12.3

Typical Nutrient Composition of Crop Residues

Product	Dry matter (%)	Analysis of dry matter (%)			
		Crude protein	TDN	Calcium	Phosphorus
Cornstalks	20–80	5	45	0.16	0.13
Cornstalk silage	35	5	50	0.16	0.13
Corn cobs	90	2.5	47	0.12	0.04
Soybean straw	87	4.3	40	0.90	0.10
Wheat straw	90	2.3	45	0.17	0.03
Grass aftermath	80	5	45	0.40	0.20

The incorporation of legume into the fescue pastures apparently adds a factor needed to balance out fescue pasture for beef cattle. These research data present the dramatic effect of forage type on beef cattle performance. Tall fescue is a hardy forage grass which will grow under rather widespread conditions. However, it does need help as a grazing forage for beef cattle.

II. FEEDING CROP RESIDUES

Typical crop residues chemical analyses are shown in Table 12.3. When beef cow daily requirements are matched with what they can obtain daily from typical crop residues (Table 12.4), it is obvious that most crop residues alone cannot

TABLE 12.4

Beef Cow Nutrient Requirements and Nutrients Supplied by Normal Daily Intakes of Crop Residues

Item	Beef cow requirements (lb/head, daily)						
	Dry matter	As-fed intake	Protein	TDN	Ca	P	Vitamin A (IU)
Dry pregnant mature cow (1000 lb)	15	—	0.88	7.8	0.03	0.03	20,000
Lactating cow (1000 lb)	22	—	2.0	12.3	0.06	0.05	40,000
Nutrients provided by crop residues							
Cornstalk silage	20	57	1.02	10.0	0.03	0.03	—
Cornstalk stacks	14	16	0.71	6.3	0.01	0.01	—
Corn cobs (ground)	18	20	0.50	8.5	0.02	0.01	—
Soybean straw	13	15	0.56	5.2	0.12	0.01	—
Wheat straw	15	16	0.35	6.8	0.03	—	—
Dead grass aftermath	15	19	0.75	6.8	0.06	0.03	—

meet total beef cow nutrient requirements. If crop residues are to be used in beef cow diets successfully, the nutrient deficiencies should be corrected (Table 12.5). Otherwise, such cows will lose weight and condition, may experience difficulty in calving and recovery, fail to return to estrus promptly after calving, and wean smaller calves.

A. Soybean Straw

Soybean straw is lower in feeding value than cornstalks and most other residues. This is because the residue consists mainly of stems. There are few if any leaves obtained by most harvesting methods. However, many of the pods are still attached to the stems after harvest, which improves the feeding value. Cows grazing soybean fields require approximately 1.5 lb of a 32% protein supplement per head daily. Once leaves and pods have been consumed, supplemental hay or grain should be fed. Studies at Iowa State University indicate that cows grazed on soybean fields with access to soybean straw stacks and protein supplement can maintain body weight for at least 100 days during winter. However, if cows are forced to consume only harvested soybean straw and protein supplement, they will lose weight and condition. If harvested soybean straw is available, it should be used only for dry cows. Feed 1–1.5 lb of 32% protein supplement along with hay free choice or 2–4 pounds of corn grain.

B. Small Grain Straw

Wheat straw is likely the most readily available in Indiana. Wheat straw can supply part of the energy needs of dry pregnant beef cows. It is best to use straw

TABLE 12.5

Supplemental Feed Recommended with Corn Residue for Beef Cows

	Supplementation suggested	
Product and situation	Protein 32% supplement (lb)	Energy
Grazing dry residue		
Dry cow, first 2 months	None	None
Dry cow, last 2 months	1.0	Hay or corn, 3 lb
Lactating cow, fall calving		
First month	1.0	None
Last 2–3 months	2.0	4 lb corn or good hay
Ensiled corn residue		
Dry cow (50 lb)	0.5	None
Lactating cow (60 lb)	2.0	3 lb corn

Fig. 12.1. Large hay packages, properly shaped, can be stored out-of-doors for up to 1 year without great nutrient loss.

in combination with high quality roughages such as grass-legume hay. Feeding 8-10 lb of grass-legume hay plus a full feed of wheat straw will provide sufficient protein and energy for mature dry cows. If less hay is fed, cows should receive 1-1.5 lb of a 32% protein supplement per head daily. (Figs. 12.1 and 12.2)

Fig. 12.2. Large hay packages provide roughage in times of heavy snow coverage.

C. Dead Grass Aftermath

The use of pasture aftermath as feed for beef cows is satisfactory as long as pasture is available and weather permits grazing. Grazing aftermath is often practiced in combination with use of field-stored hay. This works well provided the hay is not wasted. The need for supplemental protein depends upon the protein content of the hay. First-cutting hay harvested before mid-June will not require supplemental protein with pasture aftermath. However, first-cutting hay harvested later will likely be deficient in protein along with the aftermath. In this case, cows should receive 1 lb of a 32% supplement per head daily.

D. Corn Cobs

Corn cobs may be available in certain areas. In order to obtain a satisfactory intake by cows, the cobs should be ground prior to feeding. Because they are dry, ground cobs work well in complete mixed rations. A mixture of 90% cobs and 10% protein supplement fed at the rate of 16–18 lb per head daily will provide the protein and energy needs of dry cows. Mixing in 5% cane molasses will reduce dustiness and increase intake by cows. For lactating cows, a mixture of 65% ground corn cobs, 20% shelled corn, and 15% protein supplement can be full fed to provide all the energy and protein needed during lactation.

III. GRAIN SORGHUM STOVER FOR BEEF COWS*

Combine-type grain sorghum is a perennial grass which acts as an annual under Midwest conditions due to killing frosts. Since it is a perennial until frost, the stover accumulates dry matter between grain harvest and frost. Data indicate that after combine-harvest, 50–60% of the total plant dry matter of grain sorghum remains in the field, as compared with estimates of 45–50% of the corn plant remaining. Grain sorghum stover remains standing after grain harvest, lending itself to easy harvest later in the fall by either animals or a forage harvester.

Harvest of grain sorghum stover for dry stacking with a hay packaging machine does not appear possible under Corn Belt conditions unless the crop is mowed and windrowed after field-curing. The moisture concentration for packaging should be less than 40%, and grain sorghum stover grown in the Corn Belt would not be anticipated to drop that low of a moisture content. Therefore, use of such product probably is limited to either ensiling or grazing in the field in a humid environment.

In the Iowa study from which these conclusions and recommendations were

*Adapted from Iowa State University Agric. Exp. Stn., Ames, Iowa, A. S. Leaflet R203 (1974) (Anonymous).

drawn, dry, pregnant beef cows (heifers and mature cows) were grazed on grain sorghum stover for a 98-day period at a stocking rate of 1.5 acres per cow. The combined heads and tailings of the grain sorghum were collected with a Foster dump wagon and stored at the ends of the field as a reservc fccd supply. Snow cover occurred from December 20 through late January and there was intermittent snow cover during February and March. The cows received a mineral–vitamin A supplement the first 42 days, and a 32% liquid supplement was fed free-choice (lick wheel) during the last 56 days.

For the first 42 days, the mature cows gained 44 lb and the bred heifers gained 6 lb with an average gain of 26 lb per cow. Over the last 56 days, the mature cows and heifers closely held their weights for a net weight change of 1–2 lb per cow. The 98-day weight gain avarages were 46 lb for mature cows and 8 lb for the bred heifers, an average of 27 lb per cow. The average liquid supplement consumption for the last 56 days was 2.1 lb per animal.

An obvious advantage of sorghum stover was the availability of standing forage when heavy snow cover occurred during the first half of the test. At the same time, supplemental hay was needed for cows grazing cornstalks. The sorghum leaves were selectively grazed during the first period which caused stover quality to be much lower in the last grazing period. This was evidenced by the lack of weight gains, even with supplement consumption of 2.0 lb per head daily.

The field-stored dumps of combined heads did not appear to be well utilized. Even with grazing controlled by an electric fence, much waste occurred. Accurate measurements of utilization were not possible.

The grazing program in this study demonstrated that sorghum stover provided about 65 cow–days per acre. Mature cows were adequately maintained by the grazing program, but weight gains for bred heifers were inadequate and supplemental energy would be needed for normal growth and development under such a program.

When grazing grain sorghum stover, producers must be aware of possible prussic acid poisoning. Grain sorghum stover may be utilized safely by cattle after a killing frost if the cattle are kept from the stover for three days following the killing frost. The frost must also have completely killed any new shoots which may have appeared after grain harvest for cattle to safely graze the stover.

IV. CORN RESIDUE UTILIZATION BY BEEF CATTLE

Cornstalks represent the plant material remaining after the corn grain has been removed. This material makes up about 50% of the total corn plant dry matter. Moisture content of this product will vary greatly (from 20 to 75%) which affects the method of harvest and storage.

There are several methods which have been proposed and investigated for utilizing corn residue as a feedstuff. Corn residue is a cellulose and lignin feed with very little protein and a TDN value of about 40%. It is low in calcium and phosphorus but fairly high in potassium. The product is useful as a low-quality roughage, primarily for pregnant brood cows, or for maintaining stocker cattle at a low level of grain.

Admittedly, corn residue has quite a feeding potential under the proper conditions. Therefore, one must consider alternative methods for feeding it.

A. Field Grazing

1. Normally, the grazing period will be between October 1 and February 1.
2. On an average, two acres of stalks are required to carry one cow for approximately 100 days.
3. For more efficient utilization of cornstalk material, cows should be restricted to half an acre of stalks per cow for each 4- to 5-week period. A hot wire may be used to restrict the cow within the grazing area.
4. Only 15–20% of the stalks available are actually consumed by the cow.
5. If more than normal amount of corn is left in the field behind the combine, caution should be taken to prevent the cows from consuming too much of the ear corn. If this is a problem, the cows should be restricted to a smaller area (0.1 acre/cow for one week).
6. There are obvious disadvantages to this method of harvest: (a) fencing is required, (b) fall plowing is not possible, and (c) on muddy days cattle traffic may have an undesirable effect on the subsequent physical quality of the soil.

B. Combine Trash—Accumulated and Stacked

1. In this process the husk, cob, some leaves, and any bypassed grain is collected in a trailer behind the combine instead of being scattered on the ground. Once the trailer is filled, the material is dumped at random in the field or moved to one side or end of the field (Fig. 12.3).
2. When the stacks are placed at one end of the field in a concentrated area, a hot wire may be used to restrict access to a limited number of stacks or to only a portion of one stack.
3. Yield of this product may run as high as one ton dry matter per acre.
4. The advantages include (a) less fencing required, (b) frees corn fields for fall plowing, and (c) stacks usually are higher than the snowbanks when snow comes.
5. Disadvantages include (a) cost of equipment plus labor involved, (b) more management required to control (cows love to get on top the stacks and lie

Fig. 12.3 The equipment illustrated here gathers husks, cobs and bypassed grain into stacks. (Photo courtesy of BEEF Magazine.)

Fig. 12.4. Cornstalks stored in large round bales. (Photo courtesy of BEEF Magazine.)

down), and (c) cows do not eat the whole cobs so in the spring these must be scattered.

C. Whole Stalk Collected in Large Packages

1. Large machines will make packages of from 1000 pounds each and upward. (Fig. 12.4)
2. Yield of this product may run from two tons per acre and more.
3. Feeding management is similar to that for the combine-accumulated trash listed above.
4. Advantages include those listed for (B) above.
5. Disadvantages are similar to those for (B) above.

D. Ensiling the Residue

1. The material is harvested utilizing methods (B) or (C) above and hauled to a tubgrinder where it is chopped. Water is added and the material is then ensiled. This method really does not increase the feeding value of corn residue, but it preserves the quality status quo and in the grinding, also makes the entire product consumable. When fortified with vitamins, protein, and minerals, ensiled corn residue is an excellent energy feed for dry pregnant cows. The addition of some concentrates (4–5 lb per day) makes it a lactation diet.
2. The greatest disadvantage to this method is the overall cost of getting the mission accomplished. Furthermore, the harvesting of corn residue comes at a time which is unpredictable and also which is competitive with fall plowing. Not many people make corn residue silage.

Table 12.6 attempts to place a dollar evaluation on various methods of feeding corn residue. Naturally, any table is out of date before it can be published. However, it does give a point of reference that with 100 head of cows or more, ensiling of corn residue might not present an economically prohibitive alternative.

V. LEAST-COST BROOD COW DIETS USING CROP RESIDUES

Iowa State University researchers utilized the linear programming technique in determining least-cost diets for winter-feeding beef cows (Boehlje and Vetter, 1978). First, consideration was given to products which could be considered as cow feed. Naturally, corn, milo, and small grain residues are used when avail-

TABLE 12.6

Estimated Costs of Alternative Corn Residue Utilization Systems for Beef Cows, For 100-Day Wintering Period

System	Number of beef cows in herd ($ cow-day)					
	25	50	75	100	150	200
Field grazing[a]	0.245	0.245	0.245	0.245	0.245	0.245
Grazing plus husk–cob stacks[b]	0.393	0.300	0.268	0.252	0.237	0.232
Dry corn stalk stacks[c]	0.445	0.342	0.307	0.288	0.271	0.262
Corn stalk silage[d]	1.076	0.689	0.559	0.492	0.425	0.386

[a] Includes cost of maintaining fence and water supply, 5 lb of supplemental hay and 2 lb of protein supplement per cow per day. Assumed that next year's crop yields would not be reduced due to soil compaction or delayed planting because of spring plowing.

[b] Includes annual costs associated with $2800 investment in dump trailer, cost of additional labor and machine time at harvest, cost of electric fence to control access to husk–cob dumps, cost of 2 lb of protein supplement per cow per day plus cost of maintaining fence and water supply in stalk field.

[c] Includes annual costs associated with one-half of the $5200 investment in stack wagon and stack mover of tractor and labor for harvest and feeding, cost of feeding racks, cost of 2 lb of corn and 2 lb of protein supplement per day, plus value of fertilizer in stalks removal from field.

[d] Includes annual costs associated with $4650 investment in flail type forage harvester and blower–recutter and one-half of $3200 investment in forage wagons, annual costs of new silo unloader, and feed bunks, tractor, labor, and electric costs for handling feed, plus 2 lb of corn and 2 lb of protein supplement per cow day, plus value of fertilizer in stalks removal from field.

able. It has been shown, for example, that beef cows can be maintained on stacked corn or milo stover plus a balancer supplement. A 25% wastage of harvested stover dry matter was observed when it was field-fed using feeder control panels. Chopped wheat or barley straw at 65–75% of the diet plus hay will accomplish the same thing. According to Nebraska research, treating stover materials with about 4% of sodium, calcium, or ammonium hydroxide will improve their nutrient availability by about 20%.

To determine the economic potential using such crop residues as beef cow feedstuffs, least-cost daily consumption beef cow diets were developed in Iowa programs. Adjusted costs of the various feedstuffs and their respective nutrient contents are summarized in Table 12.7. These include proposed harvesting, storing, processing, and feeding costs incurred in utilizing the feedstuffs. Losses in handling and feeding also were added and included in the final cost. Naturally, no published feeding costs remain current for very long and thus must be recalculated for any given time and place. These figures, however may be utilized for comparative purposes.

These adjusted costs were used to obtain least-cost daily consumption diets for beef brood cows by using a linear programming model (Iowa State University,

TABLE 12.7

Nutrient Content and As-Fed Cost of Beef Cow Feedstuffs[a]

Item	Nutrients (%)			Cost \$/cwt[b] as-fed
	Dry matter	Crude protein	TDN	
Corn stover	70	3.36	35.0	0.642
Corn stover silage	50	2.4	25.0	0.530
Corn stover silage[c] (treated)	50	2.4	29.0	0.793
Wheat stover	90	3.69	39.6	0.947
Wheat stover[c] (treated)	70	2.87	36.4	1.105
Barley stover	90	3.69	36.4	0.947
Oat stover	90	3.96	41.4	0.947
Corn silage	35	2.94	24.5	0.921
Mixed alfalfa hay	90	13.0	48.6	2.200
32% Liquid supplement	61	32.0	43.9	5.250
Shelled corn	87	8.7	79.2	3.571
Broiler/turkey litter	80	24.0	48.0	0.500
Excreta stover silage	40	10.5	18.0	0.550

[a] From Boehlie and Vetter (1978)a.

[b] Cost adjusted for storage and feeding losses: silages, 5%; hay, 10%; stacked corn stover (field-fed, feeder control panels), 25%; and wheat, barley, or oat stovers, 5%.

[c] Chemical treatment: 3% NaOH + 1% CaOH per ton of residue dry matter. Estimated cost of \$10 per ton of dry matter treated.

1974, Anonymous). The results of the computer analysis assume that all stover and conventional feedstuffs are available. The data is summarized in Tables 12.8 and 12.9. Note that with hay at \$20 per ton and the other feedtuffs as indicated, least-cost diet consists of 24 lb corn stover plus 4.01 lb of hay on an as-fed basis. The cost per day for this combination is 19.41 cents. This combination holds pretty well until hay reaches \$40 per ton when corn silage begins to appear in the printout at 2.99 lb per day. The hay at this point is reduced to 2.51 lb per day with corn stover remaining constant. The diet thus remains constant until hay reaches \$60 per ton at which point hay is priced out of the diet. The optimal diet with hay priced over \$60 per ton consists of 24 lb of corn stover, 6.76 lb corn silage and 0.67 lb protein supplement. The cost of this feeding program is 25.2 cents per day.

The least-cost diet, assuming drier stover stack harvested and field-fed is not available, is summarized in Table 12.8 for different hay prices. Under these conditions, a preponderance of hay shows up when it is \$20 per ton. At \$50 per ton for hay, the ratio shifts to 10.85 lb corn stover silage, 17.57 lb oat straw and 0.82 lb protein supplement at a daily cost of 26.7 cents.

TABLE 12.8

Optimal Beef Cow Formulas Utilizing Dry Harvested Corn Stover

Hay price ($/ton)	Ingredients per day (lb) Corn stover	Hay	Corn silage	Supplement	Cost/day (¢)
20	24.00	4.01			19.4
25	24.00	4.01			20.4
30	24.00	4.01			21.4
35	24.00	4.01			22.4
40	24.00	2.51	2.99		23.2
45	24.00	2.51	2.99		23.8
50	24.00	2.51	2.99		24.4
55	24.00	2.51	2.99		25.1
60	24.00		6.76	0.67	25.2
65	24.00		6.76	0.67	25.2
70	24.00		6.76	0.67	25.2

TABLE 12.9

Optimal Beef Cow Formulas Utilizing Corn Stover Silage

Hay price ($/ton)	Ingredients per day (lb) Corn stover silage	Oat straw	Hay	Supplement	Cost/day (¢)
20	0.63		20.97		21.3
25	11.42	14.86	2.74		23.6
30	11.42	14.86	2.74		24.3
35	11.42	14.86	2.74		24.9
40	11.42	14.86	2.74		25.6
45	11.42	14.86	2.74		26.3
50	10.85	17.57		0.82	26.7
55	10.85	17.57		0.82	26.7
60	10.85	17.57		0.82	26.7
65	10.85	17.57		0.82	26.7
70	10.85	17.57		0.82	26.7

Because of the greatly increased cost of alkali-treated products, as related to improved utilization, such products did not appear in the formulations. In other words, our present technology—coupled with high cost—hardly warrants treating crop residues with alkali products.

VI. DRYLOT VERSUS CONVENTIONAL COW HERD MANAGEMENT SYSTEMS

Traditionally, many beef cow herds must depend on the marginal land in the farming operation for feed source, including pasture and hay, with an assist from cornstalk residue for a part of the post-corn picking season. Today as land values continue to rise and fencing costs increase, one could envision a beef cow herd in confinement 12 months out of the year. A 4-year study was conducted at Purdue University to study three beef cow herd management systems which included a complete confinement program (Perry *et al.,* 1974).

In this program, mature commercial Hereford cows bred for spring calving each year were maintained in their three respective 25-cow herds for the 4-year test. The cows in each herd were bred naturally to performance-tested Hereford bulls. Bulls were rotated among cow herds; calves were born outside in February and March. All calves were weighed, ear tagged, and tattooed at one day of age. Creep feed was not available. The cows were hot-branded initially for permanent identification.

This research was adapted to three rather obvious naturally occurring phases from calving to weaning. All cows and calves were weighed individually at the beginning and end of any particular phase. Phase I was from the time calves were born until the "permanent-pasture" herd was placed on bluegrass pasture; Phase II was the interval between the start of the permanent-pasture grazing until summer annual pasture was available for the "temporary-pasture" herd to graze; Phase III started when the temporary-pasture herd went to pasture and ended when this pasture was either depleted or killed by frost. Calves were weaned and weighed when this phase was completed.

A. Permanent Pasture Herd (Bluegrass Pasture, Cornstalks, Hay)

From May through October this herd grazed bluegrass (permanent) pasture on land which did not permit alternative cropping. For the 4-year period, bluegrass was grazed an average of 158 days during the summer grazing season and carried an average of 0.6 cow–calf units per acre. In other words, it required 1.73 acres for a cow and her calf for the 5-month grazing season. It should be pointed out

TABLE 12.10

Herd Treatments

Herd	Diet	Period
Permanent pasture	Bluegrass pasture	May thru Oct.
	Cornstalk pasture	Nov. thru Jan.
	Mixed hay	Feb. thru April
Temporary pasture	Bluegrass pasture	May thru June
	Summer annual	July thru Sept.
	Aftermath	Oct., Nov.
	Corn stover silage	Dec. thru April
Total confinement	Corn silage (60 lb per day)	May thru Sept.
	Corn silage (25 lb per day)	Oct., Nov.
	Corn silage (30 lb per day)	Dec. thru April

expanded grazing might be realized if other species of plants were incorporated into the pasture program, but this might also increase the investment cost in the pasture greatly.

Cornstlaks were grazed an average of 89 days by this herd. Finally, 1486 pounds of medium-quality mixed hay were required to round out each cow's needs for the year (Table 12.10).

B. Temporary Pasture Herd

This herd was started on sorghum–sudan grass hybrid pasture around July 1 each year. The temporary pasture seeding was on 150 bushels per acre corn land which had been plowed and disked with potash plowed down; anhydrous ammonia was applied prior to planting, and starter fertilizer was placed in the row. Seeding was made in mid-May with a corn planter in 38-inch rows and then doubled back so the resulting seeding was in 19-inch rows. The resulting growth was divided into three subpastures for rotational grazing. Each of the subpastures was clipped after cattle were removed, but adequate grazing pressure resulted in only some central stubble remaining. This pasture was grazed an average of 97 days per season by 1.25 cow–calf units per acre or 0.81 acre for each cow and calf. The calculated cost of producing an acre is relative to any current year, but it is about the same as that for producing an acre of corn.

When the summer annual was gone, this herd had access to aftermath bluegrass pasture until corn stover silage was ready. The winter feeding program consisted of corn stover silage plus a balancing 32% protein supplement. When spring permanent pasture was ready, this herd was switched to it until about July 1 when summer annual pasture was ready.

C. Total Confinement Herd

This herd was not kept under shelter, but rather was quartered on a small amount of land (3 acres) that was devoid of herbage at a population of 10 cows per acre. During the heaviest lactation (May through September), corn silage was fed at the rate of 60 lb per cow along with 2 lb of a 32% protein balancing supplement. Since the calves did not have access to a creep, it is obvious the calves tended to compete with their mothers for the feed provided. Immediately after calf weaning, the level of silage fed was dropped to 25 lb per day and supplement was reduced to 1 lb per day; as the cows approached parturition, silage feeding was increased to 30 lb per day. These latter two levels of silage feeding did not satisfy the hunger of the cattle, but over a 4-year period no ill effects appeared.

Average corn silage yield was 17 tons per acre (65% moisture); and since each cow-calf unit consumed from 7.5 to 8.0 tons of corn silage per year, two plus cow-calf units can be fed for a year from one acre's corn silage yield.

A summary of feed budget plus relative feed costs is presented in Table 12.11.

TABLE 12.11

Summary of Feed Budget and Costs for Three Systems of Managing Beef Cows (4-Year Summary)[a]

Material and amount fed	Unit cost	Cost per cow–calf unit
Permanent pasture herd		
180 days bluegrass (permanent) pasture	10¢/day	$ 18.00
89 days cornstalk pasture	5¢/day	4.45
1486 lb medium quality mixed hay	1¢/lb	14.86
Minerals		3.11
	Total	$ 40.42
Temporary pasture herd		
78 days bluegrass (permanent) pasture	10¢/day	$ 7.80
0.81 acre summer annual pasture	$60.45/acre	48.96
1.71 tons corn stover silage	$3.71/ton	6.34
136 lb 32% protein supplement	8¢/lb	10.88
Minerals		3.11
	Total	$ 77.09
Total confinement herd		
7.7 tons corn silage	$18/ton	$138.60
475 lb 32% protein supplement	8¢/lb	38.00
Minerals		3.11
	Total	$179.71

[a] Perry *et al.* (1974).

TABLE 12.12

Birth to Weaning Performance of Calves and Annual Performance of Cows Due to Treatment (4-Year Average)[a]

Item	Permanent pasture herd	Temporary pasture herd	Total confinement herd
Total number of calves	93	89	91
Birth weight of calves (lb)	75	76	75
Daily gain of calves			
Birth to May 1 (lb)	1.68	1.58	1.63
May and June (lb)	1.89	1.79	1.40
July tō October (lb)	1.51	1.71	1.81
Birth to weaning (lb)	1.66	1.71	1.64
Calf weaning weight (lb)	445	454	449
Cow net weight change (lb)	−50	+13	−7

[a] Perry *et al.* (1974).

Even though the "bottom line" favors the "permanent pasture" reared calves, nevertheless the data point up the possibility of either of the other two options when the price of feeder calves warrants it. In other words, weaned 450-pound feeder calves selling at 70 cents for steers and 60 cents for heifers could gross an average of $292.00 per head. The research also demonstrates the comparable nutritional adequacy of each of the three programs for the production of feeder calves.

Calf performance—at least weaning weights—was comparable for all three methods (Table 12.12). However, it should be noted that calves in the "total confinement" herd did not gain as rapidly as those in the other two herds during May and June. This could be explained by the fact that they were too small to compete with their dams at the feed bunk when silage and supplement were fed. By July, though, the calves "muscled" into the feed bunk and obtained a reasonably fair share of feed. Meanwhile, starting in July, calves in the "permanent pasture" herd were finding shorter droughty pastures and their rates of gain declined during that period. Most consistent calf gains were made by those in the "temporary pasture" herd where pretty good pasture was available right up to weaning time. Naturally, some of the poorer calf gains might have been leveled off had a creep diet been provided for them.

REFERENCES

Anonymous, 1974. Iowa State Univ. Agr. Exp. Sta., Ames, Iowa, A. S. Leaflet, R203.

Boehlje, M. D., and Vetter, R. L. (1978a). "Use of Crop Residues in Least Cost Rations for Beef Cows," Proc. Anim. Ind. Week, A.5. Leafl. R281. Iowa State University, Ames.

Bohhlje, M. D., and Vetter, R. L. (1978b). "Use of Cattle Excreta Stover Silage in Least Cost Beef Cow and Feedlot Rations," Proc. Anim. Ind. Week, A.5. Leafl. R282, Iowa State University, Ames.

Lechtenberg, V. L., Smith, W. H., Pettrtz, D. C. and Hawkins, G. K. (1975). "Feeding Crop Residues to Brood Cows," Beef Forage Res. Day, p. 27, Purdue University, West Lafayette, Indiana.

Perry, T. W., Peterson, R. C. and Beeson, W. M. (1974). A comparison of drylot and conventional cow herd management systems. *J. Anim. Sri.* **38,** 249.

13

Milk Production and Calf Performance

Many researchers have reported correlations ranging from 0.40 (medium) to 0.88 (high) between level of milk consumption and weight gain among beef calves. Also, it has been demonstrated that the introduction of dairy breeding into beef cows results in increased milk production and heavier calf weaning weight. However, when one introduces dairy breeding into beef cows, evaluation of the results may be confounded a bit by the fact both increased milk production and increased rate of gain are seen at the same time. Oklahoma researchers (Wyatt *et al.*, 1977) used a unique approach to determine milk level effect on calf performance. Hereford and Friesian cows were employed in a system designed to expose calves of two biological types to a low (Hereford) or high (Friesian) level of milk. Two distinct biological calf types were produced by breeding Hereford cows to Angus bulls and Friesian cows to Charolais bulls. Then a reciprocal cross-fostering scheme was used in which half the calves born to Hereford dams (A × H) were nursed by Friesian cows and, conversely, one-half the calves born to Friesian dams (C × F) were nursed by Hereford cows. Thus, within each biological type, some calves obtained a low milk level (10–12 lb per day) while others obtained a high level of milk (21–24 lb per day) based on periodic calf suckle estimates. In addition, cow–calf pairs were maintained on either tall grass native range or in a completely confined drylot.

The performance of A × H and C × F calves consuming low and high milk levels is summarized in Table 13.1.

Under range conditions, and within the A × H breeding, calves on high milk were 114 lb heavier ($p < 0.05$) at weaning, representing a 20% increase. Drylot calves of the A × H group were 97 lb (19% more) heavier at the high level of milk. Within the C × F calves, calves on high milk and on range gained 111 lb more (22% more) and those in drylot gained 122 lb more (23% more) than comparable calves on low milk.

At the high milk levels, A × H and C × F consumed similar amounts of milk,

TABLE 13.1

Milk Level and Biological Type Effect on Calf Performance[a]

Item	Breed of calf and level of milk			
	Angus × Hereford		Charolais × Friesian	
	Low milk	High milk	Low milk	High milk
Range environment				
Number of calves	13	9	9	14
Daily milk consumption (lb)	12	24	11	22
Birth weight (lb)	67	67	87	104
Weaning weight (lb)	510	614	561	689
Daily gain (lb)	1.84	2.29	1.98	2.44
Drylot				
Number of calves	9	9	9	9
Daily milk consumption (lb)	11	22	11	22
Birth weight (lb)	66	71	98	93
Weaning weight (lb)	497	594	539	656
Daily gain (lb)	1.80	2.20	1.84	2.35

[a] Wyatt *et al.* (1977).

indicating the potential growth capability had little effect on milk intake by the respective calves.

These data look exciting and suggest that the cow-calf herdsman should introduce some dairy breeding into his cow herd in order to achieve greater milking capability and thus produce heavier beef calves at weaning time. The professional purebred-showman often achieves this extra growth by use of a supplemental nurse cow to provide extra milk for a potentially valuable calf. But the commercial cow–calf herdsman will need to do some intensive figuring. The matter of increased milk production is not free. According to the best estimates set forth in National Research Council requirements, the Hereford cows in this test would need to consume 25.46 Mcal of digestible energy during lactation whereas the Holstein cows would need to consume 35.75 Mcal of digestible energy during lactation to produce the indicated milk differential. This represents a 50% increase in digestible energy for the higher producing cow. The extra 100 pounds of calf at weaning time may justify the extra energy required, but it must be calculated carefully.

I. CHOOSING A PROFITABLE COW SIZE

The data above relative to milk level effect on calf weaning size also implied an extra 40–60 lb at weaning time favoring a larger breed calf (C × F) on either the high or low milk regimen. Size of the feeder calf—or calf at weaning time—has a great bearing on dollar return to the cow-calf herdsman. Unfortunately it is not possible to recommend one specific feeder calf size or type because the "most desirable type" will change from time to time. On the other side, however, there is a demand for almost any type of feeder calf. Many factors influence the most desirable type of feeder calf and thus the type for which the feedlot operator is willing to pay a bit of a premium.

First of all, what does the feedlot operator want in terms of size? It is a well-known fact that when fed to the same compositional endpoint (USDA grade), larger, faster gaining cattle have no advantage in efficiency of feed conversion. Larger, faster gaining cattle have an advantage only if slaughtered on a weight constant basis. Therefore, desirable weight of the feeder calf may be related more to the price of feedstuffs. Thus, when corn is low in price, money is made in the long feed and thus the operator will pay a little more for lighter cattle; when feed is high, he will pay more for the yearling 700 plus feeder cattle that he can upgrade and move out quickly. Unless weather or political maneuvering enters into the feedstuffs picture, generally corn and the other grains are a "bargain."

The cow-calf herdsman likes to produce a heavy calf (high sale price) from a small cow (low inputs) because the cost of maintaining (exclusive of milk production) is related directly to her size. The influence of size of the dam on maintenance requirements is reflected in Table 13.2. The relative energy requirement as expressed as a percentage of that required by a 1000-pound cow is shown and can be visualized also as a relative land requirement per cow.

Thus, if a 1000-pound cow requires 10 acres of land on a given range, a 1600-pound cow requires 48% more land, or 14.8 acres, merely for maintenance

TABLE 13.2

Influence of Body Size of a Cow on Her 365-Day Maintenance Energy Requirement (Exclusive of Milk Production)[a]

Cow weight (lb)	TDN requirement[b] for maintenance for 365 days (lb)	Percentage of requirement for a 1000-lb cow
800	2274	84
1000	2704	100
1200	3140	116
1400	3573	132
1600	4006	148
1800	4439	164
2000	4873	180

[a] Totusek (1975).
[b] TDN, total digestible nutrients.

purposes. Further, if a $90.00 land charge is assessed for the 1000-pound cow, the 1600-pound cow would incur a land charge of $133.20. Additional energy to support a higher level of milk production, which might be the case with the larger cow, has not been included in the comparisons. On the other hand, the greater salvage value of the heavier cow partially offsets her additional maintenance cost.

The apparent weaning efficiency of a small cow may be misleading, especially if she calves a small type early-maturing calf which is apt to be discounted in the feeder calf auction ring. This problem can be alleviated somewhat by use of large growthy bulls, except that calving difficulty may become a potential threat. Furthermore, to produce the smaller dams, it is estimated a herd of 60 cows would be needed to maintain and replenish a 100-cow herd of the small, specialized cows. The economics of such a total program, including the small "by-product" steers from the herd producing the small females, must be considered.

The data in Table 13.3 set forth the combined body size and milk requirements for energy for beef cows, based on research by the Oklahoma researchers. The TDN figures might be more meaningful if visualized as feed. Good quality nonlegume hay contains about 50% TDN, so if the TDN requirements are doubled, one would have a hay requirement, as a benchmark, to compare energy requirements needed to support the four cow situations for 365 days. A 1000-pound cow giving 10 lb of milk would need the equivalent of 3.4 tons of good nonlegume hay equivalent to meet her energy needs for a year; a 1400-pound cow giving 10 lb of milk a day during lactation would require 4.3 tons of good nonlegume hay for the same period of time, or 26% more feed with no greater milk production. The same comparisons might be made on pasture land or on a combination of hay and pasture, as the case may be. Most research information

TABLE 13.3

Combined Weight and Milk Production Effect on 365-Day Energy Requirement of a Beef Cow[a]

Cow weight (lb)	Daily milk (lb)	365 day TDN (lb)	Number of cows[b]
1000	10	3,416	100
1000	20	4,112	83
1400	10	4,354	78
1400	20	5,050	68

[a] Totusek (1975).

[b] Relative number of cows which could be supported by energy required by 100 cows weighing 1000 pounds and producing 10 lb of milk daily during lactation.

TABLE 13.4

Necessary Weaning Weight of Calves for Cows Varying in Weight and Milk Production[a]

Cow weight (lb)	Daily milk (lb)	Number of cows	Necessary weaning weight[b] (lb)	Total calf produced[b] (lb)	Necessary weaning weight adjusted for cow salvage (lb)[c]
1000	10	100	470	42,300	470
1000	20	83	566	42,280	566
1400	10	78	603	42,331	569
1400	20	68	691	42,289	657

[a] Totusek (1975).

[b] Based on 470-pound calf produced by 1000-pound cow with 10 lb daily lactation and 90% calf crop for all cows.

[c] Based on the assumption that the productive life of cows will be 6 years. An additional year is assessed for the development of the replacement female, so 57 lb additional salvage is available each year from the larger cows (400 lb ÷ 7 yr = 57 lb). Since cows have a market value approximately 60% that of calves, 34 lb (57 lb × 60%) less necessary weaning weight is required for the larger cows.

indicates increased weaning weights shown in Table 13.4 are difficult to obtain with the larger cows.

II. CREEP FEEDING BEEF CALVES

A program of management which provides energy feeds other than milk, plus grass or hay, is usually defined as a creep feeding arrangement. Not all farms and ranches lend themselves to this management technique which requires additional labor, equipment, and feed. However, creep feeding usually results in increased calf gain during its suckling period (Fig. 13.1). In fact, many plain-type calves can be fattened sufficiently on a creep feeding program to sell as heavy fat calves at time of weaning, although sale of heavy fat calves is not a very common practice in the United States.

Creep-fed suckling calves will make more rapid gains if they are fed a well-balanced diet in addition to any pasture or hay they are receiving, but energy is the primary demand. Although creep feeding can be beneficial at almost any season of the year, the greatest benefit is derived when pasture or hay is poor and thus a decline in milk production by the dams would be anticipated, as well as there being less forage for the calves. Varying results may be found in the literature as to the anticipated response cow and calf operators may anticipate from creep feeding beef calves. However, data reported from the Oklahoma Agricultural Experiment Station (Anonymous, 1954), is found in Table 13.5,

Fig. 13.1. Creep feeding of calves running with their dams on pasture can produce heavier weaned calves. (Photo by J. C. Allen and Son.)

perhaps reflects a typical growth response one might anticipate from creep feeding suckling beef calves.

Cows and calves in the Oklahoma study were grazing native grass pastures at the Lake Carl Blackwell range area where adequate range was available. This trial was from April 29 through October 12. It was concluded by the researchers that it was not profitable to creep-feed Oklahoma calves born in February and March, which would be put on pasture in the spring, if they were to be sold at weaning. In other words, the increased weight of the calf due to creep feeding did not warrant the additional cost of the feed required per extra pound gain (10–12 lb).

In contrast, Purdue University (Perry, 1974) has shown the benefits of creep feeding calves where the dams and their calves are restricted to drylot—or possibly in times of extremely sparse pasture conditions (Table 13.6).

In the Purdue study, a herd of grade Hereford cows and their calves was

TABLE 13.5

Effect of Creep Feeding Calves on Pasture[a]

Item	Lot 1 Not creep fed	Lot 2 Creep fed	Lot 3 Creep fed (molasses)
Number of calves	15	15	15
Average feed for 166 days			
Corn (lb)	—	188	330
Oats (lb)	—	94	198
Cottonseed meal(lb)	—	31	66
Molasses (lb)	—	—	66
Initial weight (lb)	163	166	164
Gain for 166 days (lb)	291	317	357
Feed per extra pound gain (lb)	—	12.0	10.0
Daily gain (lb)	1.75	1.91	2.15

[a] Anonymous (1954).

confined to drylot as a part of a long-term study. During the lactation phase, from approximately May 1 through September 30, whole plant corn silage was fed at a level of 60 lb per day along with 2 lb of a complex 32% protein cattle supplement per day. The free-choice creep diet available to the calves consisted of two parts whole shelled corn to one part of whole oats by weight.

Naturally, many formulations for a creep diet could be presented. However, it should be borne in mind that milk is a pretty well-balanced combination of protein, minerals, and vitamins and that probably the first limiting constituent of milk is energy. Therefore of selecting energy feeds which are most acceptable to

TABLE ₁

Effect of Creep Feeding Calves in Drylot[a]

Item	1 (53 days)		2 (60 days)		3 (71 days)	
	No creep	Creep	No creep	Creep	No creep	Creep
Number of calves	27	27	35	34	35	35
Initial weight (lb)	301	300	312	320	350	355
Final weight (lb)	379	423	404	448	450	530
Gain per calf (lb)	78	123	92	128	100	185
Daily gain (lb)	1.47	2.30	1.54	2.15	1.41	2.46
Creep consumption per day (lb)	—	3.2	—	2.6	—	6.7
Feed per extra pound gain (lb)	—	3.9	—	4.2	—	6.4

[a] Perry *et al.* (1974).

TABLE 13.7

Grain Preference of Creep-Fed Beef Calves[a]

Item	Year: 1 (126 days)	Year: 2 (138 days)
Number of calves	48	50
Initial age (Days)	64	49
Initial weight (lb)	152	135
Final weight (lb)	380	380
Daily gain (lb)	1.80	1.78
Average daily feed, first 60 days		
Rolled corn (lb)	0.39	0.65
Rolled barley (lb)	0.44	0.43
Rolled oats (lb)	0.02	0.11
Average daily feed, total period		
Rolled corn (lb)	1.67	2.15
Rolled barley (lb)	0.81	1.00
Rolled oats (lb)	0.05	0.04

[a] Anonymous (1961).

the calf. University of Minnesota (Anonymous, 1961) researchers have shown corn is generally preferred over barley or oats (Table 13.7). When given the opportunity to choose, calves steadily showed their increasing preference for corn as they approached weaning.

TABLE 13.8

Stilbestrol Implants for Nursing Calves[a]

Item	Year 1 (147 days): Control	Year 1 (147 days): 12 mg	Year 2 (126 days): Control	Year 2 (126 days): 12 mg
Number of steers	16	14	25	27
Initial weight (lb)	272	254	190	199
Final weight (lb)	364	376	326	332
Daily gain (lb)	0.63	0.83	1.08	1.05
Increase in gain (%)		24		2

[a] Anonymous (1959).

III. IMPLANTING CALVES

The matter of implanting suckling calves with hormones is worthy of consideration. A number of trials have been conducted to evaluate this practice. Unfortunately, the results from this practice vary—sometimes increases in gain may be obtained from hormone implants whereas in others no benefit is obtained. Mississippi State University reported results which confirm the uncertainty of benefits to be derived from this practice (Table 13.8).

In one year's data, the Mississippi researchers reported a 24% increase in gain due to diethylstilbesterol (DES) implants whereas in another year there was no benefit. However, generally it is a good practice to implant calves which are destined for feeding out as slaughter cattle. Do not implant purebred calves which may be saved as seed stock. The normal level of DES implant is 12 or 15 mg within a few days after the calf is born.

Kansas State University (Corah *et al.*, 1977) reported on the benefits of double implantation of suckling calves with Ralgro. Forty-nine steer and 87 heifer Polled Hereford or crossbred calves were divided into three groups based on breed, age, and sex. Sex of the calf had no effect on response to treatment. The three treatments were (1) control group, not implanted, (2) one implant (36 mg Ralgro) during the suckling period when the calves were approximately 2½ months of age, and (3) two implants during the suckling period. Initial implant was at 44 days average age, reimplantation was approximately 70 days later (Table 13.9).

Calves given one Ralgro treatment gained 8 lb more than nonimplanted calves; calves given two treatments gained 43 lb more ($p < 0.01$) at weaning time than nonimplanted calves.

In the same Kansas report, data from 62 implant demonstrations conducted in 31 different counties in Kansas, of the 1402 implanted suckling calves (either Ralgro or DES—both gave comparable responses) (Table 13.10) an average gain of 15.1 lb more than 694 nonimplanted calves resulted. In the yearling trials (Des or Ralgro), (Table 13.11) 616 implanted yearling steers and heifers gained 20 lb more than 365 nonimplanted yearlings on summer pasture.

IV. EARLY WEANING OF BEEF CALVES

Early weaning of beef calves is possible and sometimes may be practical. Most cattlemen will agree it is most difficult to improve on a good nursing cow as a source of nutrients for the young calf as long as good pasture is available. Under certain economic conditions, it might be worthy of consideration to wean calves as early as 4 weeks—and possibly even at 2 weeks of age. Louisiana State Uni-

TABLE 13.9

One versus Two Ralgro Implants for Suckling Beef Calves[a]

Treatment	Number of calves	Age of first implant (days)	Age of second implant (days)	Days between first implant and weaning	Birth weight (lb)	Weaning weight (lb)	Gain (lb)	Treatment advantage (%)
Not implanted, control	41	—	—	—	72	373	301	—
One implant during suckling period	43	79	—	114	71	380	309	+8.4
Two implants during suckling period	43	44	114	149	71	415[b]	344**	+43.0

[a] Corah *et al*. (1977).

[b] $p < 0.01$.

TABLE 13.10

Effect of Implanting Suckling Steers and Heifers with DES or Ralgro

Area of Kansas	No. trials	No. counties	No. implanted calves	No. control calves	Average response to implants (lb)
Southeast	15	6	323	253	11.3
South Central	2	2	35	18	14.9
Northeast	9	3	180	129	14.0
Southwest	1	1	110	12	31.0
Northwest	6	11	754	282	14.7
Total	43	23	1402	694	15.1

versity (O'Neal *et al.*, 1977) researched such a program and found it was workable; the 205-day adjusted weaning weights of such calves were not significantly different from those which had nursed their mothers for the same 6½-month period. In the LSU program, early-weaned calves were confined to small indoor pens when pasture was not available and then moved to relatively small pasture plots when spring and summer pasture was available. The early-weaned calves at 2 and 4 weeks of age were given access to some grass hay and water, in addition to the diets listed in Table 13.12.

The LSU data seems to indicate that it is more advantageous to wean the calves at 4 weeks of age, but in an emergency can be weaned at 2 weeks. For about the first 60 to 90 days, the early-weaned calves should have access to Diet No. 1 (LSU diet); Diet No. 2 (no milk replacer) may be then substituted until the calves are 205 days of age. Several cases of calf scours were observed in the early-weaned calves, but such was not a problem in the calves nursing their mothers. However, this problem was alleviated by rotating the calves among clean pens

TABLE 13.11

Effect of Implanting Yearling Steers and Heifers with DES or Ralgro

Area of Kansas	No. trials	No. counties	No. implanted calves	No. control calves	Average response to implants (lb)
South Central	5	4	136	76	30.2
Northeast	5	3	192	97	19.7
Northwest	4	3	161	67	15.3
Southeast	5	4	127	125	15.5
Total	19	14	616	365	20.0

TABLE 13.12

Early-Wean Calf Formulas[a]

Ingredient	Diet 1 (%)	Diet 2 (%)
Corn	37.3	42.9
Oats	22.0	21.5
Soybean meal (49%)	22.0	32.2
Calf Manna[b]	15.4	—
Curaphos[c]	1.1	1.2
Salt	1.1	1.1
Vitamin–antibiotic mix	1.1	1.1
Total	100.0	100.0

[a] O'Neal *et al.* (1977).
[b] Albers Milling Co.
[c] Calcium, 35% and phosphorus, 14%.

combined with the use of Neo-Terramycin calf boluses (Chas. Pfizer and Co., New York) as needed.

It is difficult to assess an economic comparison of this program since feedstuffs are not the only costs involved. A great deal more labor and management are involved in calf early weaning programs.

REFERENCES

Corah, L. R., McKee, M., and Schalles, R. R. (1977). "Effect of using One vs. Two Growth Promoting Implants on the gains of Nursing Calves," Cattlemen's Day Rep., p. 13. Kansas State University, Manhattan.

O'Neal, E. N., Smart, L. I., Posey, J. T., Godbe, R. A., (1977). "Livestock Day." Louisiana State University, Baton Rouge.

Perry, T. W., Peterson, R. C. and Beeson, W. M. (1974). A comparison of drylot and conventional cow herd management systems. *J. Anim. Sci.* 38, 249.

Totusek, R. (1975). "Indiana Beef Forage Research Day," University, Totusek, R. (1975). "Indiana Beef Forage Research Day," p. 67. Purdue University, West Lafayette, Indiana.

Wyatt, R. D., Gould, M. B., Whiteman, J. V., and Totusek, R. (1977). Effect of milk level and biological type on calf growth and performance. *J. Anim. Sci.* 45, 1138.

IV

Finishing Beef Cattle

14

Starting Cattle on Feed

I. THE PROGRAM

A recommended program for handling newly purchased feeder cattle upon their arrival at their new feedlot home is presented below. Naturally it would be expected that most feedlot operators would plan to impose various modifications in this program. However, it is a quite workable program when followed in principle.

1. Isolate the new cattle so that they are not adjacent to cattle which are already "adjusted." The author has experienced an outbreak of "shipping fever" in cattle which had been on site for 6 weeks, undoubtedly via exposure to new cattle which were brough in and penned adjacently. It is therefore a good idea to have receiving pens where the isolation and adjustment phase may be managed separately.
2. Clean dry quarters should be available where the new cattle may rest following a long uncomfortable transit period.
3. Cattle should not be crowded. At least 40 square feet per animal under roof is desirable—especially during the rainy period (Figs. 14.1 and 14.2).
4. It is best to have the shed-type shelter in the receiving pens to provide the cattle the option of shelter or open air.
5. Fresh clean water is critical since cattle tend to become dehydrated on long hauls. If drainage is good, it is especially desirable to us overflowing tanks. Automatic waterers should be avoided since new cattle are usually unacquainted with using this type of equipment.
6. The timing of hormonal treatments and treatment for grubs, lice, and worms varies. The large commercial operator prefers to give the complete treatment to the new cattle immediately upon arrival at the farm; others prefer to administer the typical veterinary shots at this time, but to delay such things as implants, worming, administration of degrubbers, until the cattle have passed through the 3-week shipping fever period. It really will have about the same net

Fig. 14.1. Excellent working facilities enhance handling cattle rapidly and efficiently. (Photo courtesy of BEEF Magazine.)

Fig. 14.2. New feeder calves should have loose hay scattered about the lot for at least the first three to four days. (Photo by J. C. Allen and Son.)

Fig. 14.3. External and internal parasites should be controlled at all times. (Photo by J. C. Allen and Son.)

effect whichever way it is done so the timing will be left to the discretion of the operator.

7. The shots to be administered as a prevention for the typical feedlot diseases of cattle should vary somewhat from one locale to another. These should be determined based on the problems typical of the area and in consultation with your veterinarian.

8. Treatment for external parasites may be handled by spraying, dipping, or pouring. In certain states the practice of dipping is required. There is no hard and fast "one best method" of accomplishing this and more than one method is available. The author is convinced that a good spraying with a pressure sprayer is effective where dipping facilities are not available (see Fig. 14.3 and 14.4).

9. Warble eradication must be handled with some understanding of the geography of origin of the cattle. In other words, the use of the very effective systemic pour-ons must be at the proper time in the life cycle of the grub. If the grub is killed at the wrong time in its movement through the cow's body, a severe "protein" shock may be experienced. A qualified entomologist should be consulted to define typical warble locations in the body of the cattle of various geographic origins at various months of the year.

10. Parasites of the gastrointestinal tract may be handled at the time of arrival. At least two excellent methods are available—either subcutaneous Tramisol or oral Thibendazole will do a good job. It is important, however, to follow the directions of the manufacturer. The author does not routinely worm feeder cattle; others worm all cattle. Perhaps this is a matter of judgement and should be decided by each feedlot manager.

11. The Aureo S-700 program (feeding 350 mg each of aureomycin and sulfamethazine per head daily for from 14 to 21 days—or even as long as 28 days) is an excellent one. Many researchers, including the author, have demonstrated the beneficial effects of this program. It is currently a part of the routine program for working new cattle. Aureo S-700 is administered in the natural protein supplement.

12. Even though the cattle have been injected intramuscularly with vitamin A in the "shots" step, it is a good idea to feed 50,000 IU of vitamin A per head daily during this adjustment period. If the cattle have been injected with 2.5 million IU, no more oral vitamin A is needed following completion of the adjustment period.

The above program presents general guidelines that need to be considered in the starting program. The specific day-to-day regimen of the first 21 to 28 days is given below.

Arrival. Administer the shots, vitamin A injection, worming if necessary, hormone implant, and the required dip. Cattle should then be moved to their comfortable receiving quarters. Bunches of medium-to poor-quality hay should

Fig. 14.4 Dipping vat with hold-down rack. The hold down rack prevents the animals from jumping on top of each other. (Photo courtesy Temple Granden, Tempe, Arizona.)

be provided in hay feeders as well as scattered on the ground. A free-choice mineral mix should be available which contains phosphorus, calcium, salt, and trace minerals. In fact, a free-choice mineral mix of two parts dicalcium phosphate to one part trace mineralized salt is an excellent one. The cattle should be immediately placed on their Aureo S-700 medication which can be mixed with a protein supplement. Acutally, it is also advantageous to put a mixture of 2 lb of corn and 2 lb of natural protein supplement per head daily in the bottom of the bunk and then to put a thin layer of their hay on top of the grain mix to begin the feeding program. Additional hay can be scattered about the lot.

Day Two. The same general program is followed as for the first day—limited corn and supplement under a little hay plus other hay scattered about the lot.

Day Three. Introduce the cattle to corn silage. Mix the 2 lb of corn and 2 lb of supplement with 4 lb of corn silage on the bottom of the trough, covered with a little hay. Additional hay should be available about the lot.

Day Four. Continue to introduce the cattle to more corn silage if the feed from the previous days aliquot was used satisfactorily. Incorporate 7 or 8 lb of corn silage plus the corn and medicated supplement.

Day Six to Seven. Hopefully all the trough-fed hay can be dispensed with by this time. Feed whatever corn silage the cattle will utilize (20–25 lb) plus 2 lb of corn and 2 lb of protein supplement containing the Aureo S-700. Bowel movement should firm by now.

Day Seven to Twenty-One. This time should be used for a *gradual* introduction to a much higher, "hotter" diet. During this period cattle may have been moved to thier permanent pens.

Day Twenty-One for the Next "Several Weeks." The cattle should now be adapted to the management changes. The high-moisture corn and urea protein programs can now be started. Rumensin feeding can be considered but should be delayed until the cattle are past the risk of shipping fever syndrome.

II. STARTING NEW FEEDER CATTLE ON CORN SILAGE

Starting new cattle on feed is a major problem which faces cattle feeders. During the stress of weaning, assembling at sales barns, and at least two trucking experiences, calves will shrink at least 10%. Probably half of this is intestinal loss, but the other half is a combination of tissue loss and dehydration. Cattle thus arrive at the feedyards somewhat depleted nutritionally and need wholesome feed quite promptly.

Michigan State University (Fox *et al.*, 1977) conducted a series of trials to compare methods of getting new feeder cattle started. This section summarizes that research. The data presented are from trials varying in length from 28 to 46 days—but primarily 28 days—which corresponds to the typical feedlot adjustment period.

A. The Value of Adding Hay to Corn Silage

In this research, summarized in Table 14.1, 455-pound calves trucked nearly 2000 miles were fed silage which had been treated with Pro-Sil (to 12% crude

TABLE 14.1

Value of Added Hay in Starting New Feeder Cattle on Corn Silage[a]

Item	Corn silage only	Corn silage plus free-choice hay
Daily gain (lb)	1.93	1.72
Difference (lb)		−0.21
Daily dry matter (lb)	9.8	9.7
Dry matter per lb gain (lb)	5.2	5.8

[a] Fox *et al.* (1977). Five experiments, 252 cattle.

protein) alone, or in combination with 1 lb of soybean meal per 20 lb of silage versus feeding alfalfa–brome hay fed free choice along with the corn silage treatment. In summarizing five experiments, there was no advantage to offering hay in addition to the corn silage. In all but one of the five experiments, gains were decreased by the feeding of hay, whereas the overall depression in gain was 11%.

The apparent explanation for difference in performance is that those fed hay had similar dry matter consumption to those fed only corn silage, resulting in a lower net energy intake. Cattle having access to hay tended to eat more readily the first few days, but those fed silage adjusted quickly and their consumption increased rapidly. However, hay may be desirable and necessary for the first few days in order to attract new cattle to the feed bunk quickly.

B. Hay Plus Grain versus Corn Silage

Feeder calves (460–490 lb) were fed either corn silage treated to contain 12% crude protein or else a mixture of hay, corn, and soybean meal, such that the two diets were isocaloric as well as isonitrogenous. In all cases, those fed the hay plus corn diets had highest rates of gain, but there was very little difference in dry matter required per pound of gain (Table 14.2). Furthermore, there is some evidence suggesting cattle started on a hay diet will still require an adaptation period to adjust to the silage changeover.

C. Value of Corn Added to Corn Silage

Table 14.3 gives the results of one experiment in which a 50% silage, 50% concentrate diet was compared to an all-silage diet. Cattle fed the higher grain diet gained about 10% faster (1.89 versus 1.72 lb). However, this performance is approximately 12% less than would be expected based on the comparative energy levels of the two diets, suggesting greater benefit may be obtained from fortification of corn silage with extra corn after the cattle are started on silage.

TABLE 14.2

Hay and Corn Compared to Corn Silage for Starting New Feeder Cattle[a]

Item	Corn silage	Hay plus corn
Calves		
Daily gain (lb)	1.49	1.90
Difference (lb)		+0.41
Daily dry matter (lb)	9.4	12.2
Dry matter per lb gain (lb)	6.4	6.4
Yearlings		
Daily gain (lb)	3.85	4.23
Difference (lb)		+0.38
Daily dry matter (lb)	16.4	18.5
Dry matter per lb gain (lb)	3.9	4.8

[a] Fox *et al.* (1977). Two experiments with calves (66 head), 1 experiment with yearlings (50 head).

D. Value of Soybean Meal versus NPN-Treated Silage

Several experiments were conducted to study NPN utilization by new cattle starting on silage. Basically, three treatments were compared: (1) untreated silage supplemented with soybean meal, (2) silage which had been treated with sufficient Pro-Sil to bring the protein up to 12% and (3) treated silage (Pro-Sil-plus 1 lb of soybean meal with every 20 lb of silage consumed (Table 14.4).

In all comparisons, all cattle fed NPN-treated silage gained more slowly than cattle fed soybean meal as the source of supplemental protein. When all data were lumped together, gains were depressed by about one-third when NPN silage was fed. These results suggest NPN utilization by calves is quite poor during the time of stress, and thus probably natural protein should be utilized for new cattle. It has been recognized for many years that cattle must adapt to NPN over a period

TABLE 14.3

Value of Added Corn for Starting New Cattle on Corn Silage[a]

Item	All-silage	Silage 50%, concentrate 50%
Daily gain (lb)	1.72	1.89
Difference (lb)		+0.17
Daily dry matter (lb)	9.0	9.2
Dry matter per lb gain (lb)	5.3	5.0

[a] Fox *et al.* (1977). One experiment, 132 cattle.

TABLE 14.4

Soybean Meal versus Nonprotein Nitrogen as a Source of Protein for Starting New Cattle on Corn Silage[a]

Item	Type of silage treatment		
	Untreated plus soybean meal	NH_3 treated	Treated plus soybean meal
Calves			
Daily gain (lb)	2.03	1.35	1.87
Differ from untreated control (lb)		−0.68	−0.16
Daily dry matter (lb)	10.2	9.1	9.8
Dry matter per lb gain (lb)	5.2	7.8	5.6
Yearlings			
Daily gain (lb)	4.23	3.77	3.65
Differ from untreated control (lb)		−0.46	−0.58
Daily dry matter (lb)	16.4	14.7	16.6
Dry matter per lb gain (lb)	3.9	3.9	4.5

[a] Fox *et al.* (1977). Five calf experiments (270 head) and one yearling experiment (50 head).

of from 2 to 4 weeks, during which time gains may be depressed. It is not logical to use the stress period of newly arrived feeder cattle to attempt the NPN adaptation.

E. Mineral Programs for New Cattle

Minerals represent one of the prime considerations for new cattle. Limestone and salt are relatively inexpensive, but the phosphorus supplements are much more expensive to provide range cows and their calves. Recently weaned calves or yearling feeder cattle may arrive at the feedlot with an almost depraved appetite for phosphorus. Clean fresh minerals should be provided in a box where the cattle may have free access. A complicated mineral mix is not necessary for any beef cattle feeding situation. A combination of two parts dicalcium phosphate to one part of trace mineralized salt mixed together and offered at all times will meet practically all of the supplemental mineral needs—especially if a well-fortified protein supplement is fed.

III. FEEDING REGIMENS FOR NEW FEEDER CATTLE

Several good programs can be utilized for starting new feeder cattle. Two or three such programs will be outlined.

A. Complete Pelleted Diet

This is a bit more expensive approach, yet such diets are being manufactured and sold by feed manufacturers. This program is based on feeding medium- to poor-quality hay free choice along with the mineral mixture listed above free choice. In addition, feed 10 lb per day of a ¼-inch pelleted complete mixture whose composition is shown in the tabulation below:

	1000-lb mix
Ground corn cobs	685 lb
Soybean meal	100 lb
Molasses	100 lb
Cracked corn	100 lb
Dicalcium phosphate	10 lb
Trace mineralized salt	5 lb
Vitamin A	5 million IU
Aureomycin, terramycin, or tylosin	35 gm
Sulfamethazine	35 gm
Total	1000 lb

From the approximately 10 lb per day of the above formulation, a calf will receive 50,000 IU of vitamin A plus 350 mg each of antibiotic and sulfamethazine. This program should be continued for from 2 to 4 weeks, based on when the danger of shipping fever has passed.

B. Hay Diet

New feeder cattle tend to take to hay readily because it is so nearly like some of the dry range forage to which they have been accustomed prior to leaving range country. Good-quality hay should not be used because extra leafy alfalfa hay, for instance, would encourage looser bowels which means the calf will be defecating extra water, thus tending to aggravate its semi-dehydrated condition. Bright clean nonlegume hay that is free of mold is ideal for this version of the program. The bunks should be loaded with this type hay so that when the new calves arrive in their receiving pens, they can have free access to water, minerals, and hay. This program is exceptionally good as a receiving diet, but by day two it will need to be expanded. Introduce the calves to 2 lb of whole shelled corn plus 2 lb of a protein supplement in which the source of protein is natural, for example, from either soybean meal or cottonseed meal. The protein supplement should be specially prepared so that from 2 lb per head per day each calf will get its 50,000 IU of vitamin A plus 350 mg each of antibiotic and sulfamethazine.

The mechanics of feeding hay plus 2 lb each of shelled corn and pelleted protein supplement may have to be worked out. If you could feed hay in a tight-bottom bunk, then the corn and supplement could be placed in the bottom of the bunk and then the day's aliquot of hay placed on top of the concentrates. Thus, in order for the calves to get to the very palatable concentrates, they would have to burrow through their hay.

C. Corn Silage Program

Wherever corn silage is available, it is a most excellent basis for a conditioning program because of its high energy value and its high palatability. However, the cattle feeder should bear in mind that most new feeder cattle have never been exposed to corn silage so that problems of introduction of this type of diet should be considered. It is a good idea to utilize medium-quality hay for the first day or two since cattle will relate to hay very well. The second or third day 8–10 lb of corn silage per head should be placed in the bottom of the bunk with only limited hay on top of the silage. As the cattle begin to feed on the hay they get their first corn silage as they lick off their nose. Like the hay program listed above, a corn silage program should be fortified with 2 lb of whole shelled corn and 2 lb of natural protein supplement. The 2 lb of protein supplement should supply 50,000 IU of vitamin A plus 350 mg each of antibiotic and sulfamethazine. This medicated program should be continued for from 2 to 4 weeks or until the risk of shipping fever has passed.

D. Hay as a Rumen Regulator

Dry hay is an important product and should be the first feed for all cattle. As indicated above, you can go from hay to any other feeding program but medium-quality grass-type hays are a reliable foodstuff especially when problems are incurred. Hay seems to exert a "rumen regulating" effect. The rumens of most new arrivals are usually in poor condition and thus they are in need of re-conditioning before concentrates formulations are begun. Furthermore, the hay may be used in the open scattered on the ground for the new cattle—they will literally stumble over it. The too rapid introduction to grain, might initiate the symptoms of acidosis. The added hay will then be extremely valuable in getting the rumen back into condition.

E. Soybean Meal versus Urea as Sources of Protein for Newly Received Feeder Calves

Ohio Agricultural Research Center (Anonymous, 1975) demonstrated that the use of natural protein supplements on newly received feeder calves, as contrasted to urea supplements, is best for about the first month. Feeder calves averaging just under 400 lb initially were full-fed limestone-treated (0.5%) corn silage plus

2 lb of a fortifying supplement containing trace mineralized salt, vitamin A, aureomycin, and sulfamethazine (AS-700), and either soybean meal or urea and corn. In other words, soybean meal and urea provided the supplemental protein in various ratios of 100:0, 67:33, 33:67, and 0:100.

Feedlot performance of the calves for the first 65 days is shown in Table 14.5. Gain and efficiency of feed conversion were progressively and significantly improved when an increased proportion of the supplemental protein came from soybean meal. Steers fed increased levels of urea tended to consume less silage. Generally, the depressing effect from urea occurred the first 2 weeks of the study although the effect persisted somewhat through the 28th day. Subsequently, there was little effect on performance due to urea level in the supplement.

Since adjustment of the rumen to urea utilization undoubtedly represents considerable stress, it is recommended that urea supplements be replaced with natural protein supplements for newly received feeder calves.

South Dakota research (Embry, 1977), essentially confirmed the earlier Ohio data (Table 14.6). The South Dakota calves were fed prairie hay plus 2 lb of a 40% protein supplement containing either soybean meal or urea as a source of protein.

F. Protein Level Effect on New Feedlot Cattle

The daily intake of protein is of major importance for new feedlot cattle and is required in relatively large amounts compared to other nutrients. Protein is a nutrient for which the body is not able to store adequate reserves. Thus, unless

TABLE 14.5

Influence of Soybean Meal and Urea as Supplemental Protein Sources on the Performance of New Feeder Calves[a]

	Protein supplement			
Item	Soybean meal	2/3 soy 1/3 urea	1/3 soy 2/3 urea	Urea
Number of calves	18	18	18	18
Initial weight (lb)	376	380	399	393
Final weight (lb)	547	542	539	529
65-day gain (lb)	171	162	140	136
Daily gain (lb)	2.63	2.49	2.15	2.09
Daily feed				
Corn silage (lb)	29	28	27	27
Supplement (lb)	2	2	2	2
Dry matter/gain (lb)	4.6	4.8	5.3	5.5

[a] Anonymous (1975).

TABLE 14.6

Urea Utilization with Prairie Hay during Beef Cattle Adaptation Phase of 37 Days[a]

Item	Type of supplement: Soybean meal	4% urea
Number of cattle	32	32
Initial weight (lb)	422	422
Final weight (lb)	482	473
37-day gain (lb)	60	51
Daily gain (lb)	1.61	1.34
Daily feed (lb)	13.2	12.9

[a] Embry (1977).

intake is adequate, the body may become deficient in protein quite rapidly, especially during times of stress. Protein-deficient animals tend to be more prone to infections. South Dakota research (Embry, 1977) demonstrated the effect of adequate supplemental protein on the performance of new feeder calves.

Feeding diets containing more than 13.5% protein, on a total dry matter basis, appeared to have little benefit. The 13.5% protein diet translates over to about 12% protein on an "air-dry" or 90% dry matter basis. A diet containing 10.5% protein was not adequate. Note from Table 14.7 the level of protein in the diet did not affect daily feed consumption.

G. Preconditioning Feeder Cattle

There has been a lot written and discussed about the advantages of "preconditioning" feeder cattle—primarily feeder calves—prior to shipment for their new feedlot homes. Preconditioning in its simplest definition involves the weaning process coupled with conditioning, shots, and feed initiation. This usually involves 4 to 6 weeks. "Conditioned" calves are then shipped to feedlots or perhaps so-called backgrounding lots or even to wheat pastures for an intermediate stop.

In a preconditioning program cattle are wormed, grubbed, liced, implanted, eartagged, castrated or spayed if desired, branded, given a variety of shots—but at least an IBR (Infections bovine rhinotracheitis) shot—and injected with vitamins A and B_{12}.

Immediately upon arrival cattle should be placed on to medium-quality hay, and one of the conditioning and medicated programs listed above should be pursued—it really does not make much difference nutritionally which program is

TABLE 14.7

Levels of Protein Supplementation on Feedlot Adaptation of New Feeder Calves—29 Days[a]

Item	Percentage protein in the diet			
	10.5	13.5	16.5	19.5
Number of cattle	19	19	19	19
Initial weight (lb)	374	374	373	375
Daily gain to date				
7 days (lb)	1.9	2.8	2.5	2.7
14 days (lb)	1.9	2.5	2.3	2.7
21 days (lb)	2.4	2.7	2.6	3.0
28 days (lb)	2.2	2.4	2.3	2.5
Daily feed				
7 days (lb)	10.4	10.2	10.1	10.9
14 days (lb)	10.7	10.8	10.9	11.1
21 days (lb)	12.1	12.3	12.3	12.6
28 days (lb)	13.7	14.0	13.0	14.4

[a] Embry (1977).

followed. The key to any preconditioning program is basically to get the cattle to eat promptly.

H. Preconditioned Calves—Do They Do Better?

This is a question that is extremely difficult to treat in summary form. One can pull out isolated data which will show just about anything one desires. In other words, preconditioning data can be presented to make it look quite worthwhile whereas other data can cause its practice to be questioned.

The Purdue Veterinary and Animal Sciences group cooperated with comparable investigators from Oklahoma State University (Meyer *et al.*, 1971) in working with Oklahoma calves which were shipped to Purdue University to be fed out. Four general treatments were imposed on the Oklahoma calves prior to shipment: (1) those weaned and shipped the same day, (2) those weaned 30 days prior to shipment and placed on hay, grain, and supplement, (3) those vaccinated with BVD vaccine and IBR and bovine parainfluenza-3 vaccine approximately 45 days before shipment and then weaned on the day of shipment, and (4) those treated as in group (3) above and weaned 30 days prior to shipment and fed as listed for group (2) above.

What does it cost to precondition a calf? One might assume a medicinal cost of $5 per head plus "about" $20 per head for labor, yardage, and feed for a 30-day stay—a ballpark figure might well be $25 per head. The economics perhaps does not sound too bad until the bargaining begins as to who is going to pay the bill. Naturally, some feedlot operators are happy to pay a reasonable amount to have someone get their feeder calves prepared for immediate introduction to the feedlot—others may not. Another way of calculating the cost is on a hundred-weight basis so that on a 500-pound calf, a reasonably conservative cost would be an additional \$5 per cwt.

How do preconditioned calves perform compared to nonpreconditioned ones? Note from Table 14.8 that preconditioned or vaccinated calves started faster (see first 28-day performance); but by the end of the total feedlot life (252 days), no differences existed in feedlot performance between weaned and early-weaned or between vaccinated and nonvaccinated calves. Based on these data, it would seem doubtful that preconditioning practices could be justified on a year-to-year basis. Rather, it should be viewed as a type of disaster insurance. Similarly, the performance of vaccinated and nonvaccinated calves was similar; but in a "disaster" situation, a disease outbreak might wipe out a very large number of calves. Therefore, the vaccination procedure is insurance. In the case of vaccination in the test research it was not presumed vaccination would affect performance of the cattle on test, but naturally it was critical to obtain data on that aspect. A vaccination program saves lives and probably will have very little effect on the performance of those that survive. Naturally there will be exceptions to this observation. For example, cattle which have a rather prolonged siege of respiratory infections may be delayed in starting feedlot; occasionally such cattle may develop adhesions of the lung wall to the chest cavity.

TABLE 14.8

Feedlot Performance of Weaned versus Nonweaned and Vaccinated versus Nonvaccinated Feeder Cattle[a]

Item	Treatment			
	Weaned	Nonweaned	Vaccinated	Nonvaccinated
Number of animals	72	71	72	71
First 28 days after shipment				
Initial weight (lb)	453	436	445	443
28-day weight (lb)	534	502	521	514
Gain per animal (lb)	81	66	76	71
Daily gain (lb)	2.89	2.35	2.71	2.54
Total 252-day performance				
Initial weight (lb)	453	436	445	443
252-day weight (lb)	992	975	991	976
Gain per animal (lb)	539	539	546	533
Daily gain (lb)	2.14	2.14	2.16	2.12

[a] Hendrix and Smith (1975).

I. Backgrounding Feeder Cattle

Backgrounding (Hendrix and Smith, 1975) can be contrasted to "preconditioning" primarily by the length of time involved. Whereas a preconditioning program usually runs for 28 to 35 days, a backgrounding program may last for up to 6 or 7 months—sometimes even longer. The backgrounding phase includes the time lapse between weaning and when the cattle enter the feedlot to be finished for slaughter. Calves generally gain from 100 to 400 lb, depending upon diet fed and length of the backgrounding program. Background gain is mainly muscle and frame, with minimal fattening. Growth gains are achieved as economically as possible primarily utilizing forages. Very little grain is used in backgrounding diets.

1. Selecting Calves for Backgrounding

Not all calves are suited for a backgrounding program. Generally, calves less than 8 months of age that have been creep fed and in a fleshy condition do not fit into this program well since they lose weight and condition rapidly when fed high roughage rations. Heifer calves finish for market at weights 150–200 lb less than steers, thus they do not fit into a lengthy backgrounding program as well as steers. However, for the cow–calf producer, backgrounding the heifer calves may allow him to make a better selection of replacement heifers since the selection can be made when heifers are older. In addition, the rate of gain obtained in most backgrounding programs is optimum for developing replacement females.

Steer calves weighing 400–500 lb in thin to moderate flesh are ideally suited for most programs. These calves when grown to 700–800 lb on high roughage rations are ready for finishing and should be in demand by cattle feeders.

2. Management during Backgrounding

A successful backgrounding operation requires a high level of operator skill and management. Calves are usually subjected to stress and are highly susceptible to disease and death. A sound health program developed with the aid of a vererinarian is recommended. The program should include handling of calves before and after weaning or prior to and after shipment in the case of purchased cattle.

Vaccinations should include those for IBR, PI_3, blackleg, malignant edema, and leptospirosis. Vaccinate for BVD if it is a problem in your area. The vaccinations should be given when purchased cattle arrive or at weaning.

Castration and dehorning should be done prior to weaning if possible but delayed a week or so after new cattle are acclimated and recovered from stress. Worming and treatment for external parasites should be done separately from other practices, especially castration and dehorning, to avoid excessive stress at any one time.

With homegrown calves, weaning stress can be reduced somewhat by providing good-quality mixed hay and 2 lb of corn in a creep 2–3 weeks prior to weaning. This will accustom calves to eating dry feeds from troughs. If this is not possible or in the case of purchased cattle, provide first cutting grass–legume hay and 2 lb of corn when they are first weaned or arrive in the lot. Avoid second- and third-cutting legume hay as it has the tendency to scour calves. Corn silage plus 1.5–2.0 lb of a protein supplement also provides a good starting ration. The protein supplement should be all natural. Urea supplements are not well utilized by new cattle.

Supply 50,000 IU of vitamin A per head daily during the first 3 weeks. This may be supplied in the protein supplement or by adding 500,000 IU of vitamin A to a free-choice mineral mix. The mineral mixture should consist of 1–2 parts of dicalcium phosphate or bonemeal and 1 part trace mineralized salt. This mixture should be available free choice at all times.

Feed 350 mg of a broad-spectrum antibiotic such as Aureomycin or terramycin per head daily during the first 3–4 weeks. In many cases, combination of 350 mg of Aureomycin and 350 mg of sulfamethazine during the first 3–4 weeks helps to reduce sickness and increase rate of gain.

Adequate quantities of clean fresh water should be provided for calves. If calves are extremely stressed or dehydrated, hold them off water for 2–5 hours until they have had a chance to eat. However, at first, continuous flow of running water may help to attract calves to the stock tank.

Implant steer calves with a growth stimulant such as Synovex, or Ralgro. Heifers that may be saved for replacements should not be implanted.

3. *Backgrounding Systems*

Once claves have recovered from the weaning and/or shipping stress period (2–4 weeks), they are ready to be placed on a backgrounding program. The program will depend upon when the calves were born (spring versus fall) and the feedstuffs available. The following are some alternative methods for handling spring and fall born calves. A number of wintering rations are shown in Table 14.9 using various feeds and resulting in different rates of gain during the winter.

The methods of backgrounding spring calves weaned or purchased in the fall weighing 350–500 lb are listed below.

1. Winter pasture thrives in at least the southern part of the United States, extending up as far north as about the 38th parallel. At least the southern portions of Illinois, Missouri, and Oklahoma plus much of Kentucky, Virginia, and Arkansas on south are regions where winter perennial pastures are available. In addition, winter wheat and oats provide winter pasture in such areas as Oklahoma, Kansas, and certain portions of Texas. The utilization of wheat pasture by beef cattle is a well-established practice that is utilized by growing beef cattle.

TABLE 14.9

Suggested Winter Rations for Backgrounding 350- to 500-lb Calves

	Ration (lb/head daily)							
Feed	1	2	3	4	5	6	7	8
Grass hay (late cut)	12–14							
Ground corn cobs		10–12				7–9		
Grass-legume hay (early cut)			12–14				10–12	
Legume hay (early cut)				12–14				
Cornstalk silage (65% moist)					30–35			
Corn silage (65% moist)						10–15		30–35
Corn or sorghum grain							3–4	
Protein supp (32% protein)[a]	1	2			2	2		2
Mineral mixture[b]	FC[c]	FC	FC	FC	FC	FC	FC	FC
Expected daily gain (lb)	0.5	0.5	0.75	1.0	1.0	1.0	1.25	1.75

[a] The supplement should contain 15,000 IU vitamin A per pound.
[b] 2 parts dicalcium phosphate and 1 part trace mineralized salt.
[c] FC = free choice.

The feeding value of winter cereal grains where they can be grown is extremely high. Such pasture may be utilized until about the time the plant is ready to "joint." Pasturing later than this will be detrimental to the grain yield; yet, on the other hand, the beef potential for gains from grazing winter wheat on into June is so great that this route may be followed. Little other than free-choice minerals is needed to supplement winter wheat pasture for cattle.

2. This method involves 30 days of grazing on fall pasture or stalk fields followed by 165 days of winter feeding. After the winter feeding period, the calves would go to spring pasture in May and graze for 60 days and sold about July 1.

If on fall pasture, the calves should gain 0.75–1.0 lb per head daily. No supplemental protein or energy would be required. Provide a free-choice mixture of 2 parts dicalcium phosphate and 1 part trace mineralized salt. If calves are used to glean stalk fields, they should receive 1.5 lb of a 32% protein supplement daily plus a salt-mineral mixture free choice. They should gain about 0.75 lb per day while dropped corn is available and the leaves and shucks have not been subjected to weathering. After that, calves should be fed additional feed in drylot. The wintering ration could consist of either of the first 7 listed in Table 14.9 and would depend upon available feeds. Ration 8 which is a full-feed of corn silage is not recommended because calves will likely be carrying too much flesh to make good use of pasture the following spring.

If calves are wintered on rations 1, 2, or 3, they will have gained about 100 lb from weaning until spring. After being wintered at this level, they should gain 1.75–2.0 lb per day during the 60 days on pasture. Stocking rate on pasture should be about two head per acre. This will result in calves weighing 650–700 lb by July 1. If wintered on rations 4–7, the calves should gain 175–200 lb during the winter and gain an additional 100 lb during the 60 days on pasture. This would result in calves weighing 700–800 lb by July 1. This system would require approximately 1 ton of dry roughage or 2.5–3.0 tons of stalk silage. Supplement requirement would range from 0 to 350 lb, depending upon the ration used. Approximately 20–25 lb of salt–mineral mixture would be required.

3. This method is similar to 2, with the exception that the cattle are carried on pasture during the second summer period until October 1 rather than being sold in July. Because less pasture is available during the summer and that which is available is lower in quality, daily gains can be expected to drop. Cattle would likely gain an additional 50 lb from July 1 to October 1. In addition, because less forage is available, stocking rate should be reduced to 1.5 head per acre during this period. Thus, one-third to one-half of the cattle would need to be sold on July 1 or supplemental grain fed during the summer grazing season. Feeding 1% body weight of grain during the summer will permit stocking rates similar to those during the spring and will result in a dramatic increase in daily gain. Feeding this level of grain would be satisfactory provided that you were going to finish the cattle for slaughter. If cattle are to be sold as feeders, however, it may be advisable to omit the grain unless a premium could be obtained at sale time for having cattle started on feed.

4. This method involves 30 days of grazing on fall pasture or stalk fields as described in the previous two methods. The 165-day winter feeding period involves the use of corn silage and protein supplement (ration 8). Calves should gain 275–300 lb during the winter on this ration. Thus, calves would weigh 700–800 lb after the winter feeding period. They should go directly to a feedlot and finished for slaughter. It is likely that after this length of silage feeding, the cattle would be carrying too much flesh to go on pasture unless grain was fed. This system would require 2.5–3.0 tons of corn silage, approximately 350 lb of protein supplement, and 20–25 lb of salt–mineral mixture per head.

The methods of handling fall calves weaned or purchased in June, weighing 350–500 lb are enumerated below.

1. Since pasture is available after weaning, one alternative is to graze on pasture until October or November and sell as heavy feeder calves in the fall. Calves handled in this manner should gain 50–100 lb during the remaining grazing season provided the pasture contained some legume. Only a salt–mineral mixture would be required in addition to the pasture. If the pasture is strictly grass

without legume, 3–4 lb of grain plus 1 lb of protein supplement would improve daily gains during July and August.

2. If calves from the previous system were not sold, they would be carried over the winter as yearlings. Since they are heavier and have more frame going into the winter than spring-dropped calves, they have sufficient capacity to utilize more low-quality roughage such as pasture aftermath and stalk fields. When grazing these lower quality roughages, supplemental protein, vitamin A, and phosphorus would be required. Feeding 1.5 lb of a 32% supplement fortified with vitamin A and phosphorus plus the roughage should provide sufficient energy and protein for yearling cattle to gain 0.5 to 0.75 lb per head daily. When most of the dropped grain, shucks and leaves were gleaned from stalk fields and pasture aftermath became limited, hay or other roughage should be provided for the remainder of the winter. The ration shown in Table 14.9 could be modified by increasing the roughage levels 25–30% to meet the requirements of the heavier weight yearlings. Using a ration that provides at least 1.0 lb daily gain would result in long yearlings weighing 700–800 lb ready for sale or finishing in drylot the following spring.

REFERENCES

Anonymous (1975). "Beef Day and Cattlemen's Roundup," p. 6. Ohio State University, Columbus.

Embry, L. B. (1977). "Feeding and Management of New Feedlot Cattle," Cattle Day Rep., p. 47. South Dakota State University, Brookings.

Fox, D. G., Woody, H. D., Danner, M. L., Cools, R. J., Bates, D. B., and Lomas, L. W. (1977). Starting new feeder cattle on corn silage. *Mich. Agric. Exp. Stn., Res. Rep.* **328,** p. 110.

Hendrix, K. S., and Smith, W. H. (1975). Calf backgrounding systems and rations. *Purdue Univ. Hoosier Cow-Calf Days, 1975* p. 31.

Meyer, K. B., Beeson, W. M., and Armstrong, J. H. (1971). Observations on preconditioning of feeder cattle. *Purdue Univ. Annu. Cattle Feeders Day Rep.* p. 5.

15

Feeding Stocker Cattle

A stocker is a young animal fed to grow but not to improve its general condition of flesh. In other words, a stocker that gained 100 pounds over a given period of time would be in the same degree of flesh at the end of the period as it was when it started. There are two type of stockers—heifers grown for replacement, and steers grown for subsequent fattening. If sufficient minerals and vitamins are present, it is possible to limit the rate of gain of immature beef cattle by limiting either the protein or energy made available to them. However, it is far more desirable to limit gain by limiting the energy fed rather than limiting daily protein allotment since the latter technique conceivably might exert a permanent effect on the animal.

Stocker growth is nourished, normally with a preponderance of roughages, balanced with adequate protein, minerals, and vitamins. Unless subsequent programs will limit gains markedly, stockers should be fed to gain from 0.9 to perhaps as much as 1.5 pounds per head daily. With any less gain, the overhead cost will be prohibitive; and with any greater gain, cattle will come off the stocker program carrying too high a degree of finish. A "too fat animal" coming off a stocker program can be expected to have a lowered performance record during the finishing phase.

I. FEEDING PROGRAMS FOR STOCKERS

A. Hay Program

Many stocker programs are conducted in drylot when pasture is scarce (Fig. 15.1). Thus, the feeding of harvested roughages is most common. Naturally hay can vary so greatly in quality that it is dangerous to generalize on a specific feeding program. However, if medium-quality hay is available, a typical consumption of from 12 to 18 pounds of such hay should cause stocker calves to gain from 0.75 to 1.00 pound per day. It is quite likely there will be enough protein

Fig. 15.1. High-quality grass–legume hay fits well into drylot cattle feeding. (Photo courtesy of BEEF Magazine.)

for normal growth since "average" grass–legume hay contains upward of 12–15% protein, a growing stocker requiring only 10–11% protein. With adequate minerals (free-choice nimeral of 2 parts dicalcium phosphate: 1 part trace mineralized salt) and an intramuscular injection of 2.5 million IU of vitamin A initially, such stockers' nutrient needs should be met for nearly one pound per day gain. If it is desirable to raise the gain an additional one-third pound per day (1.2–1.4 lb total daily gain), 4 lb of concentrate per head daily would be added.

B. Cornstalk Pasture Program

Well-matured weanling calves can do pretty well on cornstalk fields as long as feed is ample. However, cornstalk leaves and husks provide primarily energy so additional protein will be needed. This can be accomplished by feeding a minimum of 3 lb of legume hay per head daily scattered on the ground—5 or 6 lb of legume haylage will accomplish the same goal. Also, a good mineral mix is needed to supply calcium, phosphorus, salt, cobalt, and zinc and possibly manganese. This program will not support much more than about 0.8 lb daily gain, but perhaps it is not too expensive.

C. Winter Range

In certain parts of the country, calves are "overwintered" on winter range or pasture to be sold as short yearling feeder cattle the next spring. Unless some supplemental protein and/or energy is provided, such calves do little more than remain stable on such a program. Naturally the performance on this program is related to pasture quality. However, this may be justified in order to hit a higher spring market than might have been realized from a fall sale of the calves.

D. Silage Programs

It is most difficult to administer a corn silage or sorghum silage that will limit the gains of calves to no more than 1.5 lb per day. Since both corn silage and sorghum silage are deficient in protein, when supplemental protein is added good-quality calves might gain nearly 2 lb per day—which is usually too much for a stocker calf. The only control then would be to limit the silage level fed which results in calves that are hungry and restless.

E. Winter Wheat Pasture

Many years ago wheat producers of the Great Plains area learned they could safely pasture off the winter growth of fall-sown wheat. In fact, the grazing process helped the wheat plants to spread out and was actually beneficial to the wheat growing program. As a result shipment of stocker cattle for many miles to overwinter on wheat pasture has become a common practice and a big business. Thrifty calves should gain from 1 to 1½ lb per head daily on good wheat pasture. Probably their only supplemental need on the wheat pasture program is free-choice minerals.

F. Summer Pasture Program

The performance one can expect from healthy stocker calves on a summer pasture program is highly correlated to pasture quality. Herein the word "quality" refers not only to the predominant pasture crops grown but also to the condition of growth of the pasture. In other words, pasture quality can range all the way from the eiptome of lush young growth of palatable grass–legume mixture down to a woody droughty vegetation that has very little growth potential even when cattle can find it and are willing to eat it. Therefore, it is most difficult to predict performance to be expected from a summer pasture program without pinpointing almost all the specifics. As a good example of this, typically in the Corn Belt or Midwest area, it is a common statement that stockers grazing Kentucky bluegrass–bromegrass mixtures will "make two-thirds of their growth on summer pasture during the first one-third of the pasture season." Thus, under these conditions it is necessary to consider whether a concentrate feeding of stocker cattle on pasture around July 1 to 15 should be started rather than staying with declining performance on pasture alone from July 15 to the time of killing frosts (November 15 ?).

II. CONTROLLED GROWTH OF STOCKERS

Earlier it was implied the growth of stockers could be controlled by the level or concentration of available energy made available to them. It is not possible to give an exact formula for predicting a specific gain because different types of cattle (genetically) may respond somewhat differently to the same type of feeding program. Even if genetics were ruled out, feedstuffs of the same name will vary considerably in quality. However, it is possible to approximate a diet formulation for cattle to gain at a predetermined rate. The National Research Council tables on predicted performance are extremely valid. If these conditions are followed, performance will be close to what the tables predict.

Several years ago researchers at the Miller–Purdue Agricultural Center conferred with Purdue animal scientists to formulate winter diets for stocker calves which would cause four groups of calves to gain 0.5 lb, 1.0 lb, 1.5 lb, and 2.0 lb, respectively, per day over a 140-day period (McVey *et al.*, 1958). The cattle used were weanling Hereford steer calves purchased from the Great Plains area. The study was repeated for four winters, starting in December and ending 140 days later, for each of the four variables (Table 15.1).

From Table 15.2, it is obvious the animal scientists, based on National Research Council recommendations, were able to formulate and predict performance of stocker calves quite accurately.

III. WINTER GAIN EFFECT ON SUMMER PASTURE GAIN

As a followup to the winter gain of stocker calves, the effect of such gain on summer pasture gains of nonyearling stocker steers was studied. In other words, following the 140-day winter studies indicated above, pasture gain studies were made each of the four summers following the previous winter studies. Beef pro-

TABLE 15.1

Diets for Causing Stocker Steer Calves to Gain at Various Rates (4 Years)[a]

Ingredient per day	Desired daily gain over 140 days			
	0.5 lb	1.0 lb	1.5 lb	2.0 lb
Ground corn cobs (lb)	9.0	7.0	5.0	—
Corn silage (lb)	—	9.0	14.5	30.5
Purdue Supplement A (lb)[b]	1.5	1.5	3.0	3.5
Actual performance by years, daily gain (lb)				
Year 1	0.58	1.06	1.69	2.14
Year 2	0.48	1.00	1.50	2.01
Year 3	0.52	0.97	1.53	1.86
Year 4	0.61	0.94	1.42	1.85
Average	0.55	0.99	1.54	1.96

[a] McVey *et al.* (1958).
[b] 32% protein supplement fortified with vitamin A, dehydrated alfalfa meal, dicalcium phosphate, salt, and cobalt.

TABLE 15.2

Suggested Winter Diets for Growing 400- to 500-lb Stockers

Feed	Diet (lb per head daily)							
	1	2	3	4	5	6	7	8
	Pounds per head daily							
Grass hay (late cut)	12–14							
Ground corn cobs		10–12				7–9		
Grass–legume hay (early cut)			12–14				10–12	
Legume hay (early cut)				12–14				
Cornstalk silage (65% moist)					30–35			
Corn silage (65% moist)						10–15		30–35
Corn grain							3–4	
Protein supplement (32% protein)[a]	1	2			2	2		2
Mineral mixture[b]	FC[c]	FC	FC	FC	FC	FC	FC	FC
Expected daily gain (lb)	0.5	0.5	0.75	1.0	1.0	1.0	1.25	1.25

[a] The supplement should contain 15,000 IU vitamin A per pound.
[b] Two parts dicalcium phosphate and one part trace mineralized salt.
[c] FC = free choice.

TABLE 15.3

Effect of Previous Winter Gain on the Subsequent Gain of Stocker Cattle on Pasture[a]

Grazing season	Number of days	Previous 140-day winter daily gain 0.5 lb	1.0 lb	1.5 lb	2.0 lb
Year 1	140	1.49	1.18	0.88	0.79
Year 2	168	1.24	1.08	0.97	0.67
Year 3	189	1.13	1.05	0.66	0.70
Year 4	140	0.98	1.01	0.80	0.50
Average		1.21	1.08	0.83	0.66

[a] McVey *et al.* (1958).

duction on alfalfa–grass pasture over the summer was studied for each group for each of four years. The average daily gains on pasture alone were affected greatly by the previous winter's daily gain (Table 15.3). In fact, cattle which had gained approximately 0.5 lb per day the previous winter gained nearly double the rate on pasture of cattle which had gained 2.0 lb per day the previous winter (1.21 versus 0.66 lb per day).

IV. PASTURE MANAGEMENT EFFECT ON STOCKER PERFORMANCE

Stocker cattle are often turned to pasture without much regard for optimal management of the pasture. But one must then ask, would it pay to fertilize the pasture lands—would it pay to rotationally graze the pasture areas? Research was conducted at the Miller-Purdue Agricultural Center (Mott *et al.*, 1948) to compare stocker beef performance on pastures which were fertilized, limed, and rotationally grazed. Also, a comparison was made on a fertilized pasture containing birdsfoot trefoil in addition to the permanent bluegrass plants. Table 15.4 shows the results of this research.

The data in this table emphasize the return to be realized from fertilizing beef stocker cattle pasture. The addition of 300 lb of 0–20–10 fertilizer, coupled with rotational grazing, resulted in a 47% increase in beef production per acre or an extra 71 pounds of beef. (Previous research showed no benefit for rotation grazing on nonfertilized permanent pasture.) The addition of the ammonium nitrate (treatment 3) did not contribute very much as only 16 pounds extra stocker beef gain was obtained over the treatment which did not include the ammonium nitrate fertilization.

TABLE 15.4

Pasture Treatment Effect on Stocker Beef Production[a]

Bluegrass pasture type	Steers per acre	Daily gain (lb)	Total gain per steer (lb)	Beef production per acre (lb)
1. Permanent, no fertilizer, continuous grazing	0.93	1.14	169	150
2. Permanent, 300 lb 0–20–10, rotation graze	1.40	1.13	159	221
3. Permanent, 300 lb 0–20–10, 200 lb ammonium nitrate, rotation graze	2.09	0.81	112	237
4. Same as (3) above, but deferred grazing (not grazed July 12–September 9)	2.55	1.61	139	346
5. Birdsfoot trefoil-bluegrass mixture, 300 lb 0–20–10, rotation graze	2.29	1.03	123	334

[a] Mott *et al.* (1948).

Deferred summer grazing—or holding cattle off pasture during the summer droughty period of mid-July to early September—resulted in a further 56% increase in total stocker beef production (treatment 4 versus treatment 2). Inclusion of birdsfoot trefoil, along with fertilizer (treatment 5), resulted in a 51% increase in stocker beef per acre over fertilized permanent pasture not containing the legume (treatment 2).

V. GRAIN FEEDING LEVELS ON PASTURE

Research at this Station (Perry *et al.*, 1972) has shown the effect of feeding various levels of concentrates to stocker cattle on pasture. This study covered four years (4 experiments) of pasture studies. As a followup to the pasture studies, all cattle were finished out to a slaughter finish to determine the effect of summer pasture gains on subsequent fattening gains. The pasture phase averaged 135 days in length (May 6 to September 8), and the finishing or drylot phase extended from an average of 52 days to 124 days. The criterion for termination of the drylot finishing phase was when the cattle reached USDA Choice and as near to 1100–1150 lb live weight as possible. The levels of concentrate feed on pasture which were compared were (1) no concentrate, (2) one-third full feed of concentrates, (3) two-thirds full feed of concentrations, and (4) a full feed of concentrates during the entire pasture period. In addition, (5) a "control group" was fed a full feed of concentrates in drylot from the start of the pasture season until they reached finish weight and grade.

The concentrates fed, both on pasture and in drylot, consisted of 8 parts ground ear corn and 1 part of a complex 32% protein supplement mixed together. The pasture crop was a legume–perennial grass mixture. Drylot phase for the pasture cattle started on September 18, when they were moved alongside the fifth group which had been in drylot throughout the entire experiment each year. From the two full-fed lots—one on pasture and one in drylot—it is possible to calculate how much the pasture contributed (or how much concentrate was saved).

From Table 15.5 note that full-fed cattle on pasture required 6.4 lb concentrate per pound gain whereas those in drylot required 7.6 lb or 1.2 lb more. Therefore, gain per steer (full feed on pasture) was 393 lb × steers per acre (4.9) × feed saved per pound of gain (compared to those in drylot, or 1.2 lb) which equals 2310 lb of concentrates.

Table 15.6 shows the effect of previous summer pasture gain of stocker cattle on their performance in drylot. Gains in the drylot finishing phase were negatively correlated ($r = -0.96$) with previous pasture gain. For each additional 10 lb the stocker steer gained on pasture, it gained 1.3 lb less in the finishing lot.

VI. STOCKER RESPONSE TO MONENSIN SODIUM ON PASTURE

In July, 1978, the Food and Drug Administration cleared monensin sodium (Rumensin) as a growth stimulant for use on cattle on pasture, however, with certain legal restrictions, namely, it must be fed in a minimum of one pound per

TABLE 15.5

Effect of Level of Grain Feeding on Pasture on Rate of Gain of Stocker Steers; May 6 to September 18, 135 days (4 years)[a]

Item	Level of concentrate fed on pasture				
	0	1/3	2/3	Full-feed	Drylot
Initial weight (lb)	638	612	625	634	616
Weight, end of pasture (lb)	759	878	933	1027	1025
Daily gain (lb)	0.95	1.84	2.36	2.94	3.03
Steers/acre pasture	1.7	2.0	2.9	4.9	—
Average daily concentrates (lb)	0	6.2	12.5	18.7	22.9
Total concentrate/animal (lb)	0	832	1694	2526	3089
Concentrates/lb gain (lb)	0	3.1	5.5	6.4	7.6

[a] Perry *et al.* (1972).

TABLE 15.6

Finishing Phase of Pasture Concentrate Level of Feeding (September 18 to Finish), 4-Year Summary[a]

Item	Level of concentrate fed previously on pasture				
	0	1/3	2/3	Full-feed	Drylot
Final weight (lb)	1113	1150	1162	1140	1107
Drylot daily gain (lb)	2.82	2.54	2.28	2.28	1.74
Daily concentrates (lb)	23.8	23.8	24.2	24.0	24.2
No. days in drylot phase	124	110	98	57	46
Concentrates/steer, both phases (lb)	2941	3456	4083	3777	4200
Total gain/steer, both phases (lb)	475	543	537	510	491
Concentrates/steer gain (lb)	6.2	6.4	7.6	7.4	8.6

[a] Perry *et al.* (1972).

animal daily at a dose range of from 50 to 200 mg Rumensin per head daily (Fig. 15.2). The matter of "one pound of feed per head daily minimum" may present some problems in administering Rumensin, although the use of a free-choice lick tank liquid supplement may fill that need with the aliquot of Rumensin suspended in the liquid supplement.

Brown (1978) summarized twenty-three pasture experiments in which Rumensin fed over an average of 125 days to calves weighing 500 lb at the start resulted in a 15% increase in gain. Normally in drylot on higher energy diets, cattle gain similarly with and without supplemental Rumensin. However, Runensin-fed cat-

Fig 15.2. Rumension feeding on pasture stimulates rate of gain of beef cattle.

tle eat about 10% less feed per day so at the same rate of gain on 10% less feed consumed, Rumensin-fed cattle are 10% more efficient. On pasture, cattle fed Rumensin consistently gain more rapidly. In the research summary referred to above, control cattle gained an average of 1.32 lb per day whereas those fed Rumensin gained 1.52 lb per day. Thus, in 125 days, Rumensin feeding resulted in an extra 25 lb of stocker beef. The propionic acid increase appeared in the twenty-three pasture studies, averaging 19.3 molar percent in control cattle and 24.2 molar percent in Rumensin-fed cattle, or a 25% increase in propionic acid produced.

It has been suggested that on a high-energy diet, cattle eat to a caloric satiety, that is, they eat until they have absorbed sufficient calories from dietary intake. Thus, a steer on a high concentrate diet meets its caloric requirements on 10% less feed when it is also fed Rumensin because of the increase in propionic acid production. On the other hand, cattle eating a high roughage (low-energy) diet would like to eat more than their digestive tract will hold in order to meet their caloric needs. Therefore, on a high roughage diet, they eat the same amount of feed whether they receive Rumensin or not. Therefore, since Rumensin provides 10% more energy, cattle fed a high roughage diet plus Rumensin eat the same and gain 10% faster.

Wherever it is possible to supply Runemsin in a minimum of one pound of feed to cattle on pasture, it is a desirable practice.

REFERENCES

Brown, H. (1978). "Elanco Products Feedlot," Tech. Manual No. 1. Eli Lilly Co., Indianapolis, Indiana.

McVey, W. M., Smith, R. E., Wheaton, H. N., and Mott, G. O. (1958). Beef production from alfalfa-grass pastures when grazed by steers wintered at four levels of nutrition. *Purdue Univ. Agric. Ext. Serv. mimeo* AY-147.

Mott, G. O., Beeson, W. M., and McVey, W. M. (1948). Grazing trials at Miller-Purdue Farm. *Purdue Univ., Agric. Exp. Stn., Agron Mimeo* AY-11A.

Perry, T. W., Huber, D. A., Mott, G. O., Rhykerd, C. L., and Taylor, R. W. (1972). Effect of level of pasture supplementation on pasture, drylot and total performance of beef cattle. II. Spring plus summer pasture. *J. Anim. Sci.* 34, 647.

16

Cattle Finishing Systems

Cattle feeding types tend to fall into one of a few categories. For instance, cattle purchased (or raised) are usually either calves or yearlings; they are of the type that will feed out to a government grade of Choice, Good, or Standard, and they are heifers or steers—and sometimes bulls. Similarly, feeding program types are few in number in which only the pattern or timing of feeding high-energy and medium- or low-energy diets is manipulated. In the Corn Belt, most cattle feeders rely on corn and whole plant corn silage as the main sources of energy; in barley country, barley is the main energy source; and in milo country, milo will supply most of the energy. Beef cattle are so well adapted to respond to almost any wholesome source of energy that proximity to a specific grain type is no real problem in locating a cattle feedlot. As programs are presented, it should be borne in mind that one energy source may be replaced by almost any other equivalent energy source without affecting feedlot performance.

A cattle feeding system is defined as a feeding enterprise for which the animal's sex, grade, and starting and finishing weight have been specified. In fact, we have learned so well just how to predict these results that it is almost possible to sell your cattle on the day you purchase them and rest assured you can come very close to that target.

Table 16.1 summarizes the most common systems of finishing cattle but does not include growing or stocker programs. Rather it takes up at a point where a wintering or summer pasture stocker program leaves off and whereby cattle are pushed the remainder of their feedlot life for a slaughter finish. The starting weight is the weight of the animal after it has arrived at the destination where it will be fed; the finishing weight is the final weight of the animal at the feedlot without actual or pencil shrink. All systems described and detailed in the following programs are not designed to include the initial health conditioning period of from 2 to 4 weeks. Admittedly, the programs are based on corn (and corn silage), but these are meant to be prototypes only in which other feedstuffs may be substituted freely on the basis of price per unit of energy.

TABLE 16.1

Feedlot Finishing Systems

USDA grade	Age	Sex	Starting weight (lb)	Finishing weight (lb)	Daily gain (lb)	Days on feed
Choice	Calf	Steer	425–550	1050	2.6	200
	Calf	Heifer	400–500	900	2.3	190
	Yearling	Steer	575–650	1150	2.8	200
	2-year-old	Steer	750	1160	2.8	150
Good	Calf	Steer	400–500	1000	2.4	240
	Yearling	Steer	600–700	1050	2.7	150
	2-year-old	Steer	750	1100	2.9	120
Standard	Yearling	Steer	700	1125	2.8	150
	2-year-old	Steer	800–900	1250	3.0	140

I. CHARACTERISTICS OF CATTLE FINISHING SYSTEMS

A. Choice Grade

1. Steer Calf

This is a quality type animal which can carry a relatively high degree of finish. It can produce one of the most popular carcasses for the chainstore trade. The Choice steer calf does not have the time nor the capacity to be subjected to very high silage rations at any time except for a very short time initially. To take advantage of maximum performance, the steer calf is implanted early with 36 mg RAL, or Synovex-S. This animal should finish at 1050 lb and should be in the feedlot for 200 days with an average daily gain of 2.6 lb (Table 16.2).

2. Heifer Calf

Ration for a heifer calf should contain a high proportion of concentrates and a low proportion of corn silage. The calf should be in the feedlot only as long as it will require her to reach United States Choice. The heifer calf should be fed MGA (0.35 mg per head daily); or implanted with 36 mg RAL, or Synovex-H. There should be a maximum weight gain of 400 lb at a rate of 2.3 lb per day for 190 days (Table 16.3).

3. Yearling Steer

This animal can utilize a fairly high proportion of corn silage in the ration early in the feeding period. However, it should be given a rather "hot" ration for at

TABLE 16.2

Detailed Feeding Programs: Choice Steer Calf (Gain 500 lb; 200 days, 2.6 lb/day)

Roughage program	First 65 days (lb/day)	Second 65 days (lb/day)	Last 60 days (lb/day)	Total requirement (lb)	Value
1. Corn silage	20.0	18.0	12.0	3190	$ 31.90
Corn[a]	5.0	7.0	16.0	1345	53.80
High urea supplement	1.0	1.0	1.0	190	15.20
Total					$100.90
Feed cost per lb gain					20.2¢
2. Corn cob					
Ground ear corn[a]	14.0	18.0	24.0	3520	$119.68
High urea supplement	1.0	1.0	1.0	190	15.20
Total					$134.88
Feed cost per lb gain					26.9¢
3. Haylage					
Haylage	20.0	15.0	12.0	2995	$ 44.92
Corn	8.0	13.0	18.0	2145	85.80
High urea supplement	0.5	0.5	0.75	110	8.80
Total					$139.52
Feed cost per lb gain					27.9¢

[a] Where high-moisture corn is to be used, increase the figures for corn by 14%.

TABLE 16.3

Detailed Feeding Program: Choice Heifer Calf (Gain 450 lb, 190 days, 2.3 lb/day)

Roughage program	First 80 days (lb/day)	Last 100 days (lb/day)	Total requirement (lb)	Value
1. Corn silage				
Corn silage	12.0	12.0	2160	$ 21.60
Corn	12.0	16.0	2560	102.40
High urea supplement	1.0	1.0	180	14.40
Total				$138.40
Feed cost per lb gain				30.8¢
2. No roughage				
16 parts whole shelled corn	17.0	19.8	3340	$133.60
1 part high urea supplement	1.0	1.2	180	14.40
Total				$148.00
Feed cost per lb gain				32.9¢

least 50 days in order to establish United States Choice grade. A 36 mg RAL implant, or Synovex-S should be used. Such steers should gain 2.8 lb per day for 200 days to a final weight of 1150 lb (Table 16.4).

4. *Two-Year-Old Steer*

Probably this steer has come from some other feedlot. This animal should be pushed hard all the way to finish. It should gain 2.8 lb per day for 150 days to a finish weight of 1160 lb (Table 16.5).

B. Good Grade

1. *Steer Calf*

This calf should be allowed to make some early economical gains by giving it a rather heavy feed of corn silage. You can afford to keep it around for a little longer time—perhaps as long as 7 or even 8 months. Implant early and reimplant at 90–100 days with hormone in order to comply with federal regulations. Feed to a final weight of 1000 lb in 240 days at a gain rate of 2.4 lb per day (Table 16.6).

TABLE 16.4

Detailed Feeding Program: Choice Yearling Steer (Gain 550 lb, 200 days, 2.8 lb/day)

Roughage program	First 65 days (lb/day)	Second 65 days (lb/day)	Third 65 days (lb/day)	Total requirement (lb)	Value
1. Corn silage	30.0	18.0	12.0	3900	$ 39.00
Corn	5.0	16.0	18.0	2535	101.40
High urea supplement	1.0	1.0	1.0	195	15.60
Total					$156.00
Feed cost per lb gain					28.3¢
2. Corn cob					
Ground ear corn	18.0	22.0	25.0	4225	$109.85
High urea supplement	1.0	1.0	1.0	195	15.60
Total					$125.45
Feed cost per lb gain					22.8¢
3. Pasture					
Pasture, 200 days pasture @ 20¢					$ 40.00
Corn, free choice, 21 lb whole shelled corn per day				4116	164.64
Total					$204.64
Feed cost per lb gain					37.2¢

TABLE 16.5

Detailed Feeding Program: Choice Two-Year-Old Steer (Gain 410 lb, 150 days, 2.8 lb/day)

Roughage program	First 90 days (lb/day)	Last 60 days (lb/day)	Total requirement (lb)	Value
1. Corn silage				
Corn silage	35.0	15.0	4050	$ 40.50
Corn	10.0	20.0	2100	84.00
High urea supplement	1.0	1.0	150	12.00
Total				$136.50
Feed cost per lb gain				33.3¢
2. Corn cobs				
Ground ear corn—a full feed, averaging 23 lb/day			3450	$117.30
High urea supplement	1.0		150	12.00
Total				$129.30
Feed cost per lb gain				31.5¢

2. *Yearling Steer*

Some silage provided early can fit into this feeding program. However, in order to keep this animal gaining rapidly, keep the ration rather "hot." Implant with hormone. It should gain 2.7 lb per day for 150 days for a gain of 400 lb (Table 16.7).

TABLE 16.6

Detailed Feeding Program: Good Steer Calves (Gain 550 lb, 240 days, 2.4 lb/day)

Roughage program	First 70 days (lb/day)	Second 70 days (lb/day)	Last 80 days (lb/day)	Total requirement (lb)	Value
1. Corn silage					
Corn silage	25.0	25.0	12.0	4460	$ 44.60
Corn	6.0	10.0	18.0	2560	102.40
High urea supplement	1.0	1.0	1.0	230	18.40
Total					$165.40
Feed cost per lb gain					30.1¢
2. Corn cobs					
Ground ear corn—a full feed, averaging 19 lb per day				4370	$148.58
High urea supplement	1.0			230	18.40
Total					$166.98
Feed cost per lb gain					30.4¢

TABLE 16.7

Detailed Feeding Program: Good Yearling Steer (Gain 400 lb, 150 days, 2.7 lb/day)

Roughage program	First 90 days (lb/day)	Last 60 days (lb/day)	Total requirement (lb)	Value
1. Corn silage				
Corn silage	35.0	12.0	3850	$ 38.50
Corn	5.0	24.0	1890	75.60
High urea supplement	1.0	1.0	150	12.00
Total				$126.10
Feed cost per lb gain				31.5¢
2. Corn cobs				
Ground ear corn—a full feed, averaging 25 lb per day			3750	$127.50
High urea supplement	1.0		150	12.00
Total				$139.50
Feed cost per lb gain				34.8¢

3. Two-Year-Old-Steer

This may be a "warmed-up" steer with some flesh already on it. This animal should be maintained on a minimum of corn silage and a full feed of corn. It should gain about 2.9 lb per day for 120 days to a final weight of 1100 lb (Table 16.8).

C. Standard Grade

1. Yearling Steer

Probably this animal has a high percentage of dairy breeding. There are two types of feeding programs which can be used—either high silage early and then a hot finishing ration as to complete the program or it can be on a hot ration for the entire management period. This animal should gain 2.8 lb per day for 150 days to a final weight of approximately 1125 lb. Hormone implants may definitely be used. (See Table 16.9.)

2. Two-Year-Old Steer

This is probably a Holstein steer. Use hormonal implant and push hard with a minimal level of roughage. Animals should gain 3.0 lb per day or better for 140 days to a final weight of 1250–1300 lb. One should tend to sell these cattle by the time clock rather than by appearance because appearance may be deceiving.

TABLE 16.8

Detailed Feeding Program: Good Two-Year-Old Steer (Gain 350 lb, 120 days, 2.9 lb/day)

Roughage program	First 80 days (lb/day)	Last 40 days (lb/day)	Total requirement (lb)	Value
1. Corn silage				
Corn silage	45.0	15.0	4200	$ 42.00
Corn	5.0	23.0	1320	52.80
High urea supplement	1.0	1.0	120	9.60
Total				$104.40
Feed cost per lb gain				29.8¢
2. Corn cobs				
Ground ear corn—a full feed, averaging 27 lb per day			3240	$110.20
High urea supplement	1.0		120	9.60
Total				$119.80
Feed cost per lb gain				34.2¢

Some of these cattle will grade US Good and a few will make US Choice (Table 16.10).

Table 16.11 presents feed costs (per pound) and Table 16.12 gives the formulas for high-urea supplements used in cattle finishing programs.

TABLE 16.9

Detailed Feeding Program: Standard Yearling Steer (Gain 425 lb, 150 days, 2.8 lb/day)

Roughage program	First 90 days (lb/day)	Second 50 days (lb/day)	Total requirement (lb)	Value
1. Corn silage				
Corn silage	50.0	12.0	5100	$ 51.00
Corn	3.0	24.0	1470	58.80
High urea supplement	1.0	1.0	150	12.00
Total				$121.80
Feed cost per lb gain				28.6¢
2. Corn cobs				
Ground ear corn, a full feed, averaging 25 lb per day			3750	$127.50
High urea supplement	1.0			12.00
Total				$139.50
Feed cost per lb gain				32.8¢

TABLE 16.10

Detailed Feeding Program: Standard Two-Year-Old Steer (Gain 400 lb, 140 days, 3.0 lb/day)

Roughage program	First 80 days (lb/day)	Last 60 days (lb/day)	Total requirement (lb)	Value
1. Corn silage	55.0	12.0	5120	$ 51.20
Corn silage	3.0	28.0	1920	76.80
High urea supplement	1.0	1.0	140	11.20
Total				$139.20
Feed cost per lb gain				34.8¢
2. Corn cobs				
Ground ear corn—a full feed, averaging 28 lb per day			39.20	$133.28
High urea supplement	1.0		140	11.20
Total				$144.48
Feed cost per lb gain				36.1¢

II. SELF-FEEDING FINISHING CATTLE

Perhaps the most common method of feeding finishing cattle is to feed them according to their appetite either once or twice daily. This method is almost necessary when high-moisture feedstuffs are a part of the diet in order to keep the feed fresh and to prevent mold and spoilage from building up in the feed trough. Usually the amount of feed presented represents that which is anticipated the cattle will finish by the time the next feeding is due. This represents self-feeding of sorts because essentially feed is available to the cattle for the greater part of the twenty-four hours; ideally the troughs are empty for an hour or so just prior to the

TABLE 16.11

Feed Costs

Item	Price (¢) per lb
No. 2 corn	4
Ground corn cobs	1
Corn silage	1
High urea supplement	8
Haylage	1.5

TABLE 16.12

Formulas for High Urea Supplements

	Percentage	lb/ton
1. Purdue dry 64% supplement		
Urea (45% N)	20.0	400
Cane molasses	14.0	280
Dehydrated alfalfa meal (17%)	51.0	1020
Dicalcium phosphate	10.5	210
Iodized salt	3.5	70
Premix[a]	1.0	20
2. Purdue liquid 64% supplement		
Liquid urea (32%)	29.0	580
Cane molasses	38.5	770
Ammoniated polyphos (10–34–0)	9.0	180
Distillers solubles (27% dry matter)	9.3	186
Salt solution (28%) salt	9.0	180
Calcium chloride	1.2	24
Sodium sulfate	1.0	20
Premix[b]	3.0	60
Total	100.0	2000

[a] Premix contributes per ton: 40 million IU vitamin A; 2500 gm zinc oxide; 8 gm cobalt carbonate; 14 lb dehydrated alfalfa meal.

[b] Premix contributes per ton: 40 million IU vitamin A; 8700 gm zinc sulfate; 19 gm cobalt carbonate; 38 lb water.

next feeding. Generally, this characterizes most cattle feeders' idea of feeding cattle once or twice daily, as the case may be.

In the case of dry feeds (85–90% dry matter), it is possible to feed cattle from self-feeders which need to be refilled only as often as the contents of the feeder are about to be depleted. Thus, only the capacity of the feeder—and the amount the cattle are consuming—dictate how often additional feed should be added to the self-feeder (Figs. 16.1–16.3). Thus, conceivably new feed is added only once or twice a week. However, the feedlot manager should inspect the feeding mechanism, because almost any self-feeder can become clogged. Under such a situation, cattle could not have access to the feed intended for them until the mechanical feeder is unplugged.

Self-feeding of a balanced diet contained approximately 20% of roughage is practical and uncomplicated. In fact, cattle which have never been fed concentrates before can be adapted to self-feeders the first day without any harmful effects if the diet is formulated properly. It is the purpose of this section to

Fig. 16.1 Finishing cattle may be self-fed grain and minerals on pasture successfully. (Photo by J. C. Allen and Son.)

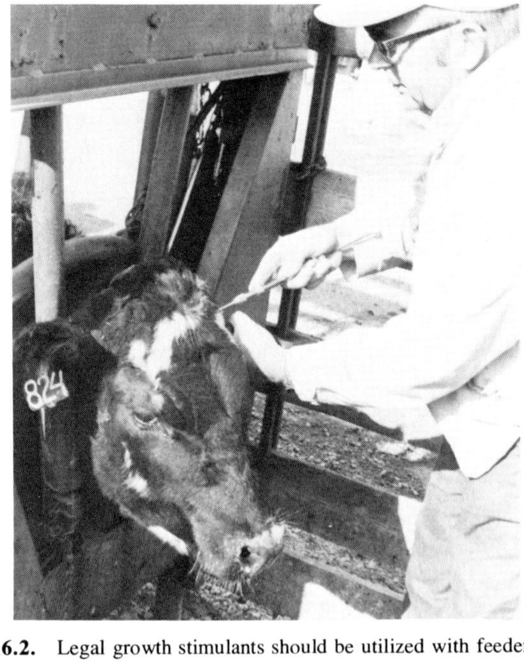

Fig. 16.2. Legal growth stimulants should be utilized with feeder cattle.

Fig. 16.3. Portable mixer trucks add flexibility to cattle feeding operations. (Photo courtesy of BEEF Magazine.)

discuss self-feeding finishing cattle and to point out the safeguards which, when built in, will make self-feeding practical.

The research data shown in Table 16.13 were from cattle which were given access to a full feed of their respective diets the very first day. There was no difficulty from looseness of bowels initially or from foundering later on in the experiment except for cattle fed the 12:1 ratio which contained 9.3% protein. Considerable looseness of bowel was observed among the cattle in that lot. Optimal performance at a reasonable consumption of protein supplement appeared to occur at a ratio of 8 parts ground ear corn to 1 part of 32% protein supplement. Undoubtedly cattle could be shifted to a leaner mixture of 10:1 or possibly 12:1 as the feeding period progressed. However, for starting fresh cattle, a mixture of 8:1 has been used as a starting point in all feeding programs for over 20 years without any problems.

Since the picker-sheller combine has made most corn cobs unavailable as a feedstuff, one may ask what can be substituted for the cob portion if self-feeding is desired. Some other roughage such as hay or straw may be chopped and substituted. Another alternative which requires somewhat more management is to mix whole shelled corn with a pelleted 32% protein supplement in a ratio of 9:1 and feed it along with free-choice hay or other roughage. Some care is needed

TABLE 16.13

Self-Feeding Various Ratios of Ground Ear Corn and a 32% Protein Supplement for Finishing Cattle[a]

Item	Ratio of ground ear corn to a 32% protein supplement				
	4:1	6:1	8:1	10:1	12:1
Percentage protein	12.3	10.9	10.1	9.6	9.3
Number of steers	12	12	12	12	12
Initial weight (lb)	633	637	630	636	636
Final weight (195 days) (lb)	1105	1143	1122	1111	1072
Daily gain (lb)	2.42	2.60	2.53	2.44	2.24
Percentage increase over 4:1	—	7.2	4.2	0.6	−7.6
Daily feed					
Ground ear corn (lb)	16.0	19.2	19.2	18.8	18.6
32% supplement (lb)	4.0	3.2	2.4	1.9	1.5
Crude protein per day (lb)	2.5	2.4	2.2	2.0	1.9
Feed dry matter per lb gain (lb)	7.2	7.5	7.4	7.4	7.8

[a] Perry *et al.* (1956).

in adapting cattle to this program, and it should require about 10 days to introduce new cattle to a full feed of the 9:1 mixture before they are allowed free access.

More recently, with the advent of higher protein supplements, it is possible to mix whole shelled corn and a pelleted 64% protein supplement together in a ratio of 16:1 and allow finishing cattle free access to a full feed after adjustment. Such adjustment means introducing them to the program gradually over 10 days, along with ad lib dry roughage to provide the scratch factor (rumen tickling).

Self-Feeding versus Hand-Feeding

How well can the feedlot manager judge the feedlot requirements as compared to allowing the steer do the job? A good feedlot operator can come pretty close, but it is the author's belief the steer can do a little bit better if it has access to a well-balanced formulation (Table 16.14). Under the conditions of this research, self-fed cattle gained one-tenth pound per day faster than hand-fed cattle. Probably the small advantage occurred from the fact the self-fed cattle were on a full feed from the first day whereas those on the hand-feeding program were brought up to a full feed of grain gradually.

Self-fed cattle have the option of regulating their own intake due to influence of their environment (such as changes in temperature, humidity, and atmospheric pressure) better than the feedlot operator is able to estimate. Perhaps hand-fed cattle do not make these adjustments as readily.

TABLE 16.14

Hand-Feeding versus Self-Feeding and Shelled Corn versus Ground Ear Corn for Finishing Steer Calves[a]

Item	Shelled corn		Ground ear corn	
	Hand fed	Self-fed	Hand fed	Self-fed
Number of steers	20	20	20	20
Initial weight (lb)	467	466	467	466
Final weight, 216 days (lb)	968	990	947	967
Daily gain	2.31	2.41	2.21	2.31
Daily feed				
Corn (lb)	12.5	13.0	12.8	15.5
32% protein supplement (lb)	2.5	2.5	2.5	2.4
Hay (lb)	4.2	3.5	3.7	3.4
Salt to control supplement consumption (oz)	—	0.5	—	2
Feed per pound gain (lb)	8.3	7.9	8.6	9.3

[a] Perry *et al.* (1960).

III. RECIPE FEEDING OF FINISHING CATTLE

Some feedlots will tend to follow a "recipe" approach to feeding finishing cattle. In this type program, purchased cattle are virtually scheduled to definite sale date, plus or minus a few days. This is particularly characteristic of larger feedlots wherein overhead cost is fairly constant. Therefore, it is critical to move

TABLE 16.15

Recipe Feeding: Diet Formulations for Feedlot Cattle[a]

Ingredient	Diet number (as is basis) (%)				
	1	2	3	4	5
Corn silage	50–60	40–50	25–30	18–23	6–8
Alfalfa pellets (sun cured or dehydrated)	5	5	7	7	7
Milo or corn	25–30	33–50	50–60	60–70	70–75
Wheat	0–5	0–5	0–10	0–10	0–10
Protein supplement (40%)	5–7	5–7	4–7	4–7	4–7
Fat	0	0.5–1	2	2	2
Salt	0.5–1	0.5–1	0.5–1	0.5–1	0.5–1
Number of days	5–10	5–10	7–21	45	60

[a] Erickson and Phar (1970).

Fig. 16.4 Even small feedlot operations can well afford a working chute with drop gates plus a squeeze chute.

cattle in and out regularly so that the next group of cattle—already scheduled for purchase—may be able to move into the "just-vacated" spaces (Figs. 16.4 and 16.5). Kansas State University researchers (Erickson and Phar, 1970) have developed such a set of "recipes" based on number of days in the feedlot. Five formulations are prepared (Table 16.15), and then each is fed for a specified number of days. The cattle are then advanced in energy content of their diet by switching to the next diet in the sequence for a given number of days. The diets set up in the table are meant to be exemplary; naturally, an infinite number of combinations could be constructed as well as appropriate substitutions made as price relationships per unit of nutrient content justified it. However, these diets represent sound formulations for a 140-day feeding period for 700-pound steers to gain about 2.75–3.00 lb per day, pay weight to pay weight.

By way of illustration, Table 16.16 follows a group of cattle through this

TABLE 16.16

Recipe Feeding: Approximate Ration in Ingredients[a]

Item	Percentage of diet	Number of pounds[b]	Feed cost/head/day, cents[c]
Diet 1 (7 days)			
Corn silage	55	14.30	14.3
Alfalfa pellets	5	1.30	6.5
Milo–corn	29	7.54	30.1
Wheat	5	1.30	5.2
Protein supplement	5	1.30	13.0
Salt	1	0.26	0.5
Total		26.00	56.6
Diet 2 (7 days)			
Corn silage	45	10.80	10.8
Alfalfa pellets	5	1.20	6.0
Milo–corn	37	8.88	35.5
Wheat	5	1.20	4.8
Protein supplement	6	1.44	14.4
Fat	1	0.24	4.8
Salt	1	0.24	0.5
Total		24.00	76.8
Diet 3 (21 days)			
Corn silage	28	8.12	8.1
Alfalfa pellets	7	2.03	10.1
Milo–corn	51	14.79	59.1
Wheat	5	1.45	5.8
Protein supplement	6	1.74	17.4
Fat	2	0.58	11.6
Salt	1	0.29	0.5
Total		29.00	112.6
Diet 4 (45 days)			
Corn silage	20	5.80	5.8
Alfalfa pellets	7	2.03	10.1
Milo–corn	60	17.40	69.6
Wheat	5	1.45	5.8
Protein supplement	5	1.45	14.5
Fat	2	0.58	11.6
Salt	1	0.29	0.5
Total		29.00	117.9
Diet 5 (60 days)			
Corn silage	7	2.10	2.1
Alfalfa pellets	7	2.10	10.5
Milo–corn	73	21.90	87.6
Wheat	5	1.50	6.0
Protein supplement	5	1.50	15.0
Fat	2	0.60	12.0
Salt	1	0.30	0.6
Total		30.00	133.8

[a] Kansas State University "recipe" feeding finishing cattle. [b] As is basis. [c] Feed costs based on the following: corn silage, $20.00 per ton; alfalfa pellets, $5.00 per cwt; milo, corn or wheat, $4.00 per cwt; protein supplement, $200.00 per ton; fat, $20.00 per cwt; salt, $2.00 per cwt.

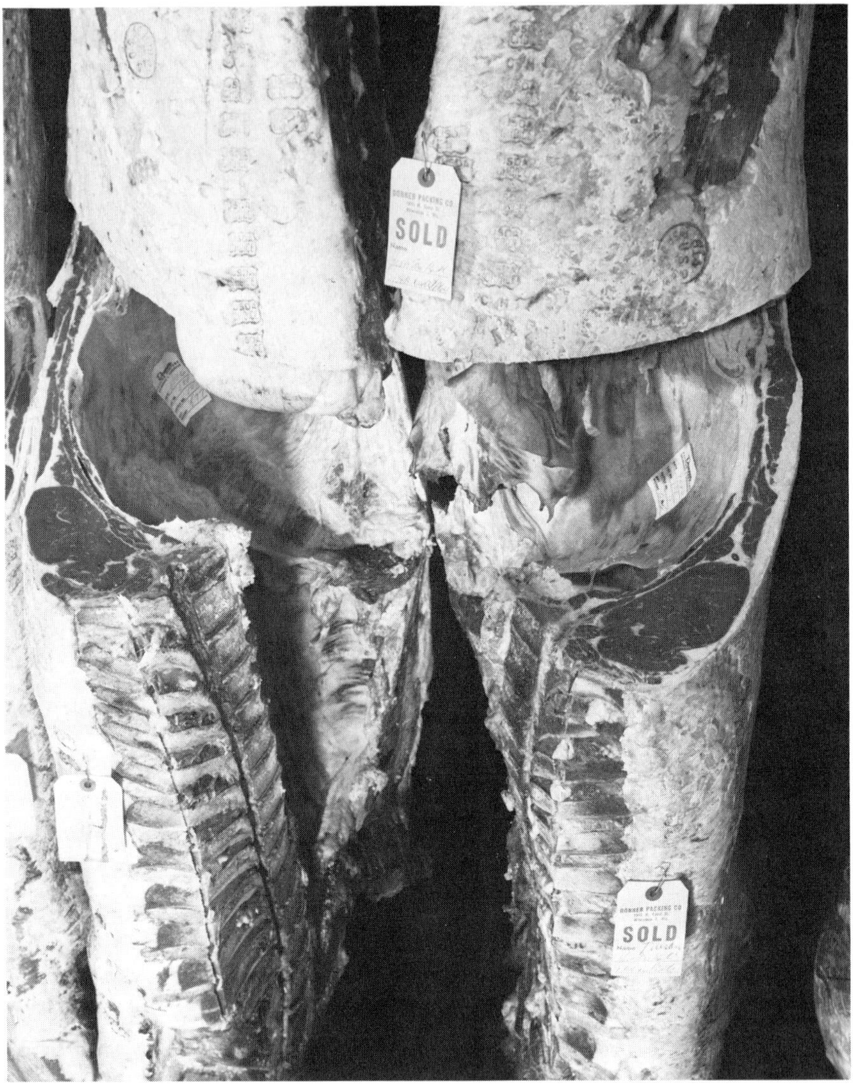

Fig. 16.5. United States Choice, yield grade 2 or 3 carcasses are the target for high quality beef. (Photo courtesy of BEEF Magazine.)

program to determine feed consumption and to project feed costs. Bear in mind these are 700-lb steers, properly hormonized, which have recovered from the shipping fever syndrome. In other words, they are healthy and prepared for the feedlot. Also, bear in mind that these Kansas recipes move up through to a fairly "hot" diet rather rapidly such that within 2 to 3 weeks the cattle are consuming 2% of their body weight daily as concentrates.

IV. DIETS FOR SHOW CALVES

One can feed show calves quite similarly to other finishing cattle, and such cattle probably will respond well if they have the inherited potential to do so. However, the basic tenet in show competition is how your animal places in its class. This section will not deal with genetics, selection of the calf, grooming, or showing. However, a formulation which puts together a combination of the very best nutrition ideas to give a showman that slight edge necessary in a tightly contested race to take the purple ribbon will be presented. This diet has been used for fitting grand champion beef, so even though it is hypothetical and scientific, it is a proved formulation. It contains some fairly expensive items per unit weight such as dried skim milk, brown sugar, and dried whole egg. However, if winning the purple ribbon is sufficient incentive, it is worth a bit more, perhaps, to pursue it to the very best of one's capability.

Since it is a rather complicated formula, the author has justified each item included in the formulation. The formulation (Table 16.17) should be pelleted into one-quarter inch pellets (smaller pellets will tend to be too hard and larger pellets will tend to be too crumbly, and lots of fine particles will accumulate). The diet should be essentially self-fed which means that it should be accessible at all times. To accomplish this, it would be best to use quantities which would not leave a great deal left over about once a day. This ensures that the feed is fresh. (There is something about "fresh" feed which is *critical.*) Thus, if the animal(s) is being given new feed once a day—or even once every two days—scoop out the leftover feed and give it to cattle not intended for the show. Practice and observation will provide the expertise necessary to weigh out within one pound, or so, what the show animal will consume of its own free will in the next 24 hours, or 48 hours, as the case may be.

To the seasoned cattle feeder, the above formula looks very complicated—and it is. He may also assume that it is no better than a standard diet used in his feedlot. However, he will really never know for sure until he tries it. The young showman who has not decided on how best to feed a show calf is more apt to devise a complex formulation than is seasoned cattle feeder.

TABLE 16.17

Complete Pelleted Formulation for Show Cattle (Growing and Finishing)

	1000-pound mix (lb)
Ground shelled corn (palatable energy)	310
Ground corn cobs (roughage)	180
Ground barley (alternate energy source—palatability)	200
Ground oats (roughage—palatability)	50
Cane molasses (aroma—palatability)	50
Dehydrated alfalfa meal (growth factors)	30
Soybean meal (high-quality protein)	35
Linseed meal (protein—hair coat effect)	60
Dried whole egg (high-quality protein—hair coat)	10
Dried skim milk (high-quality protein)	10
Ground beet pulp (palatability)	50
Brown sugar (palatability)	8
Dicalcium phosphate (calcium, phosphorus)	5
Trace mineralized salt (salt, trace minerals)	2
Aureomycin (hair coat, health)	4 gm
Vitamin A (vitamin needs)	2 Million IU
Vitamin D (vitamin needs)	200,000 IU
Vitamin B complex (vitamin needs, ?)	[a]
Total	1000

[a] Riboflavin, 2 gm; niacin, 20 gm; calcium pantothenate, 10 gm; vitamin B_{12}, 10 mg.

V. FATTENING BULLS FOR BEEF

Fattening of bulls for beef production is not a common practice in the United States or in the British Isles; yet in Continental Europe, very few bulls intended for beef production are castrated. Undoubtedly precedent and prejudice enter into the practice of castrating male beef calves intended for the feedlot in the United States.

Some of the reasons proposed by the United States cattle feeders include (1) more management problems, (2) tougher carcasses, and (3) not enough fat covering on bull carcasses to prevent deterioration or spoilage of carcasses. Admittedly, bulls older than 18 months are difficult to handle—and even dangerous. Formerly, beef production practices resulted in slaughter of cattle 2 and 3 years old whereas today most young beef are slaughtered under 2 years of age. Bulls older than 18 months will have achieved sufficient masculinity that the meat will be tough. Finally, before rapid transit and well-refrigerated trucks and trains, fat on the outside of the carcass was important in maintaining wholesome quality.

Today we are more knowledgable in the production of youthful finished bull beef prior to an age of 18 months; the meat is tender, and rapid well-refrigerated

TABLE 16.18

Bull versus Steer Performance and Carcass Value (5-Year Summary)[a]

Trait	Bulls	Steers	Bulls compared to steers (%)
Performance adjusted to 441 pounds weaning weight and 231 days weaning age			
Daily gain, postweaning			
First 28 days (lb)	2.54	1.94	+31
First 84 days (lb)	2.47	1.90	+30
First 140 days (lb)	2.32	1.88	+23
First 168 days (lb)	2.28	1.93	+18
300 days (projected) (lb)	2.03	1.76	+15
Slaughter weight (lb)	996	927	+7.5
Dressing percent (%)	62.7	62.1	
Carcass characteristics adjusted to 600 pounds carcass weight			
Weights of cuts			
Chuck (lb)	96.8	82.8	+17
Rib (lb)	30.6	30.6	
Round (lb)	65.7	63.1	
Loin (lb)	46.8	47.6	
Primal cuts (lb)	233	225	
Kidney fat (lb)	9.1	11.5	
Carcass grade	High good	Low choice	
Marbling score	3.5 (modest)	4.5 (moderate)	
Loin eye area (m^2)	12.08	10.93	+10
Backfat thickness (in)	0.54	0.82	−34
9-10-11 rib characteristics (adjusted to 11.3 pounds weight)			
Lean content (lb)	6.2	5.1	+21
Fat content (lb)	3.0	4.4	−31
Bone content (lb)	2.0	1.7	+18
Protein in rib lean portion (%)	19.5	18.3	
Fat in rib lean portion (%)	16.2	19.2	−16
Moisture in lean portion (%)	63	61	

[a] Martin *et al.* (1966).

vehicles move such beef to its destination rapidly. Young bull beef cannot be graded as regular beef but must be called "bullock," which may carry a stigma in the buyer's opinion. Thus the young bull beef of high quality still has not been accepted very well in the United States.

Continental Europeans, on the other hand, express a distinct preference for the less wasteful cuts of bull beef. Undisputed research has shown that bulls consistently gain faster, require less feed to produce a pound of beef, and produce leaner carcasses than their castrate brothers. In addition, taste panel experts are not able to separate meat from steers and young finished bull carcasses.

A. Comparison of Bulls and Steers

A good deal of research has been conducted by university experiment stations to determine the nutrient requirements that should be met in producing high quality bull beef. First, however, a comparison of performance based on university trials is in order. Bulls and steers from the same herd, and thus the same genetic background, have been compared. In such a 5-year comparison, the following observations were made:

1. Bulls gained 13% faster than steers (2.27 versus 2.01 lb per day) which resulted in an average 78 lb extra saleable animal at 18 months of age (see Table 16.18).
2. Bull carcasses contained less fat. Actual chemical analysis of the 9-10-11 rib cut from bulls contained 22% more lean meat than the comparable cut from a steer carcass.
3. Bull carcasses were less marbled and graded lower (on a steer basis) than steer carcasses. Bull carcasses averaged high Good with "small to modest amount of marbling"; steer carcasses ranged between low and average Choice with "modest to moderate amount of marbling."
4. Palatability comparisons of loin steaks from bulls and steers showed little

TABLE 16.19

Effect of Energy Level on Performance and Carcass Characteristics of Finishing Bulls[a]

	Dietary energy level		
Item	High	Medium	Low
Diet			
Corn silage (lb)	15–20	Full feed	Full feed
Corn (% body weight)	2	1	0.5
32% protein supplement (lb)	2	2	2
Daily gain (lb)	2.69	2.63	2.52
Feed per pound gain (lb)	6.72	6.55	6.16
Carcass characteristics			
Fat cover (in)	0.45	0.37	0.33
Rib eye area (in^2)	11.6	11.4	11.9
Marbling score[b] (No.)	4.9	4.3	4.0
Carcass grade[c] (No.)	12.6	11.5	9.6

[a] Martin (1974).
[b] Slight = 4, small = 5.
[c] Low choice = 12, high good = 11, average good = 10.

TABLE 16.20

Protein Level Effect on Gains of Young Weanling Bulls[a]

Time period	Dietary protein levels (%)		
	14	12	10
First 28 days	3.01	2.65	2.44
Second 28 days	2.64	2.92	2.58
Third 28 days	2.55	2.56	2.39
Fourth 28 days	2.33	2.23	2.51
Fifth 28 days	2.23	2.39	2.38
Sixth 28 days	2.01	2.21	2.17
168 days total	2.46	2.49	2.41

[a] Martin (1974). Values in daily gain (lb).

difference. Of six mechanical and panel measures of tenderness, four showed no difference and two slightly favored steer steaks.

5. Considering the four factors above, bulls produced approximately 30% more edible product per animal than steers; and the product produced by bulls was slightly less marbled and only slightly inferior in palatability and tenderness to that produced by steers.

B. Effectiveness of Diethylstilbestrol for Bulls

In a 2-year study young bulls were implanted with DES at levels of 36 and 72 mg. There was an immediate response of 0.2 lb daily gain increase for the first 84 days which then rapidly disappeared. The net effect over 154 days was an average of 8 lb extra weight per implanted bull (0.05 lb per day) over non-implanted bulls. Subsequent research has shown a greatly elevated serum testosterone level for implanted bulls, suggesting a counteracting adaptation effect by the testes of implanted bulls.

C. Energy Levels

Three levels of energy for finishing bulls were compared in a 5-year study (Martin, 1974). The high-energy diet consisted of limited corn silage (15–20 lb per day), full feed of corn (2% of body weight) and 2 lb of a 32% protein supplement; medium-energy diet limited corn to 1% of body weight plus supplement and full feed of corn silage; low-energy diet limited corn to 0.5% body weight plus supplement and corn silage (Table 16.19). On the three comparative energy levels, there were slight nonsignificant increases in rate of gain (2.52

versus 2.63 versus 2.69 lb per day) for the low, medium, and high levels of energy. Based on the energy comparisons, it would seem reasonable to feed a diet of 1% of body weight as corn plus a full feed of corn silage and balancing supplement.

D. Levels of Protein for Young Finishing Bulls

A 2-year study was conducted to compare 14, 12, and 10% of protein in the diet of finishing bulls. The results are shown by 28-day weight periods. An interesting phenomenon came from these 2 years of research which have been verified in several years research. High protein level promoted most rapid gain in the newly weaned bull calves starting at a weight of 480 lb (Table 16.20).

E. Some Generalizations on Bull Feeding

1. Feed bulls for rapid gains to reach finish by 17–18 months of age. This ensures the carcass will be desirable and decreases management problems associated with more mature bulls.
2. Do not shuffle and sort bulls among pens after they are 9 months of age. Start a group and leave them intact all the way to market. The reestablishment of the "peck order" among older bulls may result in destruction of property.
3. Pens for feeding bulls need to be sturdier than those for steers and heifers.
4. Young bulls can be fed much like heifers and steers. Be sure to use adequate levels of protein since bulls tend to lay down more protein than steers or heifers; from weaning to 750 lb feed 12% protein; from 750 lb to slaughter, 10% protein should be sufficient.

VI. COMPARATIVE PERFORMANCE OF BULLS, STEERS, AND HEIFERS FOR BEEF

Three experiments were conducted by the University of Missouri (Hendrick *et al.*, 1969) to compare the feedlot performance and quantitative and qualitative carcass characteristics of half-sib Hereford bulls, steers and heifers. In each of the three experiments, the progenies of three or four sire groups were represented evenly within sex group. After weaning, the calves were grazed up to 3 months and then placed in the feedlot. Within each sex, half the animals were killed at an "early kill" whereas the other half within each sex group was held for a "later kill" in order to bring out possible carcass differences within sex due to weight at time of slaughter. In all situations, the cattle were fed high concentrate finishing diets for a major portion of the time; within each experiment, bulls, steers, and heifers were fed the same diet.

TABLE 16.21

Carcass Characteristics of Bulls, Steers, and Heifers[a]

Item	Bulls		Steers		Heifers	
	Early kill	Late kill	Early kill	Late kill	Early kill	Late kill
Total gain (lb)	495	604	435	515	430	524
Slaughter weight (lb)	1093	1257	937	1080	897	1018
Carcass weight (lb)	656	742	570	673	543	629
Age at slaughter (days)	595	654	579	642	601	650
Carcass quality grade (No)[b]	7.3	7.8	8.2	10.3	10.3	11.2
Total retail cuts (%)	70.4	65.3	64.6	59.4	61.2	60.0
Fat trim (%)	15.9	22.1	21.2	27.7	26.4	28.9
Bone (%)	13.4	12.3	13.6	12.8	12.3	11.5
Rib eye fat content (%)	4.3	5.8	6.3	10.8	8.6	10.1
Warner–Bratzler score (lb)[c]	17.9	17.9	15.5	15.0	16.2	17.4

[a] Hedrick (1978).
[b] Good = 7, choice = 11.
[c] Warner–Bratzler shear score of pounds per one inch score.

Bulls proved to be superior ($p < 0.05$) to steers and heifers in liveweight gain and feed conversion with steers and heifers showing comparable performance.

Total weight and percentage retail cuts of the carcass consistently were greater ($p < 0.05$) for bulls than for steers and heifers and, in several instances, superior for steers over heifers (see Table 16.21). In all experiments, the percentage of ether extract (fat) in the longissimus muscle (rib eye) was less and carcass quality grade lower for bulls than for steers and heifers. Total amount and percentage of "throwaway" (trimmable) fat and intramuscular fat of all three sex groups increased as length of time on feed was extended.

Warner–Bratzler shear values and sensory panel scores indicated that steaks from bulls less than 16 months of age were comparable in tenderness to steaks from steers and heifers of similar age. However, steaks from more mature bulls were less tender. Flavor and juiciness of steaks were not affected significantly by sex condition.

VII. FEEDING HOLSTEIN STEERS

Several decades ago it was not uncommon to kill male Holstein calves not intended for herd sire replacements shortly after such calves were born. After all, what alternatives were there? Now, of course, the feeding of Holstein steers out to weights of 1000, 1200, 1500 lb—or even greater—is most common. Admittedly, a few Holstein males are raised on milk replacer formulas as white-meated veal. However, so many fortunes have been lost following such practice in the United States that it is most difficult to find very many long-time veal producers in this country.

The advantages of Holstein steer feeding include (1) extremely rapid gains of 4 pounds or more per day, (2) more rapid turnover, and (3) usually a positive margin in selling price over purchase price which is extremely rare with the so-called beef breeds. One of the obvious disadvantages is that since Holstein steers are a byproduct of the milk-producing industry, there is not much chance of an increase in numbers of them available for purchasing. In other words, the supply of Holstein steers is relatively constant from one year to the next except that the number of dairy cows in this country has declined gradually (Fig. 16.6).

Holstein steer finishing for beef production is quite new compared to finishing typical beef breeds. Whereas the beef breeds can utilize a high level of corn silage economically, Holstein steers do not fare as well on a high corn silage diet. When a cattle feeder provides a concentrated diet to the capacity of a Holstein steer, it gains rapidly and efficiently.

The University of Minnesota (Miller *et al.*, 1977) has reported research on comparative methods of handling Holstein calves from one week of age until they are marketed as finished steers at approximately 1050 lb. In the Minnesota experiments, Holstein bull calves were purchased from dairymen at approximately one week of age. Such calves were placed in individual stalls with raised

Fig. 16.6. Holstein steer feeding has developed into big business in the United States. (Photo courtesy of BEEF Magazine.)

floors and were fed milk replacer for the first 28 days. At the time the calves were 45 days of age, they were taken from their crates and moved to a confined pole building where they were fed in groups of four to six calves until the pen of cattle was marketed.

The Minnesota researchers divided the test into several periods for examining various dietary regimens. These various stages of feeding are described. The first stage of the comparison was started after the calves had been on milk replacer for about 10 days (they were now about 17 days of age) and continued for the next 5 months (141–146 days). In other words, this program overlapped the milk replacer program from the 10th day on feed through the 28th day and henceforth constituted the only diet offered the calves until the calves weighed about 300 lb or until the calves had been on the test about 5 months. In this phase, the absence of hay or the presence of 15 or 30% hay was the basis of comparison (Table 16.22). These data show that the inclusion of about 15% of high-quality hay is desirable for starting young Holstein calves. Later data will show the calves fed 30% of hay in their diet responded somewhat better in subsequent performance. Nevertheless, it seems the inclusion of from 15 to 30% high-quality hay in the

TABLE 16.22

Holstein Calf Performance during the Period From One Week of Age to Approximately 300 Pounds (141 to 146 days)[a]

	Level of hay in the diet (%)		
Item	0	15	30
Diet formulation (%)			
Rolled shelled corn	79.9	67.21	54.76
Soybean meal	17.6	15.5	13.27
Ground alfalfa hay	—	15.0	30.0
Dicalcium phosphate	1.45	1.63	1.62
Ground limestone	1.05	0.66	0.4
Vitamins A and D	+	+	+
Total	100.00	100.00	100.00
Performance			
Number of steers	67	71	73
Initial weight (lb)	94	94	97
Final weight (lb)	313	323	321
Days fed (No.)	146	141	144
Daily gain (lb)	1.49	1.62	1.56
Daily feed dry matter[b] (lb)	4.19	5.31	5.42
Feed dry matter per lb gain[b] (lb)	2.82	3.27	3.49

[a] Miller *et al.* (1977).
[b] Does not include 20 lb dry milk replacer per calf.

TABLE 16.23

Comparative Gains of Hostein Steers on High Concentrate and High Silage Diets between 300 and 600 Pounds[a]

Item	All concentrate	High silage
Number of steers	100	111
Initial weight (lb)	317	321
Final weight (lb)	592	593
Daily gain (lb)	2.42	2.12
Daily dry feed matter		
Supplement (lb)	0.89	0.89
Corn grain (lb)	10.1	2.5
Corn silage (lb)	—	8.2
Total (lb)	10.99	11.59
Feed dry matter per lb gain (lb)	4.5	5.5

[a] Miller *et al.* (1977).

diet is favorable at the stage from about 10 days of age to around 5 months of age.

Possibly a feeder might prefer to buy Holstein males (castrate) at 5 or 6 months of age. Thus, the Minnesota researchers compared feeding methods for Holstein calves starting at 5 months of age and weighing around 300 lb (average 319 lb) and extending until the steers were approaching 600 lb live weight (Table 16.23).

TABLE 16.24

Finishing Holstein Steers (600 Pounds to Market Finish) on High Concentrate or High Silage Diets[a]

Item	All concentrate	Medium silage
Number of steers	107	104
Initial weight (lb)	592	594
Final weight (lb)	1012	1002
Daily gain (lb)	2.73	2.71
Daily feed dry matter		
Supplement (lb)	0.89	0.89
Corn grain (lb)	15.5	12.7
Corn silage (lb)	—	5.1
Total (lb)	16.39	18.69
Feed dry matter per lb gain (lb)	6.0	6.9

[a] Miller *et al.* (1977).

In this research, a high concentrate diet was compared with a predominantly corn silage diet.

From 300 to 600 pounds (Table 16.23), Holstein steers fed the all-concentrate diet gained faster ($p < 0.01$) than those fed a high corn silage diet.

Many Holstein steer feeders like not to purchase steers until they weigh 600 lb or more and then start the finishing program. The Minnesota researchers compared feeding programs for this size steer (Table 16.24). Certainly in the finishing stage (600 to 1000 lb), the Holstein steer is adapted to using at least 5 lb of corn silage dry matter (about 15 lb wet corn silage) per day. Dry matter conversion figures favor the all-concentrate program but the lower cost of corn silage might tip the economical balance either way. However, Holstein steers tend to have rather large stomachs, and the consumption of an all-concentrate diet would not cause the ruminal distention (pot-belly effect) caused by the silage diet.

In conclusion, for the young Holstein steer, a level of 15% hay is effective between one week of age and 300 lb live weight; the all-concentrate diet is is preferred for the gains between 600 lb and market finish.

VIII. FEEDING CULL COWS FOR SLAUGHTER

There is a market for almost any type of healthy beef animal. The cull brood cow which has completed her usefulness as a mother cow is no exception. However, the brood cow herdsman is apt to send such cattle to the auction barn or to the stockyards for sale. Conceivably the cow–calf herdsman might consider adding some finishing feed to cull cow diets before selling them or cull cows could be assembled through auction rings and taken back to the farm to be fed a high-energy diet for a short time and then sent back through for slaughter. Cull cows, no matter what their condition and as long as they are healthy, make a lot of beef including roasts and a few steaks, (but also hamburger). In addition, the filet mignon from a cull cow is just as tender and juicy as that from a prime baby beef.

The University of Missouri researchers (Morrow *et al.*, 1978) have conducted research with cull cows. Hereford cows ranging in age from 6 to 11 years of age and which had been culled from the university herd were divided into three groups as nearly equal as possible. One group was slaughtered immediately to get estimates of carcass characteristics whereas the other two groups were fed on a heavy silage diet (60 lb per head per day) for either 28 or 42 days and then slaughtered. The cows had excellent gains but those fed only 28 days gained over 3 lb per head per day (Table 16.25) whereas those fed two weeks longer had tapered off in their gains considerably (1.89 lb per day). It would appear one would be justified in feeding cull cows a heavy feed of corn silage for up to one month, but after that time the profitability would drop off rapidly. However, this

TABLE 16.25

Feeding Cull Cows Either 28 or 42 Days on a Heavy Silage Diet[a]

Item	Length of time on silage		
	0	28 days	42 days
Number of cows	18	18	18
Initial weight (lb)	976	1005	973
Final weight (lb)	976	1090	1053
Daily gain (lb)	—	3.05	1.89
Dressing percent	48.2	48.3	50.6
Feed per day (lb)	—	58	62
Dry matter per pound gain (lb)	—	8.9	15.5
Final carcass value ($)	279.94	302.72	305.00
Pounds of silage fed (lb)	—	1725	2604
Cost of silage @ $20.00 per ton ($)	—	17.25	26.04
Yardage @ 15¢ per day ($)	—	4.20	6.30
Total cost ($)	—	21.45	32.34
Cost per pound added carcass wt[b] (¢)	—	47.6	57.8

[a] Morrow *et al.* (1978).
[b] Calculated initial carcass weights based on Lot I.

type program does offer opportunity for a person who can manage such a program.

IX. ESTRUS CONTROL IN HEIFERS: SPAYING VERSUS MGA

Occurrence of estrus in heifers is a monthly annoyance which probably contributes a great deal to the usually accepted slower rate of gain expected from heifers as compared to that from steers. Typically, a heifer may stay in estrus about 20 hours, and during that time she may be mounted a dozen or more times. Furthermore, she may eat very little during this period so that a combination of adverse conditions usually results in something less than optimum performance.

Veterinarians many years ago demonstrated that spaying, or surgical removal of the ovaries, prevented the occurrence of estrus in heifers. However, spaying has many obvious disadvantages. Often an animal will die because of surgical shock, internal hemorrhaging, or postsurgical infection. The surgical process is expensive as well as time consuming because each animal must be restrained and operated on individually. More recently a spaying procedure utilizing a vaginal approach—as contrasted to entry in front of the flank—has been developed. This technique may not cause quite the stress to the animal of the former flank approach.

Another deterrent to the spaying of heifers is the decreased performance which

TABLE 16.26

Effect of Spaying on Performance of Feedlot Heifers[a]

Item	Intact	Spayed
Average daily gain (lb)		
Trial 1	2.07	1.91
Trial 2	1.72	1.54
Trial 3	2.53	2.44
Average	2.11	1.96
Difference		−0.15
Feed per 100 lb gain (lb)		
Trial 1	775	891
Trial 2	1112	1188
Trial 3	839	869
Average	909	983
Difference		+74

[a] Trials 1 and 2, Purdue University; Trial 3, Upjohn Co., (Anonymous, 1978)

has been reported in the scientific literature (Table 16.26). In the three trials reported, both growth rate and efficiency of feed conversion were depressed by spaying heifers. Pooled results of the three spaying experiments show an average of 7–8% decrease in rate of gain and efficiency of feed conversion by heifers which had been spayed.

During the development of melengesterol acetate (MGA), over 100 feeding trials were conducted to study the reaction of heifers. A number of these made direct comparisons between MGA and no added drug for intact heifers. The average results of 47 such comparisons give MGA-treated heifers a 10% increase in rate of gain and a 6% improvement in efficiency of feed utilization (Tables 16.25 to 16.27).

TABLE 16.27

Feedlot Performance of Intact Heifers Fed MGA versus Those Not Fed the Drug[a]

Item	No drug	MGA
Daily gain (lb)	2.24	2.47
Difference (lb)		+0.23
Feed per 100 lb gain (lb)	995	930
Difference (lb)		−65

[a] 47 trials pooled, Anonymous 1978, p. 26.

REFERENCES

Anonymous (1978). Effect of spaying on performance of feedlot heifers. *Anim. Nutr. Health,* May, p. 25.

Erickson, D. B., and Phar, P. (1970). "Guidelines for Developing Commercial Feedlots in Kansas," Kansas State University, Manhattan.

Hedrick, H. B. (1978). "Carcass Characteristics of Bulls, Steers and Heifers." Charolais Bull-O-Gram, Box 95, Douglas, Arizona.

Hedrick, H. B., Thompson, G. B., and Krause, G. F. (1969). Comparison of feedlot performance and carcass characteristics of half-sib bulls, steers and heifers. *J. Anim. Sci.* 29, 687.

Martin, T. G. (1974). "Bulls for Beef Production," Cattle Feeders Day Rep. p. 3. Purdue University, West Lafayette, Indiana.

Martin, T. G., Douglass, L. W., and Garwood, V. A. (1966). Growth and carcass characteristics of Angus bulls and steers. *Indiana, Agric. Exp. Stn.,* 253.

Miller, K. P., Goodrich, R. D., and Meiske, J. C. (1977). "Feeding Programs for Holstein Steers from One Week of Age to Market," Cattle Feeders Day Rep., p. 40. University of Minnesota, St. Paul.

Morrow, R., Berger, J., Sicht, D., and Smith, M. (1978). "Performance of Cull Cows on Full Feed of Corn Silage," Beef Cattle Production and Management Report, p. 35. University of Missouri, Columbia.

Perry, T. W., Mohler, M. T., and Beeson, W. M. (1956). "Self-feeding Various Ratios of Ground Ear Corn to Supplement A for Fattening Cattle," Cattle Feeders Day Rep., Mimeo AH163. Purdue University, West Lafayette, Indiana.

Perry, T. W., Webb, D., Nickel, C. H., and Beeson, W. M. (1960). "Hand-feeding *vs* Self-feeding and Shelled Corn *vs* Ground Ear Corn for Fattening Steers," Cattle Feeders Day Rep., Mimeo AS-273. Purdue University, West Lafayette, Indiana.

17

Feedlot Disease*

A large-scale survey of nearly half a million feedlot cattle was conducted in Colorado (Jensen *et al.*, 1976) to determine causes of illness and death. The sickness rate of yearling cattle was found to be 5.1% with nearly one-fifth of these dying. Actually, about 1% of all yearling cattle in the feedlot died. Most of the cattle were 12 to 18 months of age, had originated outside Colorado, and included both steers and heifers.

The above survey showed respiratory diseases resulted in 75% of the sickness and 64% of the deaths. Pneumonia was the major disease and undoubtedly caused more economic loss than all other diseases combined (Table 17.1).

Nearly 72% of fatal cases of shipping fever pneumonia occurred during the first 45 days on feed, thus emphasizing the critical nature of getting new cattle well acclimated and started on feed. Pneumonia developed during all seasons but had a higher rate during fall and winter than during spring and summer. (See tabulation below.)

Days on feed	Incidence of fatal pneumonia (%)
1-45	72%
46-90	14%
91-141	9%
142-up	5%

Upon necropsy it was found that the tracheas and lungs contained both pathogenic and nonpathogenic organisms with the majority of bacteria isolated being from the *Pasteurella* group followed by the mycoplasmas. Infectious bovine diarrhea (red nose) virus was the most common viral agent identified.

IBR virus and pasteurellas have been recognized as probable causative agents for the shipping fever complex (Figs. 17.1 and 17.2).

*Much of the material presented herein was adapted from a paper presented by James Bailey (1977).

TABLE 17.1

Causes of Death in Colorado Feedlot Yearlings[a]

Disease	Percentage
Pneumonia	48
Brisket disease	6
Diphtheria	6
Intestinal infections	5
Riding injury	4
Bloat	3
Urinary calculi	2
Endocarditis	2
Ulcers	2
Bovine virus diarrhea	2
Pulmonary edema	1
Miscellaneous	17

[a] Jensen *et al.* (1976).

I. BULLING OR RIDING IN STEER FEEDLOTS

Bullers among steers is fairly common and represents a considerable economic loss to the cattle feeder. The main economic losses include physical injury, stress to both buller and rider, and the necessity of early isolation of the victims. Of nearly 2000 necropsies (Jensen *et al.*, 1976) performed in one year, about 4% (83) were diagnosed as riding injuries.

II. ACIDOSIS IN FEEDLOT CATTLE

Acidosis in cattle is usually characterized by loss of appetite, diarrhea, mucus in feces, dehydration, incoordination, and, often, death. Physiological symptoms include increased rumen lactic acid levels, lowered rumen pH and blood pH, dehydration and hemoconcentration, rumen stasis, rumenitis, increased osmotic pressure within the rumen, destruction of gram-negative and proliferation of gram-positive bacteria, and reduction in rumen protozoa numbers.

It is generally accepted that acidosis is caused by excessive consumption of feeds rich in readily available carbohydrates. In fact, it has been demonstrated that manually engorging rumen-fistulated animals with grain or by voluntary engorgement of starved animals with grain produced acidosis. Researchers have

*Much of the material on acidosis is adapted from published symposium proceedings on the subject (Brent, 1976; Elam, 1976; Huber, 1976; Slyter, 1976).

Fig. 17.1 New feeder cattle should be under the scrutiny of a veterinarian. (Photo courtesy of BEEF Magazine.)

been able to produce acidosis in cattle by switching them from an all-hay diet to a 90% grain diet too rapidly. The author has experienced the same situation when cattle on pasture were brought up on corn levels fed too rapidly.

A. Management Effect

Certain management factors appear to affect the onset of acidosis in cattle:

1. *Starting cattle on feed.* Often new cattle in the feedlot have never had more than token amounts of concentrates—other cattle may have much more. Therefore, background history of cattle may be helpful in determining the rate at which new cattle can be fed concentrates. Generally, however, it is best to utilize a high roughage diet for new cattle and then work them up to a full feed of concentrates rather gradually.
2. *Changing diets.* When raising the caloric or concentrate level of the diet, it is important to make such changes gradually. No increase in caloric content should be made when cattle are hungry. In fact, if feed is inaccessible, for example, due to a big snow or power and equipment failure, cattle should be fed a diet that is of lower energy content to appease the hunger. One can then proceed to the level of concentrate that was being fed prior to the interruption.
3. *Feeding high energy diets.* It is difficult to feed a high concentrate diet without experiencing some acidosis, founder, or bloat. In fact, Dr. Bart Cardon has suggested that he would like for about 1% of his finishing cattle to produce the gray off-color stool of acidotic cattle to indicate he is feeding just about the

Fig. 17.2 A yard checker looks for sick cattle from the height advantage of a "cherry picker." (Photo courtesy of BEEF Magazine.)

maximum level of concentrate possible. If some grain is replaced isocalorically with a combination of roughage plus fat there seems to be lowered acidosis incidence. Also, adding some variety to the grain portion by incorporating molasses-dried beet pulp in place of the corn, milo, or barley tends to reduce acidosis problems. Even though grain diets, in general, tend to predispose to acidosis, some grains are worse than others, wheat generally considered to be the worst, followed by corn and milo, with barley being the least predisposing.

4. *Weather and season.* The highest incidence of acidosis is observed during warmer seasons. The reason for the greater problem in summer is not known. Some keen observers have associated a higher acidosis incidence at times of weather change and the accompanying fluctuation in feed consumption.

5. *Breed differences.* Brahman-bred cattle have been incriminated as the breed most proven to develop acidosis and founder. Florida researchers (Hentges, 1970) observed that following engorgement with a high concentrate diet, blood lactate levels increased more rapidly in Brahmans than in Herefords or Angus cattle.

B. Physiological Effects of Acidosis

The lactic acid accumulation in the rumen and subsequent absorption initiate several physiological conditions which establish the acidosis syndrome.

1. *Stasis of rumen motility.* As the pH of the rumen declines to near 5, rumen contractions decrease and eventually cease altogether. The reason for this phenomenon is not understood.

2. *Diarrhea and dehydration.* A reduction in total body water of 8% of the body weight has been reported in acidotic sheep (Huber, 1971). The reduction in body water was shared by plasma, interstitial, and intercellular fluid compartments. Rumen contents became excessively wet, indicating at least a part of the body water lost had entered the rumen. Fecal fluid loss via diarrhea is large and occurs when rumen motility is depressed. Algeo (1973) switched cattle abruptly from a high roughage to a high concentrate formulation. Sixteen of 75 cattle involved died, and the survivors had weight losses of from 100 to 130 lb.

3. *Systemic acidosis.* Acute acidosis in ruminants is the result of excessive consuption of fermentable carbohydrates which causes a reduction in pH due to the production of large quantities of volatile and nonvolatile fatty acids. Furthermore, roughages have a much stronger buffering capability than is true for concentrates. It appears to occur most readily when glucose accumulates in conjunction with a ruminal pH of 5 or less. At this lower pH, amylase of the ruminal ingesta is available causing even more glucose to be liberated. During acidosis, the cellulose-digesting bacteria and protozoa are reduced greatly. It has been suggested that these organisms give stability to the rumen environment under normal conditions.

Unfortunately, animals which recover from acidosis may be plagued by other feedlot ailments. Laminitis often follows lactic acidosis; and even though there seems to be a close correlation, the biochemical relationship between the two conditions remains obscure. Laminitis is characterized by hyperemia, hemorrhage, and thrombosis with edema in surrounding tissues. There may also be severe vascular changes and extensive fibrous tissue formations in the cornium and, thus, the often elongated hoof. Because histamine is found in the rumen fluid of lactic acidotic animals, some would believe that histamine is the agent which triggers the laminitis syndrome.

Rumenitis and liver abscesses tend to closely follow acidosis conditions. With the advent of "no roughage" diets in the 1950's, damaged livers and rumens dramatically increased. Rumenitis and liver abscesses appear inseparable, since rumenitis (a consequence of acidosis) perhaps allows microorganisms to enter the portal circulation. Why acidosis predisposes rumenitis is not clear, and many theories exist on the subject. However, since the liver represents as much as 5% of the carcass weight, damage to the liver can be extremely expensive. It has been suggested by the early work of the author and others that the feeding of low levels of antibiotics (70 mg per head daily) is of considerable benefit in decreasing the number of diseased livers in beef cattle.

Polioencephalomalacia (PEM) is a condition in feedlot cattle also which seems to have its origin in lactic acidosis. In the PEM condition, cattle become dull and blindness often ensues. Muscular tremors also occur as the disease progresses. Animals tend to press on fixed objects with their heads. After a few days they may die in a coma. PEM was attributed to a wide variety of causes but was finally shown to respond to large intravenous doses of thiamin (Davies *et al.*, 1965). Thiaminase, which destroys thiamin, was later identified in the rumen fluid of PEM cattle (Edwin *et al.*, 1968). Lactic acidosis appears to establish rumen conditions conducive to PEM development. As lactic acidosis develops, the rumen population changes from predominantly gram-negative species to predominantly gram-positive with an increase in bacilli. The pH drops to near ideal for thiaminase 1, and histamine levels increase.

III. SHIPPING FEVER

Shipping fever of cattle is a syndrome characterized by elevated body temperature, dypsnea, and pneumonia. It is a disease quite similar to the common flu in humans and is triggered by the stresses associated with handling and shipping cattle. Possibly because of the added stress of the weaning procedure, light calves appear to be more susceptible to shipping fever than do yearling feeder cattle. The shipping fever syndrome is probably the greatest loss to the cattle feeding industry of any common disease complex known, except where outbreaks of hoof and mouth disease might occur. Some estimates for losses due to shipping fever in cattle will run as high as 25 million dollars per year.

Many factors contribute to the avoidance of the shipping fever syndrome and include (1) good nutrition—especially a well-balanced protein supplement, adequate energy and minerals, and higher levels of vitamin A in the range of

TABLE 17.2

Summary of Research Utilizing Aureo S-700 for the First 29 Days for New Feeder Calves

Location	Number of trials	Number of control cattle	Average initial weight (lb)	Daily gain			Feed per lb gain		
				Control (lb)	AS-700[a] (lb)	Improvement (%)	Control (lb)	AS-700 (lb)	Improvement (%)
Cyanamid									
Kansas	6	80	346	2.06	2.51	22	6.3	5.4	14
University									
Purdue	4	136	466	1.91	2.46	29	5.6	4.1	27
South Dakota	4	136	407	1.05	1.33	27	6.3	5.7	10
Texas	2	40	427	2.10	2.52	20	6.3	5.7	10
Iowa	2	76	666	1.24	1.38	11	13.6	11.2	17
Arizona	4	108	547	2.47	3.44	39	7.4	5.5	26
Kansas	5	136	372	2.05	2.26	10	6.8	5.9	14
Summary	27	712	437	1.89	2.33	23	7.9	6.3	19

[a] 350 mg each of chlortetracycline and sulfamethazine per head daily.

50,000 IU per head daily, (2) proper care and management including a comfortable, dry, draft-free, quiet place to recuperate, and (3) therapeutic levels of antibiotics and sulfa for the first 14 to 28 days on the farm. Naturally this program is not intended to preclude the usual therapy and prevention prescribed by the veterinarian, but rather it is designed to complement the veterinarian's program.

Several universities and research stations have conducted extensive research and have proven the value of the high level antibiotic–sulfa program treatment of new cattle. In four experiments at Purdue University, the feeding of a combination of 350 mg chlortetracycline (Aureomycin) and 350 mg sulfamethazine (the combination is known by the trade name, AS-700) daily to newly arrived feeder cattle resulted in a 29% increase in daily gain and a 27% improvement in the efficiency of feed conversion over the 28 day adjustment period. Calves administered the AS-700 program in their feed weighed an average of 16 pounds more than control calves not fed Aureo-S-700 (Table 17.2).

REFERENCES

Algeo, J. W. (1973). Feedlot losses in cattle. *Beef Mag*. November, p. 23.

Bailey, J. H. (1977). "Survey of Disease Problems in Large Feedlots," Cattle Day Rep., p. 88. South Dakota State University Brookings.

Brent, B. E. (1976). Relationship of acidosis to other feedlot ailments. *J. Anim. Sci.* **43,** 930.

Davies, E. T., Pill, A. H., Collings, D. F., and Venn, J. A. (1965). Cerebrocortical necrosis in calves. *Vet. Rec.* **77,** 290.

Edwin, E. E., Lewis, G., and Allcroft, R. (1968). Thiaminases and cerebrocortical necrosis. *Vet. Rec.* **83,** 417.

Elam, C. J. (1976) Acidosis in feedlot cattle: Practical observations. *J. Anim. Sci.* **43,** 898.

Hentges, J. T. (1970). Factors affecting feedlot performance of beef cattle. *Fla., Agric. Exp. Stn., Bull.* **AN 70-8.**

Huber, T. L. (1971). Effect of acute indigestion on compartmental water volumes and osmolality in sheep. *Am. J. Vet. Res.* **32,** 887.

Huber, T. L. (1976). Physiological effects of acidosis on feedlot cattle. *J. Anim. Sci.* **43,** 902.

Jensen, R., Pierson, R. E., Braddy, P. M., Saari, D. A., Lauerman, L. H., England, J. J., Horton, D. P., and McChesney, A. E. (1976). Diseases of yearling feedlot cattle in Colorado. *J. Am. Vet. Med. Assoc.* **169,** 497.

Slyter, L. L. (1976). Influence of acidosis on rumen function. *J. Anim. Sci.* **43,** 910.

18

Economics of Cattle Feeding

I. THE CATTLE FUTURES MARKET*

There are three major categories of people who use the futures markets—(a) speculator, (b) hedger, and (c) observer.

The *speculator* is the person who accepts the risk of a price change for a given commodity in the hope of making a profit. He would most likely ask the question: "How can I benefit from price fluctuations?" When he buys or sells a contract, he is doing so because he believes price movements will be in his favor. He expects contract prices to go up when he buys and down when he sells.

The *hedger* is the person who uses the futures market to minimize the risk of a price change (Fig. 18.1). He wants to shift this risk to someone else and, as he does so, settles for a somewhat fixed price. He would most likely ask the question: "How can I protect myself from up and down movements of price?" When he buys or sells a contract, he is more interested in having similar price movements for his cash commodity and the futures contract than he is in the direction of these price changes.

The *observer* is the person who uses the futures market as a guideline of things to come. He does not buy or sell futures contracts. He would probably ask the question: "Is there any way I can use the futures market without active involvement?" Although he does not use the futures market as an exact price predictor, he uses it as an indicator of direction of change or as a barometer of the reaction of the cattle industry to changing conditions (Fig. 18.2).

Why doesn't everyone employ the services of the futures market if this is a method of locking in at least minimal profit? Some typical answers to that question include: (a) "It's too complicated and I don't understand it," (b) "Someone I know was burned once and that's enough," (c) "I can't afford to use the futures market because it costs too much," or (d) "I don't want to shift

*Much of the material used in this section adapted by permission from a paper presented by Murra (1977), Extension economist.

Fig. 18.1 Selling of livestock on the futures may permit a certain built-in "safety" in the feeding business. (Photo courtesy of BEEF Magazine.)

Fig. 18.2 Feeder calves in an auction ring. (Photo courtesy of BEEF Magazine.)

price risk because I'm a born gambler and I want a shot at fatter profits, and I know I might get some lumps, also.''

Of the excuses listed, only the last one is really acceptable. Granted, the first three may apply to some people, but these excuses can usually be overcome if the interest is there.

A. An Example of Cattle Hedging

The following example illustrates some of the actions which might be taken by a cattle feeder who hedges his cattle. The example, while illustrating the procedures, is oversimplified. Also, the example is that of a perfect hedge, one where price movements of the cash and futures markets are the same. This is unlikely in a real-world situation. Also, only the fed cattle are hedged. Both grain and feeder cattle contracts can be used to ''lock-in'' these input prices.

Assume that on October 15 a cattle feeder evaluates his situation as follows:

1. He can buy 600 pound steers at $60.00 per hundredweight.
2. He knows he can add 500 pounds in 180 days (by April 15) at a cost of 40¢ per pound of gain.
3. The current price for an April contract is $52.00.
4. Price forecasts for April range from $45.00 to $55.00 per hundredweight.

On the basis of the above information, the feeder decides to buy the cattle, aim for the April market and hedge his cattle. Scheme 18.1 illustrates the results of his enterprise, both if prices went up and if they went down.

The producer would have been better off by not hedging in the situation where a price increase was assumed. However, if prices had decreased, he would have been better off with the hedge.

B. What You Should Know as a Hedger

There are several items which the hedger must know. These include (a) production costs, (b) limits and requirements of the contracts, (c) basis, (d) a broker and banker who understand hedging, (e) knowledge of factors affecting the market, and (f) how the contract affects the hedger.

1. Costs

Knowledge of production costs is always important, whether or not one uses the futures market. However, the importance is accented when hedging on the futures market because the final product price is established more firmly than when the product is not hedged. The example used earlier might best illustrate the point. There, the cattle feeder could buy 600-pound feeders for $60 per hundredweight. Also, he could use the futures market to lock-in a price of $55 for his 1100-pound steers. Should the producer feed the cattle and should he

Scheme 18.1

On Cash Market	*On Futures Market*
Oct. 15 Buy 600-lb Choice steers at $60.00/cwt. = $360.00	Oct. 15 Sell April fed cattle futures contract at $50.00/cwt.
Oct. thru Apr. Add 500-lb at $40.00/cwt. = $200.00	Oct. thru Apr. Hold futures contract but continue to evaluate it to see if the hedge should be lifted
Total cost = $560.00 or $50.90/cwt.	

IF MARKET GOES UP

Apr. 15 Sell 1100-lb steers at $55.00 = net above costs of $4.10/cwt. or $45.10/head	Apr. 15 (or before) Buy April contract at $50.00 = net loss of $0.90/cwt. or $9.90/head

$45.10 Gain on cash market
9.90 Loss on futures market
Net = $35.20/head

IF MARKET GOES DOWN

Apr. 15 Sell 1100-lb steers at $48.00 = loss over costs of $2.90/cwt. or $31.90/head	Apr. 15 (or before) Buy April contract at $45.00 = net gain of $5.00/cwt. or $55.00/head

$31.90 Loss on cash market
55.00 Gain on futures market
Net = $23.10/head

hedge on the futures market? In order to answer these questions, the cattle feeder must know his production costs. Costs greater than $190 per head would result in a loss, while lower costs would result in a profit. Once again, the cattle feeder should know his production costs even if he does not hedge. But, if a final product price is not locked in, there may be a greater tendency to "hope" for a price to cover all costs. Cattlemen seem to be eternal optimists—they like to believe the "high price" rumors. When a final price can be established prior to production, costs seem to become more important in the decision-making evaluation.

2. Limits and Requirements

Futures contracts are written in very specific language. Products involved are described carefully, with a discount schedule noted for delivery of products which do not meet specifications.

Often, the hedger's live cattle do not completely meet all of the contract specifications. Thus, the price risk is not entirely shifted. How much is shifted depends on how close the contract specifications are met and, if different, how many differences there are in prices for the different categories.

3. *Basis*

The difference between the cash price at any location and the futures price on any futures exchange is known as the basis. For example, if 1100-pound choice fed steers in Sioux City are $50 and a futures contract for fed steers on the Chicago Mercantile Exchange is selling for $52, the basis is −$2.00. For storeable commodities, the futures price is usually higher than the current cash price. However, for cattle, this is not always the case. The basis depends on: (a) total supply and demand of the commodity, (b) supply and demand of substitute commodities, (c) geographical disparity in supply and demand, (d) transportation problems and prices, (e) storage available, (f) quality factors and condition capacities, and (g) the demand for futures contracts.

Since many of the above factors vary by location, the basis must be computed for each location. For example, the "net" or localized price for a fed steer in location A might be only $50 while in location B it might be $53 if the futures contract is priced at $52. Granted, the above example is an extreme, but it is possible for one location to have a much different basis than another, even to the extent where one is positive and the other is negative.

Computation of the basis for fed cattle in South Dakota yields a complex set of patterns. Generally, the establishment of a firm, localized net price is not as precise as in the case for grain. For example, in 1976, the Omaha cash-Chicago nearby futures contract basis ranged from +$1.50 in January to −$3.58 in May. For the same months and locations in 1977, the basis was +$.54 and −$1.34, respectively. Thus, the localized prices in 1977 were closer to the Chicago futures price than was the case in 1976. The variation in basis by the amounts noted results in a wider range of "locked-in" prices than if the basis was more stable. Even then, if the basis does not vary as much as the cash price, some risk can be shifted.

4. *Broker and Banker*

The hedger must know a good broker (knowledgable and trustworthy) and a good banker (also knowledgable and trustworthy). The broker is important because he is the man who buys or sells the contracts and also has current advise on "what to do." The banker is important because he supplies the credit, probably for both the cash commodity enterprise (feedlot) and the futures market costs (margins and brokerage fees).

5. *Market Factors*

Since both the cash price and futures price are affected by market conditions, changes in these conditions and the effect of these changes must be monitored.

What may have been a good hedge today may not be tomorrow. What may have been profitable under one set of circumstances may not be under another. Therefore, the feedlot operator must continually evaluate the market to see if he should hedge, remove a hedge, or do nothing. This evaluation is not something he should not already be doing. Rather, if he uses the futures market, he may be required to evaluate on a more continuing, structured basis.

6. Effect on the Hedger

The hedger must analyze the impact of placing a hedge on both his operation and his general well being. Such questions as: (a) do I like risks, (b) can I afford a big loss, (c) does the futures market make me more or less comfortable, (d) is credit easier or harder to obtain, and (e) do I feel safer.

C. What To Watch For

The futures market is not necessarily all good or all bad. It can be a useful tool. Some of the major factors to keep in mind when deciding whether to use the futures market are discussed in this section.

1. It's Not What You Know, But Who

As noted earlier, the hedger must know a good broker and a good banker. A vast amount of knowledge about the futures market can be wasted if you do not have a good broker. He must both give sound advice and also activate your requests *as you want them* activated. Also, if your banker does not understand or trust the futures market, he may discourage its use. Sometimes this discouragement comes in the form of "no credit."

2. There Ain't No Such Thing as a Free Lunch

What must be given up by the hedger in order to get price protection? After all, no one expects to get "somethin for nothin." First, there is a broker's fee. Although this fee is not large and since a contract usually involves about 38 head, the cost is about $1.00 per head. Second, margin money is required. Even when you sell a contract, you do not get money. Rather, you must put up earnest money. An interest charge should be made against this earnest money. Usually, the margin requirement for a bonafide hedger is less than 10% of the contract value. For a 40,000 pound contract at $50 per hundredweight, a margin of less than $1,600 would probably be required. Interest on $1,600 at 10% would be about $160 per year or, if a contract were held three months, about $1.00 per head. Finally, the act of shifting price risk usually removes not only the potential for large losses due to price changes, it also removes the potential for landslide profits due to price changes. The band of possible profits or losses is narrowed. To some, the loss of windfall profits may be the highest cost of those mentioned. They enjoy the risk—and the potential gain or loss which goes along with it.

3. *Nobody's Perfect*

Even though the cash and futures prices generally move in the same direction, there are exceptions. In addition, the magnitude and/or timing of the move, even if in the same direction, are usually not the same. Thus, even though one can say that the futures market can be used to shift price risk, the above imperfections prevent this from being fully accomplished. But, if a portion of the risk is shifted, some of the goal is attained.

As we noted earlier, the basis is a key to the use of the futures market. If the basis were always completely predictable, the perfect hedge could be more nearly attained. Once again, however, if the basis varies less than the cash price, some risk usually can be shifted.

4. *Once a Loser, Always a Loser*

The futures market is not a gimmick which will automatically remove losses which go to the inefficient operator or which would occur in an unprofitable market situation. In fact, the futures market, by "locking-in" a price for the final product, can just as easily lock-in a loss as it can a profit. At times, locking-in a loss may be advantageous, especially if it prevents a much larger loss. Merely going into the futures market will not change an unprofitable operation into a profitable one. In fact, there may be instances when it may be to an inefficient producer's advantage not to use the futures market. That is, if he does not lock-in a price in the futures market, there is still the possibility that an extremely high price can bail him out. Of course, the extremely large losses resulting from very low prices are also possible.

5. *Do the Opposite*

When using the futures market as a hedge, the principal rule is "do the opposite" in the futures market than you do in the cash market. For example, when buying feeder cattle to put on feed, sell a futures contract. When selling the cattle (or before), buy back the futures contract for the same product and month sold. If the opposite transaction is not followed, but rather the same position is taken in both markets, the result is a cash side which is not hedged plus ownership of a speculative contract on the futures side.

The opposite transactions provide the hedge because, generally, the cash and futures markets move in the same direction. Losses in one market are offset by gains in the other, with the net result a shift in some of the price risk.

6. Apples Are Not Oranges

The quality specifications of a fed cattle contract, as noted earlier, are very specific. Specific weights, quality grades, yields, and sex are noted along with the price discounts which are made if these specifications are not met. Many cattle feeders do not feed cattle which meet these rigid specifications, and economically, they probably shouldn't. But, this means that if the futures market is used, the hedge will be on something different than the cash commodity. The greater the difference there is, the poorer the hedge is in terms of shifting price risk, especially since all grades, weights, and sexes are not priced the same. The extent of these price differences is the key. It may not be as bad as comparing applies and oranges, but certainly an 800 lb, yield grade 4, good-quality grade heifer is not the same as a 1,100 lb, yield grade 2, choice-quality grade steer.

7. A Pinch of Salt

There is an old saying that if a pinch of salt is good, a cup is better. Many people get burned in the futures market because they apply the above concept to their futures market activities. They believe that if it makes good dollar sense to hedge all or most of their cattle, it makes more sense to "hedge" several times what they actually have. Remember, the hedge is an attempt to shift price risk. It is not possible to shift price risk on something you do not own or expect to own. Going into the futures market with a volume greater than that on the cash side means an acceptance of price risk—i.e., a speculator. The speculator's goal is profit. He does not need or want cattle to do that. It is extremely important to recognize the difference.

D. Using the Futures Market

The futures market's active participants are the hedger and speculator. If neither of these roles is feasible or acceptable, the cattle feeder can still use the futures market. Basically, there are two ways he can do this: (1) Use the futures market as a guide to what prices are expected to be. (2) Use the futures market as a barometer in measuring the trade's reaction to livestock marketing information as it develops.

The futures market is not designed to be a price predictor. But, since there must be both a buyer and seller of a futures contract, some people are, in effect, predicting a price increase (the buyers) and others are predicting a price decrease (the sellers). In net, the predictions balance out so that usually at least the direction, if not the magnitude, of the expected price changes can be discerned. Again, although not perfect, it may be better than what many producers are currently using.

The use of the futures market as a barometer can also be beneficial. Basically, the futures market reacts to changes in supply and demand conditions. Since traders often have many dollars at stake, they are very current on these conditions and, when conditions change, the traders react accordingly. Just how much reaction takes place (how much prices go up or down) may give an indication to the nontrader as to what will happen. Because the futures market may be more psychological than the cash market, the barometer is not always completely trustworthy. But again, it may be better than the barometer currently used by many cattle feeders.

E. Basic Specifications for Fed Cattle Contracts

1. Par Delivery Unit

A par delivery unit is 40,000 lb of USDA yield grade 1, 2, 3, or 4 choice-quality grade live steers, averaging between 1050 and 1200 lb with no individual steer weighing more than 100 lb above or below the average weight for the unit. Not more than eight head of estimated yield grade 4 choice steers shall be permitted in a par delivery unit. No individual animal weighing less than 950 lb or more than 1300 lb shall be deliverable.

Par delivery units containing steers with an average weight between 1050 and 1125.5 lb shall have an estimated average hot yield of 62%. Par delivery units containing steers with an average weight between 1125.6 and 1200 lb shall have an estimated average hot yield of 63%.

All cattle contained in a delivery unit shall be healthy. Cattle which are unmerchantable, such as crippled, sick, obviously damaged or bruised, or which for any reason do not appear to be in satisfactory condition to withstand shipment by rail or truck shall be excluded. No cattle showing a predominance of dairy breeding or showing a prominent hump on the forepart of the body shall be deliverable. Such determination shall be made by the grader and shall be binding on all parties.

2. Weight Deviations

Steers weighing from 100 to 200 lb over or under the average weight of the steers in the delivery unit shall be deliverable at an allowance of 3¢ per pound provided that no individual animal weighing less than 950 lb or more than 1300 lb shall be deliverable. For purposes of computing such allowance, the weight of the over or under weight animals shall be considered the same as the average weight per head of the delivered unit.

Steers weighing more than 200 lb over or under the average weight of the load are not acceptable. The judgment of the grader as to the number of such overweight or underweight cattle in the delivery unit shall be final and shall be so certified on the grading certificate.

3. *Yield Deviations*

Delivery units with an estimated average hot yield under par shall be acceptable with an allowance of one-half cent per pound for each one-half percent or less by which the estimated yield is under par. Units with an estimated average hot yield of less than 60% shall not be deliverable.

4. *Yield Grade Deviations*

Estimated yield grade 4 Choice quality steers, up to and including eight head, are not deliverable at par.

All Good quality grade, yield grade 4 steers are deliverable at 3¢ per pound allowance for yield grade plus the quality allowance.

If nine or more steers of yield grade 4 (Good and Choice quality grade) are contained in the delivery unit, all yield grade 4 cattle in excess of eight up through a maximum of eighteen head are deliverable at a 3¢ per pound allowance.

For purposes of computing such allowance, the weight of such yield grade 4 steers shall be considered the same as the average weight per head of the delivered unit.

Any delivery unit containing more than eighteen head of cattle with an estimated yield grade of 4 shall not be deliverable. Cattle with an estimated yield grade of 5 shall not be deliverable.

5. *Quality Grade Deviations*

Delivery units containing not more than eight head of USDA Good grade steers may be substituted at a 3¢ per pound allowance for each Good grade steer. For the purpose of computing such allowance, the weight of such Good grade steers shall be considered the same as the average weight per head of the delivered unit.

6. *Quantity Deviations*

Variations in quantity of a delivery unit not in excess of 5% of 40,000 lb shall be permitted at the time of delivery, with appropriate adjustment to reflect delivered weight but with no further penalty.

7. *Delivery Points*

A par delivery of live cattle shall be made from approved livestock yards in Peoria, Illinois, Joliet, Illinois; Omaha, Nebraska; and Sioux City, Iowa.

II. CUSTOM FEEDYARDS—WHAT ARE THEY AND HOW DO THEY WORK?

A custom feedyard is one that feeds cattle for customers for a fee. It furnishes room and board for someone else's cattle, much like a motel. Thus, one can actually become a cattle feeder without doing a bit of the work. This method then permits absentee ownership of cattle sort of like a parent sending their children off to college for the university to oversee. Usually the large commercial feedlot is so efficient and since certain overhead fixed costs are spread over much larger numbers of cattle, it possibly can feed a private individual's cattle more economically than he can do himself.

Services provided by the custom feedyard are many. Cattle are fed twice or three times per day; the feed is processed; all cattle are inspected every day; sick animals are removed to protected "hospital pens" for treatment; marketing is optional but usually the commercial feedlot operator has a better "feel" for the market; and the commercial feedlot can arrange for financing.

The commercial feedlot renders a variety of services for which it must be paid. Normally the owner of the cattle receives a billing at the end of each month. However, the last month's billing is deducted from the sale price so that the owner does not have to come up with that billing. Thus, on a group of cattle on feed for 150 days (five months), the owner will have to provide cash for four months' feed with the fifth month's feed bill being deducted from the sale price. The owner of the custom feedlot is in the business of selling feed as well as feeding cattle, so he may charge a markup of at least 10% on the cost of feed. Thus, feed costing the commercial feedlot owner $80.00 per ton might be charged to the owner of the cattle at $88.00 or $90.00 per ton. In addition, usually there is charged a per head per day pen charge of 8 to 15¢. Thus, a steer in the lot for 150 days might cost the owner $15.00 pen charge plus feed cost. Furthermore, there will be a processing charge on incoming cattle covering dehorning, branding, dipping, castrating, worming, vaccinating, pregnancy checking, and implanting with growth promotants. Naturally this will vary depending on the services rendered, but may run as high as 6 or $7 per head. Cattle which become ill during the feedlot period may be removed for personal attention, and only medicinals utilized will be billed to the owner. At the end, if the commercial feedlot owner markets the cattle, he may charge from $2 to $3 per head just as a commission company would do.

Financing arrangements have become fairly uniform in the cattle industry. Lending institutions located in cattle feeding country are especially attuned to cattle feeding capital requirements and are in business to do just that. Naturally there are no set standards as to how much they will lend, but it should not be difficult to obtain up to 75% of the cost of cattle and feed. One approach to this is to require the cattle owner to put down 25–30% of the purchase cost of the cattle with the lending institution supplying the other part. Then, as the monthly feed and any bills for illness come due, the lending institution and the owner pay their proportionate part to the commercial feedlot owner. Of course, the cattle will increase in size and in their feed consumption, and potential market value may change—up or down—each month. Unless potential market value fluctuates markedly, the lending institution and the owner will pay their proportionate monthly feed bills. However, if the potential market is declining, the owner will have to pay a larger proportion of the monthly feed bill. Lending institutions tend to feel more at ease when the cattle are hedged to assure a definite sale price. Furthermore, policies of lending will vary a great deal.

What is the minimum number of cattle a person should have in a commercial feedlot? Usually this amounts to owning one pen of cattle. The typical cattle pen has a minimum capacity of 50 or 100 head and may be even higher. It is desirable that one owner not mix his cattle with that of another because bookkeeping and identification of cattle that may die become more difficult. The exception to this is investment clubs in which several people invest smaller quantities of money in an investment fund, and the manager of the fund then approaches the commercial feedlot operator to feed one or multiples of lots of cattle. The normal feeding period is from 120 to 200 days and is dependent upon initial weight, type of cattle, and type of diet fed.

What is a normal investment per head for commercially fed cattle? In other words, what would it cost for a person to put his cattle in a commercial feedlot? The following tabulation presents a set of figures which will change from day to day, but nevertheless may present "ballpark" figures for comparative purposes:

	Cost per head	100 head	30% owner's money	70% borrowed money
Cost of 600-lb steer @ 65¢	$390.00	$39,000.00	$11,700.00	$27,300.00
Gain cost, 500-lb @ 45¢	225.00	22,500.00	6,750.00	15,750.00
Interest on 70% borrowed money @ 11%				
$390.00 @ 11% for 5 months = $17.88				
Monthly bill ($45.00) prorated for 4 months = 3.27				
Total interest	21.15	2,115.00		
	$636.15	$63,615.00	$18,450.00	$43,050.00

At market time, then, the 1100-lb steer must net $57.83 per hundred pounds for the owner to "break even" on his investment. These figures are not indicative of any potential profit and loss opportunity but merely represent a model to which any given set of figures may be applied.

Who utilizes the commercial feedlots? One is the nonfarm investor who wishes to invest in the cattle feeding business just like he would invest in the stock market, for example. The tax advantage once so attractive on pre-purchased feed has lost much of its tax shelter advantage. So this type of investor must look at cattle feeding primarily on a sound investment basis.

Second, the farmer who is a cattle feeder himself wishes to expand his investment in cattle feeding but does not wish to expand his own facilities. In other words, he has faith in the cattle feeding business and wishes to invest more heavily than his own facilities will permit. One benefit of this method is that the monthly feed bill from the commercial feedyard keeps him aware of the danger of keeping his home feedlot cattle on feed too long.

A third type of commercial feedlot customer is the rancher who wants to hold on to his own calves and maybe even add to that number by having the commercial feedlot feed them to market finish for him.

Finding the right commercial feedlot and bargaining for the proper contract takes time and effort. Although most commercial feedlots operate on the same basic principles, there are at least as many specific contract types as there are different commercial feedlots. It is a good idea for the potential customer to get referrals and to study customer close-out statements. Since no two pens of cattle will perform alike, several close-outs should be studied, looking at beginning weights, number of days in the feedlot, daily gain, daily feed, feed per pound of gain, and, finally, net payoff.

III. PREDICTING PERFORMANCE AND FEED REQUIREMENTS

Normally a cattle feeder calculates rate of gain and efficiency of feed conversion based on sale weight minus purchase weight, divided by number of days in the lot (for daily gain) and total pounds of feed consumed (for feed efficiency). Naturally this represents the "payoff" data on which the cattle feeder's profit is based. However, this does not permit the cattle feeder to make interim assessments of "how he is doing" midway or whether he can afford to keep "nearly finished" cattle for additional increments of time. Kansas State University scientists (Brent *et al.*, 1978) have prepared a mathematical model to compute rate of gain, daily feed intake and efficiency of feed conversion on finishing steers starting at a weight of 650 lb and fed one of ten diet types. The diet types range from all-corn silage to all-concentrates (Table 18.1). The series of tables presented permits calculation of costs at any given time plus calculating the cost of each additional increment of gain. The cost of each pound of gain is composed of

TABLE 18.1

Ration Types Fed Finishing Steers in a Mathematical Model of Feedlot Performance[a]

Diet no.	Ingredient, % dry matter basis: Corn silage	Cracked corn	Protein supplement[b]	Net energy, therms/100 lb: Ne_m	Ne_p
1	90.4	0.0	9.6	71.8	45.8
2	80.4	9.7	9.9	74.9	48.0
3	70.4	19.4	10.2	78.1	50.1
4	60.4	29.1	10.5	81.3	52.3
5	50.4	38.8	10.8	84.4	54.3
6	40.3	48.6	11.1	87.6	56.6
7	30.2	58.4	11.4	90.8	58.8
8	20.2	68.1	11.7	93.9	63.1
9	10.0	77.9	12.0	97.1	63.1
10	0.0	87.6	12.4	98.4	65.3

[a] Brent *et al.* (1978).

[b] Varied composition of soybean meal, limestone, dicalcium phosphate, salt, trace minerals, and vitamins to assure adequate balance.

direct feed costs plus such fixed costs as interest and yardage. Thus, if a steer were feed a longer period of time on a low cost diet, total cost might be higher, because of increased fixed costs, than if it gained more rapidly on a more expensive diet.

Let us follow the calculation for a typical diet (Table 18.1). Diet number 6 contains (on a dry matter basis 40.3% corn silage, 48.6% cracker corn, and 11.1% protein supplement:

Ingredient	Dry matter level (%)	Dry matter content (%)	Proportions, on as is basis (%)
Corn silage	40.3	35	62.65
Cracked corn	48.6	87	30.40
Protein supplement	11.1	87	6.95
Total	100.0		100.00

If we assume corn silage to be $20.00 per ton and it contains 35% dry matter, cost per pound of dry matter is 2.85¢; corn costing $2.24 per bushel and containing 87% dry matter has a dry matter cost of 4.60¢; protein supplement costing $200 per ton and containing 87% dry matter has a dry matter cost of 11.49¢. Diet number 6, then, costs 4.66 cents per pound of dry matter. Table 18.2 shows that an 800-pound steer requires 6.6 pounds of dry matter from diet number 6 per pound of gain. Thus, at that weight, feed cost per pound of gain is 30.75 cents. Contrast this cost of gain for a 900-pound steer. At this point, 7.2 pounds of diet number 6 dry feed (33.55 cents) is required per pound of gain; an 1100-pound steer requires 8.8 pounds of diet number 6 dry feed (41.00 cents) per pound of gain. No interest costs are added to the cost of the dietary ingredients.

If the steer were purchased at 600 pounds for 60 cents per pound with money borrowed at 9.5% interest, the steer uses 9.37 cents per day in interest. Assuming yardage cost of 10 cents per day, daily fixed costs are 19.37 cents per day. From Table 18.3, an 800-pound steer fed diet number 6 should gain 2.99 pounds per day, so fixed costs per pound of gain are 6.47 cents. The total (feed cost plus fixed

TABLE 18.2

Daily Dry Matter Intake (lb) Computed from a Steer Performance Model

	Diet number[a]									
Steer weight (lb)	1	2	3	4	5	6	7	8	9	10
650	15.6	16.3	17.2	18.4	19.3	20.0	19.6	18.8	17.4	14.5
675	15.9	16.3	17.5	18.5	19.3	19.9	19.6	18.9	17.4	14.6
700	16.0	16.5	17.3	18.3	19.3	19.9	19.6	18.9	17.3	14.6
725	16.0	16.6	17.5	18.4	19.4	19.8	19.5	18.8	17.5	14.4
750	16.1	16.8	17.3	18.4	19.3	19.7	19.5	18.8	17.4	14.4
775	16.4	16.7	17.6	18.2	19.2	19.9	19.7	19.1	17.4	14.4
800	16.5	16.8	17.3	18.3	19.0	19.7	19.5	19.0	17.4	14.4
825	16.7	16.9	17.5	18.5	19.1	19.8	19.4	18.9	17.3	14.8
850	16.9	17.0	17.6	18.6	19.0	19.7	19.6	18.8	17.2	14.8
875	16.9	17.0	17.4	18.3	19.1	19.5	19.5	19.0	17.4	14.8
900	16.9	17.2	17.7	18.5	19.3	19.7	19.6	18.8	17.6	14.7
925	17.4	17.3	17.7	18.6	19.1	19.7	19.7	18.7	17.7	14.7
950	17.4	17.4	17.7	18.5	19.0	19.8	19.8	18.9	17.9	14.9
975	17.5	17.3	17.8	18.6	19.0	19.7	19.6	19.0	18.0	15.1
1000	17.7	17.4	17.8	18.5	18.9	19.7	19.7	18.7	18.2	14.9
1025	18.0	17.6	17.8	18.4	19.1	19.4	19.5	18.8	18.0	15.0
1050	18.0	17.7	17.8	18.4	19.1	19.3	19.3	19.1	17.6	14.9
1075	18.0	17.8	18.0	18.4	19.0	19.3	19.4	19.0	17.4	14.9
1100	18.9	18.0	18.1	18.6	19.0	19.4	19.6	18.9	17.4	15.0
1125	18.8	17.7	18.1	18.5	19.0	19.4	19.5	19.0	17.6	15.3
1150	19.1	18.0	18.2	18.5	18.8	19.5	19.6	18.7	17.5	15.3
1175	19.0	18.1	18.4	18.7	18.9	19.2	19.4	18.8	17.5	15.2
1200	19.1	18.3	18.4	18.8	19.0	19.2	19.3	18.7	17.7	15.1

[a] See Table 18.1

TABLE 18.3

Daily Weight Gain (lb) at Various Body Weights Computed from a Steer Performance Model

Steer weight (lb)	Diet number[a]									
	1	2	3	4	5	6	7	8	9	10
650	2.11	2.44	2.77	2.97	3.17	3.39	3.56	3.76	3.70	3.72
675	2.07	2.40	2.73	2.93	3.12	3.32	3.50	3.70	3.63	3.65
700	2.02	2.35	2.66	2.86	3.06	3.26	3.43	3.63	3.54	3.56
725	1.98	2.31	2.61	2.79	2.99	3.19	3.37	3.54	3.50	3.50
750	1.94	2.27	2.55	2.75	2.93	3.12	3.30	3.48	3.41	3.43
775	1.91	2.20	2.51	2.68	2.86	3.06	3.23	3.41	3.34	3.34
800	1.87	2.16	2.44	2.62	2.79	2.99	3.15	3.34	3.28	3.28
825	1.83	2.11	2.40	2.57	2.73	2.95	3.08	3.26	3.21	3.21
850	1.78	2.07	2.35	2.51	2.68	2.86	3.01	3.19	3.12	3.15
875	1.74	2.02	2.29	2.44	2.61	2.79	2.95	3.12	3.06	3.08
900	1.69	1.98	2.24	2.40	2.57	2.74	2.88	3.04	2.99	3.01
925	1.67	1.94	2.18	2.35	2.51	2.66	2.82	2.97	2.90	2.93
950	1.63	1.89	2.13	2.29	2.44	2.60	2.75	2.90	2.84	2.86
975	1.58	1.82	2.07	2.24	2.38	2.53	2.68	2.84	2.77	2.79
1000	1.54	1.78	2.02	2.18	2.31	2.46	2.62	2.75	2.71	2.71
1025	1.50	1.74	1.98	2.11	2.27	2.40	2.53	2.68	2.64	2.64
1050	1.45	1.69	1.91	2.07	2.20	2.33	2.44	2.62	2.55	2.57
1075	1.41	1.65	1.87	2.00	2.13	2.27	2.40	2.53	2.49	2.49
1100	1.39	1.61	1.83	1.96	2.07	2.20	2.33	2.46	2.42	2.42
1125	1.34	1.54	1.76	1.89	2.00	2.13	2.27	2.40	2.35	2.35
1150	1.30	1.50	1.72	1.83	1.94	2.07	2.20	2.31	2.27	2.29
1175	1.25	1.45	1.67	1.78	1.89	2.00	2.13	2.24	2.20	2.20
1200	1.21	1.41	1.63	1.74	1.83	1.94	2.05	2.18	2.13	2.13

[a] See Table 18.1

cost) then is 37.22 cents per pound of gain for an 800-pound steer consuming diet number 6.

It is interesting to note from Table 18.4 how each additional 25 pounds gain require more feed dry matter per unit gain than the previous one. For example, an 800-pound steer consuming diet number 6 requires 6.6 pounds dry matter per pound gain whereas a 900-pound steer on the same diet requires 7.2 pounds dry matter per pound gain, or 9% more feed. This should enter into the decision making of how long cattle should be kept on feed.

TABLE 18.4

Feed Efficiency (Units Dry Feed Per Unit of Body Weight Gain) Computed from a Steer Performance Model

Steer weight (lb)	1	2	3	4	5	6	7	8	9	10
650	7.4	6.7	6.2	6.2	6.1	5.9	5.5	5.0	4.7	3.9
675	7.7	6.8	6.4	6.3	6.2	6.0	5.6	5.1	4.8	4.0
700	7.9	7.0	6.5	6.4	6.3	6.1	5.7	5.2	4.9	4.1
725	8.1	7.2	6.7	6.6	6.5	6.2	5.8	5.3	5.0	4.1
750	8.3	7.4	6.8	6.7	6.6	6.3	5.9	5.4	5.1	4.2
775	8.6	7.6	7.0	6.8	6.7	6.5	6.1	5.6	5.2	4.3
800	8.8	7.8	7.1	7.0	6.8	6.6	6.2	5.7	5.3	4.4
825	9.1	8.0	7.3	7.2	7.0	6.7	6.3	5.8	5.4	4.6
850	9.5	8.2	7.5	7.4	7.1	6.9	6.5	5.9	5.5	4.7
875	9.7	8.4	7.6	7.5	7.3	7.0	6.6	6.1	5.7	4.8
900	10.0	8.7	7.9	7.7	7.5	7.2	6.8	6.2	5.9	4.9
925	10.4	8.9	8.1	7.9	7.6	7.4	7.0	6.3	6.1	5.0
950	10.7	9.2	8.3	8.1	7.8	7.6	7.2	6.5	6.3	5.2
975	11.1	9.5	8.6	8.3	8.0	7.8	7.3	6.7	6.5	5.4
1000	11.6	9.8	8.8	8.5	8.2	8.0	7.5	6.8	6.7	5.5
1025	12.0	10.1	9.0	8.7	8.4	8.1	7.7	7.0	6.8	5.7
1050	12.4	10.5	9.3	8.9	8.7	8.3	7.9	7.3	6.9	5.8
1075	12.8	10.8	9.6	9.2	8.9	8.5	8.1	7.5	7.0	6.0
1100	13.6	11.2	9.9	9.5	9.2	8.8	8.4	7.7	7.2	6.2
1125	14.0	11.5	10.3	9.8	9.5	9.1	8.6	7.9	7.5	6.5
1150	14.7	12.0	10.6	10.1	9.7	9.4	8.9	8.1	7.7	6.7
1175	15.2	12.5	11.0	10.5	10.0	9.6	9.1	8.4	8.0	6.9
1200	15.8	13.0	11.3	10.8	10.4	9.9	9.4	8.6	8.3	7.1

REFERENCES

Brent, B. E. *et al.* (1978). "Cattle Feeders Day Report," p. 90. Kansas State University, Manhattan.
Murra, G. E. (1977). "Cattle Day Proceedings." South Dakota State University, Brookings.

19

Environmental and Housing Effect on Feedlot Cattle Performance*

Cattle feeding housing ranges all the way from no housing, no windbreak, and no paving underfoot to the most sophisticated enclosed environment complete with heat and humidity-sensitive fans. Naturally each additional dollar spent on equipment must be deducted from gross profit unless its use contributes a saving in expenses. The purpose of this chapter is to summarize research data designed to sort out the pros and cons of various types of equipment which contribute to altering the environment of finishing beef cattle. Most of the data presented is based on research conducted in the North Central region of the United States, but undoubtedly these effects may be adjusted and applied to many other sections of the world where cattle may be fed.

I. HOUSING EFFECTS

A. Housing versus No Housing

A further breakdown is needed under this subheading, namely, the effects of winter housing and of summer housing. Iowa State University has conducted a number of trials to determine housing effect on winter performance. The reader should understand that Iowa winters can be extremely severe when temperature, humidity, and wind velocity all are taken into account. In yearling steer comparison, average daily gain was depressed from 9 to 18% and feed cost was increased from 7 to 21% for cattle which had no shelter (Self, 1964, 1965; Self and Hoffman, 1967). Michigan researchers (Greathouse and Henderson, 1968) compared calves wintered outside with no housing, windbreak, or bedding against calves wintered inside, bedded, and completely under roof. Daily winter gains

*Much of the material used in this section is from a handout presented by Henderson and Geasler (1968).

TABLE 19.1

Summary of Thirteen Trials of No Shelter Effects on Winter Gain of Feeder Cattle

Station	Type cattle	Date	Decrease daily gain (%)	Increase in feed cost (%)
Iowa	Yearling	1962	18	21
Iowa	Yearling	1963	15	17
Iowa	Yearling	1965	15	15
Iowa	Yearling	1966	11	8
Iowa	Yearling	1967	13	14
Iowa	Yearling	1968	9	7
Ohio	Calf	1959	2[a]	8[a]
Michigan	Calf	1968	14	23
Michigan	Calf	1968	10	14
Kansas	Yearling	1961	22	28
South Dakota	Calf	1968	4[a]	4[a]
Connecticut	Calf	1963	7[a]	12[a]
Saskatchewan, Canada	Yearling	1958	15	14
Average of 13 trials			12	14

[a] No shelter groups had bedding in experiments indicated.

for the outside group were depressed 14% and feed cost increased 23%. However, when bedding and housing costs were added to feed costs for the inside group, their total costs were greater.

The data presented in Table 19.1 are quite consistent, showing a uniform drop in daily gain of 12% and increase in feed cost of 14% for no shelter fed cattle.

In summer comparisons, eight studies are summarized (Table 19.2). Perfor-

TABLE 19.2

Summary of Eight Trials of No Shelter Effects on Summer Gains of Feeder Cattle

Station	Type cattle	Date	Decrease in daily gain (%)	Increase in feed cost (%)
Iowa	Yearling	1962	6	4
Iowa	Yearling	1964	7	7
Iowa	Yearling	1965	7	5
Iowa	Yearling	1966	4	1
Iowa	Yearling	1967	3	3
Kansas	Yearling	1961	7	8
South Dakota	Yearling	1967	2[a]	5[a]
Ohio	Yearling	1964	7	−2
Average of 8 trials			5	4

[a] Bedding provided for no shelter group.

mance favoring shelter for summer-fed cattle was much less dramatic with an average of 5% increase in gain and 4% decrease in feed cost. It is doubtful that the advantage for summer shelter will pay for the cost of providing summer shelter.

B. Comparison of Type of Housing

A number of trials have been conducted to compare performance of cattle fed in 40% and 100% covered lots. Michigan researchers (Henderson and Newland, 1965, 1966a,b, 1967; Henderson *et al.*, 1965, 1967a,b) conducted three winter trials for this type comparison and reported no difference in gain (2.50 lb per day) or in feed required per pound of gain (7.36 lb). The same researchers found no difference in performance in summer trials for 40% and 100% shelter over cattle finishing feedlots. Ohio workers (Roller *et al.*, 1960) conducted five studies and reported the results by 56-day cold and warm periods. During the cold period, cattle under 40% shelter gained 4% slower (2.03 versus 2.11 lb per day) and had a 12% higher feed requirement per unit gain than cattle under 100% shelter. However, this reversed during the 56-day warm period with cattle under 100% shelter gaining 8% slower and requiring 3% more feed per unit of gain than cattle with only 40% shelter.

C. Feedlot Surfaces

Data on this subject are not too plentiful, but Iowa State University (Self, 1964, 1965) has reported three winter trials and three summer trials which are summarized in Table 19.3. Differences in performance due to feedlot surface type were negligible for both winter and summer comparisons. This seems to be

TABLE 19.3

Feedlot Surface Type Effect on Cattle Performance[a]

Item	Surface type		
	Concrete	Part concrete	Earth
Winter Research, 3 trials			
Daily gain (99 to 151 days) (lb)	2.52	2.54	2.60
Feed required per unit gain (lb)	11.3	11.3	11.0
Summer Research, 3 trials			
Daily gain (142 days) (lb)	2.72	2.69	2.70
Feed required per unit gain (lb)	9.5	9.6	9.6

[a] Self (1964, 1965).

the general results reported by researchers who have investigated the subject. However, the author conducted research on the subject and found no difference in cattle performance during two summer comparisons; but during one extremely muddy summer, performance favored cattle whose lot was partially paved. It seèmed the cattle in the all earthern lot preferred not to wade through the mud to get to the feed bunks.

1. Slatted versus Bedded Floors

Ohio Experiment Station (Klosterman, *et al.*, 1964, 1965) results were summarized over a 2-year period, and no difference in performance was obtained between cattle housed on steel slatted floors as compared to those on bedded concrete floors. The Ohio researchers assessed a bedding cost per cwt gain of about 43 cents for bedded cattle. Michigan research data (Henderson *et al.*, 1967b) confirmed the Ohio data by which gains of steer calves (2.61 versus 2.67 lb per day) and feed per pound gain (7.1 versus 6.6) were similar for those on slatted versus those on bedded floor (Fig. 19.1 and 19.2)

2. Slatted versus Concrete Floors without Bedding

Michigan workers conducted two winter trials to compare these conditions. Differences in performance were small in both trials, favoring the unbedded

Fig. 19.1 Slatted feedlot floors outdoors decrease space and bedding requirements. (Photo by J. C. Allen and Son)

Fig. 19.2. Finishing beef cattle on slats, under roof, require from 20 to 23 square feet of floor space per head. (Photo by J. C. Allen and Son)

concrete floor group in the first and favoring the slatted floor group in the second experiment, indicating cattle should perform equally well on either type floor.

II. ENVIRONMENTAL EFFECTS

A. Effect of Summer Shade

Summer temperature and summer humidity in combination can be stressful for feedlot cattle. Thus, the value of shade has been researched. Kansas and Georgia data on feedlot cattle shown in Table 19.4 illustrates that shade may be beneficial under many summer conditions.

Research workers at the Imperial Valley Field Station, California (Ittner and Kelley, 1951; Ittner *et al.*, 1954) suggest the primary function of shade is to reduce the heat intake of cattle subjected to the absorption of the radiant energy of the sun. They suggest shades should be 10–12 feet above the ground and provide 60 square feet of shade per head of cattle.

Temperatures at the Yuma, Arizona Station (Nelms and Roubicek, 1975) were even more severe than in California and may remain in excess of 100°F for

TABLE 19.4

Summer Shade Effect on Yearling Feedlot Cattle Performance (Kansas and Georgia)

Item	No shade	Shade
Kansas[a] (3 trials)		
Daily gain (lb)	1.99	2.12[c]
Feed per unit gain (lb)	10.1	9.8
Georgia[b] (4 trials)		
Daily gain (lb)	2.17	2.21
Feed per unit gain (lb)	10.2	10.1

[a] From Boren *et al.* (1961).
[b] From McCormick *et al.* (1963).
[c] $p < 0.05$.

one-third of the year. Gains made by cattle provided with shade at the Yuma Station are significantly higher than for those not provided with shade.

It has been suggested that when temperatures rise above 80°F, some sort of shade should be provided because this condition may exert more feedlot stress than temperatures below freezing.

B. Feedlot Lighting

Most large feedlots provide night lighting. Probably the cost is justified because it seems to have a tranquil effect on the cattle. Kansas researchers (Boren *et al.*, 1965; Smith *et al.*, 1964) reported on two trials to compare the performance of cattle which did or did not have night lighting. The lighting arrangement consisted of three 25-watt incandescent lamps spaced 8 feet apart, suspended under reflectors, 7 feet above the feed bunk. A photoelectric cell turned the lights on at dusk and off at dawn. It was concluded that the presence of night light had no effect on feedlot cattle performance.

C. Animal Density

Three trials at Michigan State University (Henderson and Newland, 1965, 1966a,b, 1967; Henderson *et al.*, 1965, 1967a,b) have compared a broad range of square footage allotments of floor space for finishing both steers and heifers fed in bedded lots completely under roof but open on the south. Following completion of the third trial, the Michigan researchers concluded the square footage requirements of feedlot cattle are a function of body weight with minimum requirements being 2 square feet of bedded area under roof per 100 pounds body weight during winter months and increased to 3 square feet during summer months (Table 19.5).

TABLE 19.5

Effect of Animal Density on Feedlot Performance of Beef Cattle

	Square feet per animal							
Item	20	25	30	35	40	45	50	55
February to June (Michigan)								
Initial weight (lb)		704		704		705		704
Final weight (lb)		981		973		1003		981
Daily gain (lb)		2.41		2.34		2.59		2.41
January to July (Michigan)								
Initial weight (lb)	430	430	429	429				
Final weight (lb)	946	958	986	1000				
Daily gain (lb)	2.43	2.48	2.62	2.68				
April to August (Michigan)								
Initial weight (lb)	637	637	657	650				
Final weight (lb)	881	899	931	928				
Daily gain (lb)	1.81	1.94	2.02	2.06				

REFERENCES

Boren, F. W., Smith, E. F., Hodges, T. O., Larson, G. H., and Cox, R. (1961). Shade for feedlot cattle. *Kansas State Univ., Agric. Exp. Str., Tech. Bull.* **120.**

Boren, F. W., Lipper, R. I., Smith, E. F., and Richardson, D. (1965). The values of feedlot lighting. *Kansas State Univ., Agric. Exp. Str., Bull.* **483.**

Greathouse, T. R., and Henderson, H. E. (1968). "Outside Lots with no Shelter vs Inside Lots 100% Under Roof for Wintering Steer calves," Beef Cattle Day Rep. AH-BC-672. Michigan State University, East Lansing.

Henderson, H. E., and Geasler, M. P. (1968). "Effect of Environment and Housing on the Performance of Feedlot Cattle under Midwest Conditions," Midwest Sect., Am. Soc. Anim. Sci. (handout).

Henderson, H. E., and Newland, H. W. (1965). "Corn silage, Haylage, Vitamins and Housing for Finishing Steer Calves," Beef Cattle Day Rep. AH-BC-642. Michigan State University, East Lansing.

Henderson, H. E., and Newland, H. W. (1966a). "Effect of Animal Density on Feedlot Performance," Beef Cattle Day Rep. AH-BC-566. Michigan State University, East Lansing.

Henderson, H. E., and Newland, H. W. (1966b). "Alfalfa Haylage with Varying Moisture, Concentrate and Protein Levels for Finishing Yearling Steers," Beef Cattle Day Rep. AH-BC-652. Michigan State University, East Lansing.

Henderson, H. E., and Newland, H. W. (1967). "Alfalfa Hay, Haylage and Corn Silage with Varying Moisture and Concentrate Levels for Finishing Yearling Steers," Beef Cattle Day Rep. AH-BC-662. Michigan State University, East Lansing.

Henderson, H. E., Newland, H. W., and Esmay, M. L. (1965). "Square Footage Requirements, Controlled Atmosphere and Slotted Floors for Feedlot Cattle," Beef Cattle Day Rep. AH-BC-643. Michigan State University, East Lansing.

Henderson, H. E., Geasler, M., and Hawkins, D. (1967a). "Feedlot Housing and Environment Systems Compared," Beef Cattle Day Rep. AH-BC-663. Michigan State University, East Lansing.

Henderson, H. E., Geasler, M., and Hawkins, D. (1967b). "Effect of Animal Density on Feedlot Performance," Beef Cattle Day Rep. AH-BC-667. Michigan State University, East Lansing.

Ittner, N. R., and Kelly, C. F. (1951). Cattleshades. *J. Anim. Sci.* **10,** 184.

Ittner, N. R., Bond, T. E., and Kelly, C. F. (1954). Increasing summer gains of livestock with cool water, concentrate roughage, wire corrals and adequate shade. *J. Anim. Sci.* **13,** 867.

Klosterman, E. W., Roller, W. L., and Cahill, V. R. (1964). "Steel Slotted Floors for Fattening Cattle," Cattle Day Rep. Ohio Agric. Exp. Stn., Wooster.

Klosterman, E. W., Johnson, L. J., and Roller, W. L. (1965). "Fattening Cattle on Steel Slotted Floors," Cattle Day Rep. Ohio Agric. Exp. Stn., Wooster.

McCormick, W. C., Givens, R. L., and Southwell, B. L. (1963). Effects of shade on rate of growth and fattening of beef steers. *Univ. Ga., Coastal Plains Stn. Rep., Tech. Bull.* **NS-27.**

Nelms, G., and Roubicek, C. B. (1957). The effect of shade, type of shade and stilbestrol on the performance of summer-fed Hereford steers and heifers in Southwest Arizona. *Univ. Ariz., Agric. Exp. Str., Tech. Pap.* **474.**

Roller, W. L., Johnson, W. H., and Klosterman, E. W. (1960). A Comparison of Housing Conditions for Fattening Beef Animals," Cattle Day Rep. Ohio Agric. Exp. Stn., Wooster.

Self, H. L. (1964). Feedlot management research, annual report. *Iowa, Agric. Exp. Str., Bull.* **OEF64-31.**

Self, H. L., (1965). Feedlot management research, annual report. *Iowa, Agric. Exp. Str., Bull.* **OEF65-31.**

Self, H. L., and Hoffman, M. P. (1967). Feedlot management research, annual report. *Iowa, Agric. Exp. Str., Bull* **OEF67-31.**

Smith, E. F., Robertson, K. E., Lipper, R., Richardson, D., and Boren, F. W. (1964). The value of feedlot lighting. *Kansas State Univ., Agric. Exp. Str., Bull.* **473.**

Appendix I

Additives and Implants in Beef Cattle Feeding

The use of feed additives and other growth-promoting compounds in beef cattle feeding has been a common practice for several years. This practice has been based upon research proving that use of these products will provide an economic return to the cattleman through improved animal performance, improved utilization of feed and reduced incidence of disease and other disorders.

Antibiotics were among the first additives to be developed. Results from the continuous low level (35 to 100 mg/head/day) feeding of antibiotics have been an improvement in daily gain and feed efficiency of 2–5% in cattle fed high-energy rations along with a significant reduction in the incidence of liver abscesses. The improvement in gain and feed efficiency has been somewhat greater for growing cattle fed lower energy rations especially during the early phases of the feeding period when higher levels (350 mg/head/day) of antibiotic feeding has reduced the incidence of shipping fever.

Hormone or hormonelike compounds, some of which are fed and others implanted, such as DES, Ralgro, Synovex, and MGA have consistently improved gains in feedlot cattle 10–15% and improved feed efficiency 7–10%.

Rumensin (monensin sodium) a more recent feed additive results in an improvement of 10% in feed efficiency when fed with high grain rations. When fed with lower energy, high roughage rations, an improvement in weight gains as well as feed efficiency has been observed.

Other additives have been developed and are available for the control of both internal and external parasites in beef cattle.

Thus, many products are available for use by the cattleman to aid in his beef feeding and management program. However, only through the proper use of these compounds can the cattleman be assured of their continued availability and the expected improvements in those economic traits influencing beef production.

To aid in proper drug use, the following table was prepared listing compounds that are available for use in beef cattle feeds and as implants, their purpose for use, level of use and restrictions. The list is not complete but is intended to serve as a guide. As new research data becomes available some compounds must be deleted and others added along with revising of restrictions on particular drugs. In all cases, the producer should follow directions and withdrawal times given on the drug label.

TABLE AI Use of Feed Additives, Drugs, and Implants in Beef Cattle[a]

Compound	Level approved for use	Purpose of usage	Precautions and restrictions
A. Compounds used for nutritional purposes			
Bacitracin	35 mg/head/day	Improved growth rate and feed efficiency	
Bacitracin MD	70 mg/head/day	Reduce liver abscesses	
	250 mg/head/day continuously for 5 days out of each 30 days	Reduce liver abscesses	
Bacitracin, Zinc	35–70 mg/head/day	Improved growth rate and feed efficiency	Use only in dry feeds
Chlortetracycline (Aureomycin)	0.1 mg/lb body weight/day for calves up to 250 lb	Improved growth rate and feed efficiency	
	25–70 mg/head/day	Improved growth rate and feed efficiency	
	70 mg/head/day for cattle less than 700 lb; 100 mg/head/day for cattle over 700 lb	Improved growth rate and feed efficiency and for reduction of liver abscesses	
Erythromycin (Gallimycin)	37 mg/head/day	Improved growth rate and feed efficiency	
Monensin (Rumensin)	5–30 gm/ton of complete or 50–360 mg/head/day	Improved feed efficiency	Feed only to cattle being fed in confinement for slaughter
Oxytetracycline (Terramycin)	75 mg/head/day	Improved growth rate and feed efficiency, bloat reduction, and reduction of liver abscesses	
Tylosin (Tylan)	8–10 gm/ton of complete feed or 60–90 mg/head/day	Reduce liver abscesses	
B. Compounds used for medicinal purposes			
Ammonium chloride	Range cattle 0.75–1.25 oz/head/day Fattened cattle 1.0–1.5 oz/head/day	To reduce incidence of urinary calculi (waterbelly)	
Amprolium (Amprol)	227 mg/100 lb of body weight/day for 21 days	Prevention of coccidiosis	Withdraw 24 hours before slaughter
	454 mg/100 lb of body weight/day for 5 days	Treatment of coccidiosis	Withdraw 24 hours before slaughter

Chlortetracycline	100 mg/head/day for cattle over 700 lb	Reduction of bacterial diarrhea and prevention of footrot	
(Aureomycin)	350 mg/head/day	Prevention of bacterial pneumonia and shipping fever complex	Discontinue use 48 hours before slaughter
Chlortetracycline plus sulfa-methazine (Aureo S 700)	350 mg of each/head/day	Feed for 28 days to maintain weight gains in presence of shipping fever	Discontinue use 7 days prior to slaughter
Decoquinate	22.7 mg/100 lb body weight	Control of coccidiosis	Do not feed to breeding animals
Ethylenediamine dihydriodide (EDDI)	50 mg/head/day in feed or salt continuously	Prevention of footrot and lumpy jaw	
	400–500 mg/head/day for 2–3 weeks	Treatment of footrot and lumpy jaw	Treat with caution because of animal susceptibility to iodides
Neomycin base (Neomycin sulfate)	70–140 gm/ton of complete feed	Treatment of bacterial enteritis	Discontinue treatment for at least 30 days prior to slaughter
Oxytetracycline (Terramycin)	0.1–0.5 mg/lb of body weight/day	Treatment of bacterial diarrhea	Withdraw 5 days before slaughter when fed at the rate of 2 gm/head or above
	0.5–2.0 gm/head/day	Prevention and treatment of shipping fever	
Poloxalene[b] (Bloat guard)	1.0–2.0 gm/100 lb body weight/day	Prevention of legume and wheat pasture bloat	
C. Compounds for use in controlling internal parasites			
Coumaphos (Baymix)	0.0002 lb/100 lb body weight for 6 days	Control of gastrointestinal roundworms	Do not feed to animals under 3 months of age or to sick or stressed animals or in conjunction with feeds containing phenothiazine
Levamisole hydrochloride (Tramisol)	Level varies with animal weight (see manufacturers recommendations)	Control of gastrointestinal worms and lungworms	Do not slaughter for 48 hours after treatment
Phenothiazine	Level varies with animal weight (see manufacturers recommendations)	Control of stomach worms	

(continued)

TABLE AI (*continued*)

Compound	Level approved for use	Purpose of usage	Precautions and restictions
Thiabendazole (Thibenzole)	3–5 gm/100 lb body weight	Control of gastrointestinal roundworms	Do not treat within 3 days of slaughter
D. Compounds for use in controlling external parasites			
Methoprene (Altosid)	0.005–0.01% per 100 lb of body weight per month	Control of horn flies	
Phenothiazine (Bar Fly Phenothiazine)	0.25 gm/100 lb body weight in feed, salt, or mineral fed continuously	Control of horn flies and face flies	
Rabon	70 mg/100 lb body weight/day	Control of face flies, horn flies, horse flies, and stable flies	
Ronnel (Trolene, MoorMan's Rid-Ezy)	Level varies with length of use (see manufacturers recommendations)	Control of grubs and horn flies	Withdraw 10 days prior to slaughter
E. Hormone and hormone-like compounds used for growth promotion			
Oral usage			
Melengestrol acetate[c] (MGA)	0.25–0.50 mg/head/day	Improved growth rate and feed efficiency and suppression of heat in heifers feed for slaughter	Discontinue use 48 hours prior to slaughter
Implants			
Ralgro Calves and feedlot cattle—birth to slaughter	36 mg (3 pellets)	Improved growth rate and feed efficiency	Must be implanted minimum of 65 days prior to slaughter
Synovex S and H Calves and feedlot cattle over 400 lb	220 mg (8 pellets)	Improved growth rate and feed efficiency	Must be implanted minimum of 60 days prior to slaughter

[a] Information on compounds other than implants abstracted from 1978 Feed Additive Compendium, Miller Publishing Company, Minneapolis, Minnesota.
[b] May also be used in combination with chlortetracycline or neomycin base.
[c] Can also be used in combination with the following antibiotics: Bacitracin MD, zinc bacitracin, chlortetracycline, and oxytetracycline.

Appendix II

How Much Can I Afford to Pay for Feeder Cattle?*

How much can you afford to pay for feeder cattle? It is always good business to carefully estimate the prospects for making a profit before buying feeder animals.

Although you can not predict accurately what the market price will be when you're ready to sell, you can determine the price necessary to recover the cost of the feeder animal plus feed and other costs of production. This "break-even" selling price compared to future cattle price predictions will help you decide whether you should buy a particular group of feeders or take alternative actions.

This publication provides data and techniques for arriving at productions costs and break-even selling prices for seven alternative cattle feeding systems. Table AII.1 describes the seven systems considered; Tables AII.2 and 3 present the necessary returns and costs for each system; and Tables AII.5 through 11 list the selling prices needed to cover the cost of the animals, feed and other variable costs for different combinations of corn and feeder cattle prices. Table 4 presents information for determining the price of corn silage for alternative prices of corn.

The seven feeding systems presented here are for illustration purposes only. Table AII.12 is a budget form which has been provided to help you estimate the production costs and the necessary selling prices for your particular feeding program.

*Appendix by David Petritz, Agricultural Economics Department, and Kern Hendrix, Animal Sciences Department, Cooperative Extension Service, Purdue University, West Lafayette, Indiana.

TABLE AII.1

Approximate Feed Requirements and Rates of Gain for Seven Alternative Cattle Feeding Programs (Average Farm Conditions)[a]

Item	Choice steer calves		Choice heifer calves		Choice yearling steers		Dairy steers fed to good
	High grain	High silage	High grain	High silage	High grain	High silage	High silage
System number	1	2	3	4	5	6	7
Purchase weight[b]	450 lb	450 lb	400 lb	400 lb	700 lb	700 lb	450 lb
Market weight[b]	1050 lb	1050 lb	900 lb	900 lb	1100 lb	1100 lb	1100 lb
Gain (pay weight to pay weight)	600 lb	600 lb	500 lb	500 lb	400 lb	400 lb	650 lb
Days on farm	250 days	270 days	230 days	250 days	150 days	170 days	270 days
Daily gain [c]	2.66 lb	2.46 lb	2.39 lb	2.20 lb	3.13 lb	2.76 lb	2.65 lb
Feed requirements[d]							
Corn	60 bu	34 bu	52 bu	29 bu	50 bu	27 bu	41 bu
Corn silage[e]	1.25 tons	3.5 tons	1.15 tons	3.1 tons	0.75 tons	2.7 tons	4.00 tons
Supplement (40%)	375 lb	405 lb	345 lb	375 lb	225 lb	255 lb	405 lb
Hay[f]	200 lb	200 lb	200 lb	200 lb	150 lb	150 lb	200 lb
Salt and mineral	25 lb	27 lb	23 lb	25 lb	20 lb	23 lb	35 lb

[a] With high silage programs, the first one-half of gain is put on with a full feed of corn silage and protein supplement and the last one-half of gain put on with a full feed of corn grain, limited silage (10–15 lb per day) plus protein supplement. With high grain programs, the total gain, once cattle are on feed, is put on with a full feed of corn grain, limited silage, and protein supplement. All cattle are assumed to be of average frame size with calves in moderate condition and yearlings below moderate condition when started on feed. Cattle of larger or smaller frame size would be slaughtered at weights 100 to 150 lb heavier or lighter, respectively, and days on feed and feed requirements should be adjusted accordingly. All cattle are assumed to be implanted at optimum intervals with a growth stimulant. If a growth stimulant is not used, reduce daily gains 10% and increase days on feed accordingly. If Rumensin is fed at recommended levels, grain and silage requirements may be reduced approximately 10%. If lot conditions are muddy, reduce daily gains 10–15% and increase feed requirements 10–15%. [b] Equal to pay weight. Beginning feedlot-weight is equal to pay weight minus 5.5% shrink for steer calves, heifer calves and dairy steers and 4.5% shrink for yearling steers. [c] Equal to payweight, which is final feedlot weight minus 3.5% shrink. [d] Daily gain necessary to achieve pay weight to pay weight gain and account for shrinkage of cattle. [e] Feed requirements given on as fed basis: corn grain, 87% dry matter; corn silage, 35% dry matter; hay and supplement, 90% dry matter. [f] Legume hay or haylage may be substituted into these rations. Five hundred pounds of legume hay or 1000 pounds of haylage (17% protein) plus 7 bushels of corn grain can replace 1 ton of corn silage and 125 pounds of protein supplement.

TABLE AII.2

Approximate Production Costs and Necessary Returns for Seven Alternative Cattle Feeding Programs

Item	Choice steer calves		Choice heifer calves		Choice yearling calves		Dairy steers fed to good
	High grain	High silage	High grain	High silage	High grain	High silage	High silage
System number	1	2	3	4	5	6	7
Market weight	1050 lb	1050 lb	900 lb	900 lb	1100 lb	1100 lb	1100 lb
A. Cost per head							
1 and 2 450 lb @ 70¢	$315.00	$315.00					
3 and 4 400 lb @ 60¢			$240.00	$240.00			
5 and 6 700 lb @ 62½¢					$437.50	$437.50	
7 450 lb @ 60¢							$270.00
B. Cost of gain per head							
Feed (corn, $2.10/bu; corn silage, $17.35/ton; supplement, $10/cwt; salt and mineral, $10/cwt; hay, $50/ton)	$192.70	$180.30	$170.95	$159.70	$146.25	$135.10	$204.50
Other variable costs (deaths, vet., interest, taxes, gas, oil, repairs, market costs)	$ 71.65	$ 73.35	$ 58.83	$ 60.10	$ 57.25	$ 59.80	$ 70.28
Fixed costs (labor, buildings, equipment, and machinery depreciation)[a]	$ 64.00	$ 64.00	$ 55.00	$ 55.00	$ 33.00	$ 33.00	$ 64.00
Total costs (A and B)	$643.35	$632.65	$524.78	$514.80	$674.00	$665.40	$608.78
C. Selling price per cwt to return							
Feed and other variable costs	$ 55.20	$ 54.15	$ 52.20	$ 51.10	$ 58.25	$ 57.50	$ 49.55
All costs (including fixed)	$ 61.25	$ 60.25	$ 58.30	$ 57.20	$ 61.25	$ 60.50	$ 55.35

[a] These values are adjusted for average age of buildings and equipment and are equal to 75% annual fixed costs for new facilities. Labor wage rate used was $4.00 per hour.

TABLE AII.3

Breakdown of Variable Costs for Seven Alternative Cattle Feeding Programs (Dollars per Head)

Item	Choice steer calves		Choice heifer calves		Choice yearling steers		Dairy steers fed to good
	High grain	High silage	High grain	High silage	High grain	High silage	High silage
Veterinary and medicine	$ 7.00	$ 7.00	$ 7.00	$ 7.00	$ 5.00	$ 5.00	$ 7.00
Death loss[a]	6.30	6.30	4.60	4.60	4.25	4.25	4.80
Marketing costs[b]	15.00	15.00	13.50	13.50	17.00	17.00	15.00
Power, fuel, repairs	8.00	8.00	7.00	7.00	5.50	5.50	8.00
Interest, insurance, taxes on feed and cattle[c]	33.36	35.05	24.73	26.00	23.50	26.05	33.48
Miscellaneous	2.00	2.00	2.00	2.00	2.00	2.00	2.00
Total variable costs	$71.66	$73.35	$58.83	$60.10	$57.25	$59.80	$70.28

[a] 2% of purchase value for calves and 1% for yearlings.

[b] Includes commission cost of purchasing feeder animal and hauling to farm, and commission cost of marketing fed animal and hauling to market.

[c] Equal to 10.0% (0.10) times the following: purchase price of animal plus value of corn fed plus value of corn silage fed plus value of hay fed plus one-half of the value of supplement, salt and mineral fed. This value is then multiplied by the fraction of the year that is in the feeding period.

TABLE AII.4

Approximate Value per Ton of Corn Silage at Various Dry Matter Contents, at Various Prices of Corn Grain

Average corn price (No. 2 basis)	Value per ton of corn silage when percentage dry matter of corn silage is[a] 30%	35%	40%
$1.70	$13.25	$14.80	$16.35
1.80	13.80	15.45	17.05
1.90	14.35	16.05	17.80
2.00	14.90	16.70	18.50
2.10	15.45	17.35	19.25
2.20	16.00	17.95	19.95
2.30	16.55	18.60	20.70
2.40	17.10	19.25	21.40
2.50	17.65	19.90	22.15
2.60	18.15	20.50	22.85
2.70	18.70	21.15	23.60
2.80	19.25	21.80	24.30
2.90	19.80	22.40	25.05
3.00	20.35	23.05	25.75
3.10	20.90	23.70	26.50
3.20	21.45	24.30	27.20
3.30	21.90	24.90	27.90

[a] Value per ton of corn silage is based on following amounts of corn grain for different dry matter contents: 5.45 bushels per ton at 30% dry matter; 6.35 bushels per ton at 35% dry matter; and 7.25 bushels at 40% dry matter.

TABLE AII.5

Necessary per-Hundredweight Selling Prices at Various Corn Prices for the Choice Steer Feeding Program on High Grain Ration (System #1) to Return Purchase Cost of Animal, Feed Cost, and Other Variable Costs

Average corn price (No. 2 basis)	Necessary per-cwt selling price[a] when per-cwt purchase price for choice steer calves is									
	$50.00	$55.00	$60.00	$65.00	$70.00	$75.00	$80.00	$85.00	$90.00	$95.00
$1.70	$43.16	$45.49	$47.82	$50.15	$52.48	$54.81	$57.14	$59.47	$61.80	$64.13
1.80	43.84	46.17	48.50	50.83	53.16	55.49	57.82	60.15	62.48	64.81
1.90	44.52	46.85	49.18	51.51	53.84	56.17	58.50	60.83	63.16	65.49
2.00	45.20	47.53	49.86	52.19	54.52	56.85	59.18	61.51	63.84	66.17
2.10	45.88	48.21	50.54	52.87	55.20	57.53	59.86	62.19	64.52	66.85
2.20	46.56	48.89	51.22	53.55	55.88	58.21	60.54	62.87	65.20	67.53
2.30	47.24	49.57	51.90	54.23	56.56	58.89	61.22	63.55	65.88	68.21
2.40	47.92	50.25	52.58	54.91	57.24	59.57	61.90	64.23	66.56	68.89
2.50	48.60	50.93	53.26	55.59	57.92	60.25	62.58	64.91	67.24	69.57
2.60	49.28	51.61	53.94	56.27	58.60	60.93	63.26	65.59	67.92	70.25
2.70	49.96	52.29	54.62	56.95	59.28	61.61	63.94	66.27	68.60	70.93
2.80	50.64	52.97	55.30	57.63	59.96	62.29	64.62	66.95	69.28	71.61
2.90	51.32	53.65	55.98	58.31	60.64	62.97	65.30	67.63	69.96	72.29
3.00	52.00	54.33	56.66	58.99	61.32	63.65	65.98	68.31	70.64	72.97
3.10	52.68	55.01	57.34	59.67	62.00	64.33	66.66	68.99	71.32	73.65
3.20	53.36	55.69	58.02	60.35	62.68	65.01	67.34	69.67	72.00	74.33
3.30	54.04	56.37	58.70	61.03	63.36	65.69	68.02	70.35	72.68	75.01

[a] Important: Based on $10/cwt for protein supplement. Necessary selling price is changed by 37¢/cwt for each $1 change in the price of 100 pounds of protein supplement. Add $6.10/cwt to necessary selling price to account for fixed costs. Fixed costs are those that go on whether cattle are fed or not. If there are no alternative uses for the fixed resources, you may find that you should ignore fixed costs when you bid on feeder cattle. In such a circumstance, you "sell" the *fixed* labor and buildings for less than $6.10/cwt rather than permit them to be wasted.

TABLE AII.6

Necessary per-Hundredweight Selling Prices at Various Corn Prices for the Choice Steer Calves Feeding Program on High Silage Ration (System #2) to Return Purchase Cost of Animal, Feed Cost, and Other Variable Costs

Average corn price (No. 2 basis)	Necessary per-cwt selling price[a] when per-cwt purchase price for choice steer calves is									
	$50.00	$55.00	$60.00	$65.00	$70.00	$75.00	$80.00	$85.00	$90.00	$95.00
$1.70	$42.63	$44.97	$47.31	$49.65	$51.99	$54.33	$56.67	$59.01	$61.35	$63.69
1.80	43.17	45.51	47.85	50.19	52.53	54.87	57.21	59.55	61.89	64.23
1.90	43.71	46.05	48.39	50.73	53.07	55.41	57.75	60.09	62.43	64.77
2.00	44.25	46.59	48.93	51.27	53.61	55.95	58.29	60.63	62.97	65.31
2.10	44.79	47.13	49.47	51.81	54.15	56.49	58.83	61.17	63.51	65.85
2.20	45.33	47.67	50.01	52.35	54.69	57.03	59.37	61.71	64.05	66.39
2.30	45.87	48.21	50.55	52.89	55.23	57.57	59.91	62.25	64.59	66.93
2.40	46.41	48.75	51.09	53.43	55.77	58.11	60.45	62.79	65.13	67.47
2.50	46.95	49.29	51.63	53.97	56.31	58.65	60.99	63.33	65.67	68.01
2.60	47.49	49.83	52.17	54.51	56.85	59.19	61.53	63.87	66.21	68.55
2.70	48.03	50.37	52.71	55.05	57.39	59.73	62.07	64.41	66.75	69.09
2.80	48.57	50.91	53.25	55.59	57.93	60.27	62.61	64.95	67.29	69.63
2.90	49.11	51.45	53.79	56.13	58.47	60.81	63.15	65.49	67.83	70.17
3.00	49.65	51.99	54.33	56.67	59.01	61.35	63.69	66.03	68.37	70.71
3.10	50.19	52.53	54.87	57.21	59.55	61.89	64.23	66.57	68.91	71.25
3.20	50.73	53.07	55.41	57.75	60.09	62.43	64.77	67.11	69.45	71.79
3.30	51.27	53.61	55.95	58.29	60.63	62.97	65.31	67.65	69.99	72.33

[a] Important: Based on $10/100 pounds for protein supplement. Necessary selling price is changed by 40¢/cwt for each $1 change in the price of 100 pounds protein supplement. Add $6.10/cwt to necessary selling price to account for fixed costs. Fixed costs are those that go on whether cattle are fed or not. If there are no alternative uses for the fixed resources, you may find that you should ignore fixed costs when you bid on feeder cattle. In such a circumstance, you "sell" the *fixed* labor and buildings for less than $6.10/cwt rather than permit them to be wasted.

TABLE AII.7

Necessary per-Hundredweight Selling Prices at Various Corn Prices for the Choice Heifer Feeding Program on High Grain Ration (System #3) to Return Purchase Cost of Animal, Feed Cost, and Other Variable Costs

	Necessary per-cwt selling price[a] when per-cwt purchase price for choice heifer calves is									
Average corn price (No. 2 basis)	$45.00	$50.00	$55.00	$60.00	$65.00	$70.00	$75.00	$80.00	$85.00	$90.00
$1.70	$42.21	$44.62	$47.03	$49.44	$51.85	$54.26	$56.67	$59.08	$61.49	$63.90
1.80	42.90	45.31	47.72	50.13	52.54	54.95	57.36	59.77	62.18	64.59
1.90	43.59	46.00	48.41	50.82	53.23	55.64	58.05	60.46	62.87	65.28
2.00	44.28	46.69	49.10	51.51	53.92	56.33	58.74	61.15	63.56	65.97
2.10	44.97	47.38	49.79	52.20	54.61	57.02	59.43	61.84	64.25	66.66
2.20	45.66	48.07	50.48	52.89	55.30	57.71	60.12	62.53	64.94	67.35
2.30	46.35	48.76	51.17	53.58	55.99	58.40	60.81	63.22	65.63	68.04
2.40	47.04	49.45	51.86	54.27	56.68	59.09	61.50	63.91	66.32	68.73
2.50	47.73	50.14	52.55	54.96	57.37	59.78	62.19	64.60	67.01	69.42
2.60	48.42	50.83	53.24	55.65	58.06	60.47	62.88	65.29	67.70	70.11
2.70	49.11	51.52	53.93	56.34	58.75	61.16	63.57	65.98	68.39	70.80
2.80	49.80	52.21	54.62	57.03	59.44	61.85	64.26	66.67	69.08	71.49
2.90	50.49	52.90	55.31	57.72	60.13	62.54	64.95	67.36	69.77	72.18
3.00	51.18	53.59	56.00	58.41	60.82	63.23	65.64	68.05	70.46	72.87
3.10	51.87	54.28	56.69	59.10	61.51	63.92	66.33	68.74	71.15	73.56
3.20	52.56	54.97	57.38	59.79	62.20	64.61	67.02	69.43	71.84	74.25
3.30	53.25	55.66	58.07	60.48	62.89	65.30	67.71	70.12	72.53	74.94

[a] Important: Based on $10/100 pounds for protein supplement. Necessary selling price is changed by 40¢/cwt for each $1 change in the price of 100 pounds protein supplement. Add $6.11/cwt to necessary selling price to account for fixed costs. Fixed costs are those that go on whether cattle are fed or not. If there are no alternative uses for fixed resources, you may find that you should ignore fixed costs when you bid on feeder cattle. In such a circumstance, you "sell" the *fixed* labor and buildings for less than $6.11/cwt rather than permit them to be wasted.

TABLE AII.8

Necessary per-Hundredweight Selling Prices at Various Corn Prices for the Choice Heifer Feeding Program on High Silage Ration (System #4) to Return Purchase Cost of Animal, Feed Cost, and Other Variable Costs

Average corn price (No. 2 basis)	Necessary per-cwt selling price[a] when per-cwt purchase price for choice heifer calves is									
	$45.00	$50.00	$55.00	$60.00	$65.00	$70.00	$75.00	$80.00	$85.00	$90.00
$1.70	$41.64	$44.06	$46.48	$48.90	$51.34	$53.78	$56.22	$58.66	$61.10	$63.65
1.80	42.19	44.61	47.03	49.45	51.89	54.33	56.77	59.21	61.65	64.20
1.90	42.74	45.16	47.58	50.00	52.44	54.88	57.32	59.76	62.20	64.75
2.00	43.29	45.71	48.13	50.55	52.99	55.43	57.87	60.31	62.75	65.30
2.10	43.84	42.26	48.68	51.10	53.54	55.98	58.42	60.86	63.30	65.85
2.20	44.39	46.81	49.23	51.65	54.09	56.53	58.97	61.41	63.85	66.40
2.30	44.94	47.36	49.78	52.20	54.64	57.08	59.52	61.96	64.40	66.95
2.40	45.49	47.91	50.33	52.75	55.19	57.63	60.07	62.51	64.95	67.50
2.50	46.04	48.46	50.88	53.30	55.74	58.18	60.62	63.06	65.50	68.05
2.60	46.59	49.01	51.43	53.85	56.29	58.73	61.17	63.61	66.05	68.60
2.70	47.14	49.56	51.98	54.40	56.84	59.28	61.72	64.16	66.60	69.15
2.80	47.69	50.11	52.53	54.95	57.39	59.83	62.27	64.71	67.15	69.70
2.90	48.24	50.66	53.08	55.50	57.94	60.38	62.82	65.26	67.70	70.25
3.00	48.79	51.21	53.63	56.05	58.49	60.93	63.37	65.81	68.25	70.80
3.10	49.34	51.76	54.18	56.60	59.04	61.48	63.92	66.36	68.80	71.35
3.20	49.89	52.31	54.73	57.15	59.59	62.03	64.47	66.91	69.35	71.90
3.30	50.44	52.86	55.28	57.70	60.14	62.58	65.02	67.46	69.90	72.45

[a] Important: Based on $10/100 pounds for protein supplement. Necessary selling price is changed by 43¢/cwt for each $1 change in the price of 100 pounds protein supplement. Add $6.11/cwt to necessary selling price to account for fixed costs. Fixed costs are those that go on whether cattle are fed or not. If there are no alternative uses for the fixed resources, you may find that you should ignore fixed costs when you bid on feeder cattle. In such a circumstance, you "sell" the *fixed* labor and buildings for less than $6.11/cwt rather than permit them to be wasted.

TABLE AII.9

Necessary per-Hundredweight Selling Prices at Various Corn Prices for the Choice Yearling Steer Feeding Program on High Grain Ration (System #5) to Return Purchase Cost of Animal, Feed Cost, and Other Variable Costs

Average corn price (No. 2 basis)	Necessary per-cwt selling price[a] when per-cwt purchase price for choice yearling steers is									
	$57.50	$60.00	$62.50	$65.00	$67.50	$70.00	$72.50	$75.00	$77.50	$80.00
$1.70	$52.87	$54.54	$56.21	$57.88	$59.55	$61.22	$62.89	$64.56	$66.23	$67.90
1.80	53.38	55.05	56.72	58.39	60.06	61.73	63.40	65.07	66.74	68.41
1.90	53.89	55.56	57.23	58.90	60.57	62.24	63.91	65.58	67.25	68.92
2.00	54.40	56.07	57.74	59.41	61.08	62.75	64.42	66.09	67.76	69.43
2.10	54.91	56.58	58.25	59.92	61.59	63.26	64.93	66.60	68.27	69.94
2.20	55.42	57.09	58.76	60.43	62.10	63.77	65.44	67.11	68.78	70.45
2.30	55.93	57.60	59.27	60.94	62.61	64.28	65.95	67.62	69.29	70.96
2.40	56.43	58.11	59.78	61.45	63.12	64.79	66.46	68.13	69.80	71.47
2.50	56.95	58.62	60.29	61.96	63.63	65.30	66.97	68.64	70.31	71.98
2.60	57.46	59.13	60.80	62.47	64.14	65.81	67.48	69.15	70.82	72.49
2.70	57.97	59.64	61.31	62.98	64.65	66.32	67.99	69.66	71.33	73.00
2.80	58.48	60.15	61.82	63.49	65.16	66.83	68.50	70.17	71.84	73.51
2.90	58.99	60.66	62.33	64.00	65.67	67.34	69.01	70.68	72.35	74.02
3.00	59.50	61.17	62.84	64.51	66.18	67.85	69.52	71.19	72.86	74.53
3.10	60.01	61.68	63.35	65.02	66.69	68.36	70.03	71.70	73.37	75.04
3.20	60.52	62.19	63.86	65.53	67.20	68.87	70.54	72.21	73.88	75.55
3.30	61.03	62.70	64.37	66.04	67.71	69.38	71.05	72.72	74.39	76.06

[a] Important: Based on $10/100 pounds for protein supplement. Necessary selling price is changed by 21¢ cwt for each $1 change in tne price of 100 pounds protein supplement. Add $3.00/cwt to necessary selling price to account for fixed costs. Fixed costs are those that go on whether cattle are fed or not. If there are no alternative uses for the fixed resources, you may find that you should ignore fixed costs when you bid on feeder cattle. In such a circumstance, you "sell" the *fixed* labor and buildings for less than $3.00/cwt rather than permit them to be wasted.

TABLE AII.10

Necessary per-Hundredweight Selling Prices at Various Corn Prices for the Choice Yearling Steer Feeding Program on High Silage Ration (System #6) to Return Purchase Cost of Animal, Feed Cost, and Other Variable Costs

Average corn price (No. 2 basis)	Necessary per-cwt selling price[a] when per-cwt purchase price for choice yearling steers is									
	$57.50	$60.00	$62.50	$65.00	$67.50	$70.00	$72.50	$75.00	$77.50	$80.00
$1.70	$52.54	$54.22	$55.90	$57.58	$59.26	$60.94	$62.62	$64.30	$65.98	$67.66
1.80	52.94	54.62	56.30	57.98	59.66	61.34	63.02	64.70	66.38	68.06
1.90	53.34	55.02	56.70	58.38	60.06	61.74	63.42	65.10	66.78	68.46
2.00	53.74	55.42	57.10	58.78	60.46	62.14	63.82	65.50	67.18	68.86
2.10	54.14	55.82	57.50	59.18	60.86	62.54	64.22	65.90	67.58	69.26
2.20	54.54	56.22	57.90	59.58	61.26	62.94	64.62	66.30	67.98	69.66
2.30	54.94	56.62	58.30	59.98	61.66	63.34	65.02	66.70	68.38	70.06
2.40	55.34	57.02	58.70	60.38	62.06	63.74	65.42	67.10	68.78	70.46
2.50	55.74	57.42	59.10	60.78	62.46	64.14	65.82	67.50	69.18	70.86
2.60	56.14	57.82	59.50	61.18	62.86	64.54	66.22	67.90	69.58	71.26
2.70	56.54	58.22	59.90	61.58	63.26	64.94	66.62	68.30	69.98	71.66
2.80	56.94	58.62	60.30	61.98	63.66	65.34	67.02	68.70	70.38	72.06
2.90	57.34	59.02	60.70	62.38	64.06	65.74	67.42	69.10	70.78	72.46
3.00	57.74	59.42	61.10	62.78	64.46	66.14	67.82	69.50	71.18	72.86
3.10	58.14	59.82	61.50	63.18	64.86	66.54	68.22	69.90	71.58	73.26
3.20	58.54	60.22	61.90	63.58	65.26	66.94	68.62	70.30	71.98	73.66
3.30	58.94	60.62	62.30	63.98	65.66	67.34	69.02	70.70	72.38	74.06

[a] Important: Based on $10/100 pounds for protein supplement. Necessary selling price is changed by 24¢/cwt for each $1 change in the price of 100 pounds protein supplement. Add $3.00/cwt to necessary selling price to account for fixed costs. Fixed costs are those that go on whether cattle are fed or not. If there are no alternative uses for the fixed resources, you may find that you should ignore fixed costs when you bid on feeder cattle. In such a circumstance, you "sell" the *fixed* labor and buildings for less than $3.00/cwt rather than permit them to be wasted.

TABLE AII.11

Necessary per-Hundredweight Selling Prices at Various Corn Prices for the Dairy Steer Calf Feeding Program on High Silage Rations (System #7) to Return Purchase Cost of Animal, Feed Cost, and Other Variable Costs

Average corn price (No. 2 basis)	Necessary per-cwt selling price[a] when per cwt purchase price for dairy steer calves is									
	$40.00	$45.00	$50.00	$55.00	$60.00	$65.00	$70.00	$75.00	$80.00	$85.00
$1.70	38.11	40.35	42.59	44.83	47.07	49.31	51.55	53.79	56.03	58.27
1.80	38.73	40.97	43.21	45.45	47.69	49.93	52.17	54.41	56.65	58.89
1.90	39.35	41.59	43.83	46.07	48.31	50.55	52.79	55.03	57.27	59.51
2.00	39.97	42.21	44.45	46.69	48.93	51.17	53.41	55.65	57.89	60.13
2.10	40.59	42.83	45.07	47.31	49.55	51.79	54.03	56.27	58.51	60.75
2.20	41.21	43.45	45.69	47.93	50.17	52.41	54.65	56.89	59.13	61.37
2.30	41.83	44.07	46.31	48.55	50.79	53.03	55.27	57.51	59.75	61.99
2.40	42.45	44.69	46.93	49.17	51.41	53.65	55.89	58.13	60.37	62.61
2.50	43.07	45.31	47.55	49.79	52.03	54.27	56.51	58.75	60.99	63.23
2.60	43.69	45.93	48.17	50.41	52.65	54.89	57.13	59.37	61.61	63.85
2.70	44.31	46.55	48.79	51.03	53.27	55.51	57.75	59.99	62.23	64.47
2.80	44.93	47.17	49.41	51.65	53.89	56.13	58.37	60.61	62.85	65.09
2.90	45.55	47.79	50.03	52.27	54.51	56.75	58.99	61.23	63.47	65.71
3.00	46.17	48.41	50.65	52.89	55.13	57.37	59.61	61.85	64.09	66.33
3.10	46.79	49.03	51.27	53.51	55.75	57.99	60.23	62.47	64.71	66.95
3.20	47.41	49.65	51.89	54.13	56.37	58.61	60.85	63.09	65.33	67.57
3.30	48.03	50.27	52.51	54.75	56.99	59.23	61.47	63.71	65.95	68.19

[a] Important: Based on $10/100 pounds for protein supplement. Necessary selling price is changed by 38¢/cwt for each $1 change in the price of 100 pounds protein supplement. Add $5.82/cwt to necessary selling price to account for fixed costs. Fixed costs are those that go on whether cattle are fed or not. If there are no alternative uses for the fixed resources, you may find that you should ignore fixed costs when you bid on feeder cattle. In such a circumstance, you "sell" the *fixed* labor and buildings for less than $5.82/cwt rather than permit them to be wasted.

TABLE AII.12

Budget Form for Estimating Your Own Break-Even Selling Price for Cattle

	Costs per head		Costs per head	
1. Purchase price ______ lb @ ______ /lb =		______		______
2. Feed cost				
a. Corn ______ bu @ ______ /bu =	______		______	
b. Corn silage ______ tons @ ______ /ton =	______		______	
c. Supplement ______ lb @ ______ /lb =	______		______	
d. Other ______ @ ______ /	______		______	
e. Other ______ @ ______ /	______		______	
Total feed cost		______		______
3. Other variable costs				
a. Veterinary and medicine	______		______	
b. Death loss (2% of Line 1 for calves, 1% for yearlings)	______		______	
c. Marketing, commission charges, trucking expense[a]	______		______	
d. Power, fuel, equipment repair[b]	______		______	
e. Interest, insurance, taxes on feed and cattle[c]	______		______	
f. Miscellaneous	______		______	
Total variable cost		______		______
4. Fixed costs				
a. Buildings and equipment[d]	______		______	
b. Labor ______ hr @ ______ /hr[e]	______		______	
Total fixed cost		______		______

(*continued*)

TABLE AII.12 (*continued*)

	Costs per head		Costs per head	
5. Total of all costs (1 + 2 + 3 + 4)				______
6. Necessary selling price to return	Total cost per head	Necessary selling price per cwt[f]	Total cost per head	Necessary selling price per cwt[f]
	______	______	______	______
Cost of feeder and feed costs (1 + 2)	______	______	______	______
All variable costs (1 + 2 + 3)	______	______	______	______
All costs (1 + 2 + 3 + 4)	______	______	______	______

[a] Include commission cost of purchasing feeder animal and hauling to farm, and commission cost of marketing fed animal and hauling to market.

[b] Power, fuel and equipment repair costs are estimated to be 5–$8 per head.

[c] Equal to 10.0% (0.10) times the following: purchase price of steer plus value of corn fed plus value of corn silage fed plus one-half of the value of supplement, salt, and mineral fed. This value is then multiplied by the fraction of the year that is in the feeding period.

[d] Equal to 14% of the current investment in shelter, silage storage, and feed handling and other equipment associated with the cattle feeding enterprise. It is assumed that storage for corn is included in the price of corn.

[e] Estimated to be 3–6 hours per head typically with long-fed cattle requiring the larger amount.

[f] Divide costs per head by market weight (pay weight) per head.

Appendix III

Some Current Specifications for Beef Cattle Equipment

Housed resting space, (Ft^2/Head)
  Mature cows: 25 to 30
  Calves to 600 lb: 15 to 20
  Feeders (600 lb to market): 20 to 25
Shade
  20 to 25 ft^2 per head
  10 ft minimum height
  North–South orientation
  Provided over bunk, in open shed, or under separate shades
Additional space

Total recommended feedlot space usually exceeds the total required for feeding, resting, and shade. Year around access must be provided to the above three areas and the working area. Total additional space depends upon drainage and snow drifting (whether some space will not be available for portions of the year) tradition, etc. The following figures should be compared with local conditions and experience. The approximate additional space (add up to 50% for wet or severe climates) ft^2/head is shown below in the tabulation:

	Shed or loafing	Paved feeding	Yard
Cows	25–40	—	150–250
Yearlings	20–30	20–30	100–175
Calves	15–25	15–25	75–125

Topography

Uniform 4–6% slope away from winter prevailing winds, usually south or east. Slopes under 4% need to be paved. Slopes over 10% may erode in unpaved lots. A slight north or west slope can usually be graded to a 4% south or east slope. Consider moving from present lot site to obtain desirable slope.

Drainage

Complete land grading before construction starts. Slope away from buildings and feeding lines. Use surfaced roads and diversions or terraces as drainage ways to intercept uphill water. Provide roof gutters and drains on sheds. Crown unsurfaced roads.

Lot Pavement

Minimum paving: Heavy traffic areas, around waterers, and along feed lines. Additional pavement may pay for itself if cattle would otherwise be in mud for extended periods. Generally, concrete has been the most satisfactory paving material. Wood float or broom finish to prevent slipping, on 4″ concrete over tampe or undisturbed soil, or gravel fill. Pavement along bunks and around waterers: at least 5–7′ wide, with 1/2″ per foot slope minimum. Apron will be nearly self-cleaning with slope of 1″ per foot.

Feed bunk design—beef cattle

Apron

1/2″ per foot slope, minimum

1″ per foot slope will be nearly self-cleaning

At least 5′ wide

6′ to 8′ wide with step, with paved space for passage of animals and tractor–scraper

12′ wide if area below will be muddy or drifted with snow part of the year

Bunk dimensions (excluding fence line bunks)

Widths: 48″ of eating space if fed from both sides
if bunk is divided by mechanical feeder, 54″ to 60″

Height: of bunk floor above apron: 4″ to 6″ where apron can be kept scraped; 12″ to 14″ if frozen mud, snow, etc. will accumulate

Throat Height: Up to 18″ for calves, 22″ for feeders and mature cows
Increase to 30″ only if hogs will run with cattle

Roof

Narrow roof 5–6′ above apron for protection of bunk and minimum shade. Wide roof high enough to clear cleaning equipment provides shelter for bunk and cattle and summer shade. Wide roof may cause additional snow drifting and prevent thawing in northern regions.

Feeding space

Feeding space is determined by the size of the animal and the number of animals which must eat or drink at one time.

18–22″—calves to 600 lb

22–26″—600 lb to market

26–30″—mature cows

For feeds always available

4–6″—hay or silage

3–4″—grain or supplement

6″—grain and silage

Watering: 40 head per available water space in drylot, 3–4″ water space per head on pasture.

Working corrals

Holding pen, 20 ft^2 per head

Crowding pen, 15 ft^2 per head

150 ft^2 total minimum

Working alley, 10–12 ft wide

18–30 ft long

Loading chute, 30–42 inches wide

Expansion

Plan for expansion. Allow for additional building and lot space, bunk length, and storages. Feed processing and handling facilities should be adaptable to a large enterprise.

Shadows

In freezing climates, avoid continuous shadows in lots. Frozen manure and snow accumulate on the north side of buildings, silos, and bunks.

Buildings

Shed open one side should have the long, open side facing away from prevailing winter winds. The closed side should have doors or other provisions for summer ventilation. Floors should be at least 6″ above outside grade if floor is paved for feeding, at least 12″ if unpaved, and should be sloped to the front.

Storage Processing

Plan for expansion. Locate to avoid shadows in lot, convenient to feeding area, and away from cattle traffic patterns. Provide all-weather access. Avoid interfering with summer breezes. Consider snow drifting.

Feeding lines

Orient up and down slope or provide drainage diversion. To avoid snow drifting and give sun exposure on both sides, orient lines N–S, or NE–SW, or NW–SE. On east or west slopes, orient bunks for equal exposure (approx. N–S). In severe climates, bunk up and down slope in moderate climates.

Manure

Provide temporary stockpile area unless removal can be done at the same time as cleaning. Avoid excess drainage water, but place at lower end of slope.

Appendix IV

Nutrient Requirements of Beef Cattle

TABLE AIV.1

Nutrient Requirements for Growing–Finishing Steer Calves and Yearlings (Daily Nutrients per Animal)[a]

Weight (lb)	Daily gain (lb)	Minimum dry matter consumption (lb)	Roughage (%)	Total protein (lb)	TDN (lb)	Ca (gm)	P (gm)	Vitamin A (thousands) (IU)
200	0	4.6	100	0.22	2.6	4	4	5
	1.1	6.4	70–80	0.53	4.0	14	11	6
	1.5	6.0	50–60	0.88	4.4	19	13	6
	2.0	6.2	25–30	1.02	4.6	24	16	7
	2.4	6.0	15	1.08	5.1	28	19	7
331	0	6.2	100	0.51	3.5	5	5	6
	1.1	8.8	70–80	0.97	5.5	14	12	9
	1.5	8.6	50–60	1.08	6.0	18	14	9
	2.0	8.4	25–30	1.19	6.6	23	17	9
	2.4	8.2	15	1.28	6.8	28	20	9
441	0	7.7	100	0.66	4.2	6	6	8
	1.1	12.8	80–90	1.26	7.5	14	13	12
	1.5	12.6	70–80	1.34	7.9	18	16	13
	2.0	10.8	35–45	1.34	8.2	23	18	13
	2.4	10.1	15	1.39	8.6	27	20	13
551	0	9.7	100	0.77	5.1	8	8	9
	1.5	12.8	55–65	1.37	8.8	18	16	14
	2.0	13.7	45–50	1.52	9.9	22	19	14
	2.4	13.2	20–25	1.61	10.4	26	21	14
	2.9	13.2	15	1.67	11.5	30	23	14
661	0	10.4	100	0.88	5.7	9	9	10
	2.0	17.9	55–65	1.78	11.9	22	19	16
	2.4	16.8	20–25	1.81	12.3	25	22	16
	2.9	15.6	15	1.83	13.2	29	23	16
	3.1	16.1	15	1.92	13.7	31	25	16
772	0	11.7	100	1.01	6.4	10	10	12
	2.0	17.6	45–55	1.76	12.8	20	18	18
	2.4	17.6	20–25	1.83	13.7	23	20	18
	2.9	17.6	15	1.92	15.0	26	22	18
	3.1	18.1	15	1.98	15.4	28	24	18
882	0	13.0	100	1.12	7.3	11	11	13
	2.2	20.7	45–55	1.92	15.0	21	20	19
	2.6	18.7	20–25	1.92	15.4	23	21	19
	2.9	19.0	15	1.98	16.1	25	22	19
	3.1	19.8	15	2.07	17.0	26	23	19
992	0	14.1	100	1.19	—	12	12	14
	2.2	22.7	45–55	2.11	16.3	20	20	20
	2.6	22.5	20–25	2.14	17.4	23	22	20
	2.9	20.5	15	2.14	17.6	24	23	20
	3.1	21.6	15	2.16	18.5	25	23	20
1102	0	15.4	100	1.32	8.4	13	13	15
	2.0	23.1	45–55	2.09	16.5	19	19	23
	2.4	22.9	20–25	2.11	17.8	20	20	23
	2.6	21.2	15	2.11	18.1	21	21	23
	2.9	22.0	15	2.14	19.2	22	22	23

[a] Modified from "Nutrient Requirements of Beef Cattle," 5th rev. ed., National Academy of Sciences, Washington, D.C., 1976, Table 1. With permission.

TABLE AIV.2

Nutrient Requirements for Growing–Finishing Heifer Calves and Yearlings (Daily Nutrients per Animal)[a]

Weight (lb)	Daily gain (lb)	Minimum dry matter consumption (lb)	Roughage (%)	Total protein (lb)	TDN (lb)	Ca (gm)	P (gm)	Vitamin A (thousands) (IU)
220	0	4.6	100	0.40	2.6	4	4	5
	1.1	6.6	70–80	0.81	4.2	14	11	6
	1.5	6.4	50–60	0.92	4.6	19	14	6
	2.0	6.6	25–30	0.75	5.1	24	17	7
	2.4	6.6	<15	1.17	5.5	29	19	7
331	0	6.2	100	0.53	3.5	5	5	6
	1.1	9.0	70–80	0.99	5.7	14	12	9
	1.5	8.8	50–60	1.10	6.2	18	14	9
	2.0	8.8	25–30	1.19	6.8	23	17	9
	2.4	8.8	<15	1.32	7.5	28	20	9
441	0	7.7	100	0.66	4.2	6	6	8
	0.7	11.9	100	1.08	6.6	10	10	12
	1.1	13.2	80–90	1.28	7.7	14	13	13
	1.5	13.2	70–80	1.34	8.4	18	16	13
	2.0	11.7	35–45	1.37	8.8	22	17	13
	2.4	11.0	<15	1.41	9.5	25	19	13
551	0	9.0	100	0.77	5.1	7	7	9
	0.7	14.1	100	1.26	7.8	12	12	14
	1.1	14.3	80–90	1.36	8.6	13	13	14
	1.5	12.8	55–65	1.36	9.1	17	15	14
	2.0	13.0	35–45	1.43	10.1	21	17	14
	2.4	14.3	20–25	1.63	11.5	25	20	14
	2.6	13.9	<15	1.65	11.9	27	21	14
661	0	10.4	100	0.88	5.7	9	9	10
	0.7	16.3	100	1.39	8.4	13	13	16
	1.1	16.3	80–90	1.48	9.9	14	14	16
	1.5	14.6	55–65	1.48	10.4	16	15	16
	2.0	15.0	35–45	1.54	11.5	19	17	16
	2.4	16.5	20–25	1.72	13.2	23	20	16
	2.6	15.9	<15	1.74	13.7	24	20	16
772	0	11.7	100	1.01	6.4	10	10	12
	0.7	18.1	100	1.52	10.0	15	15	18
	1.1	18.3	80–90	1.61	11.2	15	15	18
	1.5	17.4	55–65	1.61	11.9	15	15	18
	2.0	17.9	35–45	1.70	13.2	17	17	18
	2.4	18.3	20–25	1.78	14.5	20	19	18
	2.6	17.9	<15	1.78	15.2	21	20	18
882	0	13.0	100	1.12	7.3	11	11	13
	0.7	20.0	100	1.67	11.1	16	16	19
	1.1	18.7	70–80	1.72	11.9	15	15	19
	1.5	19.2	55–65	1.74	13.2	16	16	19
	2.0	18.5	20–25	1.74	14.3	17	17	19
	2.4	18.3	15	1.78	15.9	19	18	19
992	0	14.1	100	1.21	7.9	12	12	14
	0.4	19.2	100	1.63	10.6	16	16	19
	1.1	20.5	70–80	1.76	13.0	17	17	20
	1.8	20.1	35–45	1.81	15.0	16	16	20
	2.2	18.7	15	1.83	16.3	19	19	20

[a] Modified from "Nutrient Requirements of Beef Cattle," 5th rev. ed., National Academy of Sciences, Washington, D.C., 1976, Table 1A. With permission.

TABLE AIV.3

Nutrient Requirements for Beef Cattle Breeding Herd (Daily Nutrients per Animal)[a]

Weight (lb)	Daily gain (lb)	Minimum dry matter consumption (lb)	Roughage (%)	Total protein (lb)	TDN (lb)	Ca (gm)	P (gm)	Vitamin A (thousands) (IU)
Pregnant yearling heifers—last 3–4 months of pregnancy								
716	0.9	14.5	100	1.28	7.7	15	15	19
	1.3	18.7	100	1.65	9.9	18	18	23
	1.8	20.7	85–100	1.87	12.3	22	20	26
772	0.9	15.2	100	1.34	8.1	15	15	19
	1.3	19.6	100	1.72	10.3	19	19	25
	1.8	22.0	85–100	1.94	12.9	22	21	28
827	0.9	15.9	100	1.39	8.4	15	15	20
	1.3	20.5	100	1.78	10.8	19	19	26
	1.8	24.2	85–100	2.11	13.5	22	22	31
882	0.9	16.5	100	1.43	8.7	16	16	21
	1.3	21.4	100	1.85	11.3	19	19	27
	1.8	25.6	85–100	2.22	14.0	22	22	33
937	0.9	17.2	100	1.52	9.0	16	16	22
	1.3	22.3	100	1.94	11.7	19	19	28
	1.8	26.7	85–100	2.31	14.6	22	22	34
Dry pregnant mature cows—middle third of pregnancy								
772		12.2	100	0.70	6.6	10	10	15
882		13.4	100	0.79	7.3	11	11	17
992		14.8	100	0.86	7.9	12	12	19
1102		15.9	100	0.93	8.6	13	13	20
1213		17.0	100	0.99	9.2	14	14	22
1323		18.3	100	1.08	9.8	15	15	23
1433		19.4	100	1.14	10.4	16	16	25
Dry pregnant mature cows—last third of pregnancy								
772	0.9	13.9	100	0.42	8.0	12	12	19
882	0.9	15.4	100	0.97	8.7	14	14	21
992	0.9	16.5	100	1.06	9.4	15	15	23
1102	0.9	17.9	100	1.12	10.0	15	15	24
1213	0.9	19.0	100	1.19	10.7	16	16	26
1323	0.9	20.3	100	1.26	11.2	17	17	27
1433	0.9	22.4	100	1.32	11.9	18	18	29
Cows nursing calves—average milking ability, first 3–4 months postpartum								
772		18.1	100	1.65	9.7	24	24	19
882		19.4	100	1.78	10.4	25	25	21
992		20.5	100	1.89	11.0	26	26	23
1102		21.6	100	1.98	11.7	27	27	24
1213		23.1	100	2.14	12.3	28	28	26
1323		24.2	100	2.22	13.0	28	28	27
1433		25.1	100	2.31	13.7	29	29	29

TABLE AIV.3 (*continued*)

Weight (lb)	Daily gain (lb)	Minimum dry matter consumption (lb)	Roughage (%)	Total protein (lb)	TDN (lb)	Ca (gm)	P (gm)	Vitamin A (thousands) (IU)
Cows nursing calves—superior milking ability, first 3–4 months postpartum								
772		22.4	100	2.44	12.8	45	40	32
882		23.8	100	2.58	13.5	45	41	34
992		24.9	100	2.71	14.1	45	42	36
1102		26.0	100	2.84	14.8	46	43	38
1213		27.3	100	2.97	15.4	46	44	41
1323		28.4	100	3.10	16.1	46	44	43
1433		29.5	100	3.22	16.8	47	45	45
Bulls, growth and maintenance (moderate activity)								
661	2.2	19.4	70–75	1.98	12.3	27	23	34
882	2.0	24.2	70–75	2.27	15.4	23	23	43
1102	1.5	26.9	80–85	2.35	16.5	22	22	48
1323	1.1	26.4	80–85	2.25	16.1	22	22	48
1543	0.7	28.4	90–100	2.38	17.0	23	23	50
1764	0	23.1	100	1.96	12.8	19	19	41
1984	0	25.1	100	2.18	13.9	21	21	44
2205	0	27.3	100	2.31	15.2	22	22	48

[a] Modified from "Nutrient Requirements of Beef Cattle," 5th rev. ed., National Academy of Sciences, Washington, D.C., 1976, Table 1A. With permission.

TABLE AIV.4

Nutrient Requirements for Growing–Finishing Steer Calves and Yearlings (Nutrient Concentration in Diet Dry Matter)[a,b]

Weight (lb)	Daily gain (lb)	Minimum dry matter consumption (lb)	Roughage (%)	Total protein (%)	TDN (%)	Ca (%)	P (%)
220	0	4.6	100	8.7	55	0.18	0.18
	1.1	6.4	70–80	12.4	62	0.48	0.38
	1.5	6.0	50–60	14.8	70	0.70	0.48
	2.0	6.2	25–30	16.4	77	0.86	0.57
	2.4	6.0	<15	18.2	86	1.04	0.70
331	0	6.2	100	8.7	55	0.18	0.18
	1.1	8.8	70–80	11.0	62	0.35	0.32
	1.5	8.6	50–60	12.6	70	0.46	0.36
	2.0	8.4	25–30	14.1	77	0.61	0.45
	2.4	8.2	<15	15.6	86	0.76	0.54
441	0	7.7	100	8.5	55	0.18	0.18
	1.1	12.8	80–90	9.9	58	0.24	0.22
	1.5	12.6	70–80	10.8	64	0.32	0.28
	2.0	10.8	35–45	12.3	75	0.47	0.37
	2.4	10.1	<15	13.6	86	0.59	0.43
551	0	9.7	100	8.5	55	0.18	0.18
	1.5	12.8	55–65	10.7	70	0.31	0.28
	2.0	13.7	45–50	11.1	72	0.35	0.31
	2.4	13.2	20–25	12.1	77	0.43	0.35
	2.9	13.2	<15	12.7	86	0.50	0.38
661	0	10.4	100	8.6	55	0.18	0.18
	2.0	17.9	55–65	10.0	70	0.27	0.23
	2.4	16.8	20–25	10.8	77	0.33	0.29
	2.9	15.6	<15	11.7	83	0.41	0.32
	3.1	16.1	<15	11.9	86	0.42	0.34
772	0	11.7	100	8.5	55	0.18	0.18
	2.0	17.6	45–55	10.0	72	0.25	0.22
	2.4	17.6	20–25	10.4	80	0.29	0.25
	2.9	17.6	<15	10.8	83	0.32	0.28
	3.1	18.1	<15	10.9	86	0.34	0.29
882	0	13.0	100	8.5	55	0.18	0.18
	2.2	20.7	45–55	9.4	72	0.22	0.21
	2.6	18.7	20–25	10.2	80	0.27	0.25
	2.9	19.0	<15	10.4	86	0.29	0.26
	3.1	19.8	<15	10.5	86	0.29	0.26
992	0	14.1	100	8.5	55	0.18	0.18
	2.2	22.7	45–55	9.3	72	0.19	0.19
	2.6	22.5	20–25	9.5	80	0.23	0.22
	2.9	20.5	<15	10.4	86	0.26	0.25
	3.1	21.6	<15	10.0	86	0.26	0.23
1102	0	15.4	100	8.5	55	0.18	0.18
	2.0	23.1	45–55	9.1	72	0.18	0.18
	2.4	22.9	20–25	9.2	80	0.19	0.19
	2.6	21.2	<15	10.0	86	0.22	0.22
	2.9	22.0	<15	9.7	86	0.22	0.22

[a] Modified from "Nutrient Requirements of Beef Cattle," 5th rev. ed., National Academy of Sciences, Washington, D.C., 1976, Table 2. With permission.

[b] The concentration of vitamin A in all diets for growing and finishing steer calves and yearlings is 1000 IU/lb dry diet.

TABLE AIV.5

Nutrient Requirements for Growing–Finishing Heifer Calves and Yearlings (Nutrient Concentration in Diet Dry Matter)[a,b]

Weight (lb)	Daily gain (lb)	Minimum dry matter consumption (lb)	Roughage (%)	Total protein (%)	TDN (%)	Ca (%)	P (%)
220	0	4.6	100	8.7	55	0.18	0.18
	1.1	6.6	70–80	12.4	61	0.47	0.37
	1.5	6.4	50–60	14.4	69	0.66	0.48
	2.0	6.6	25–30	15.9	77	0.80	0.57
	2.4	6.6	<15	17.8	86	0.97	0.63
331	0	6.2	100	8.7	55	0.18	0.18
	1.1	9.0	70–80	11.0	61	0.34	0.29
	1.5	8.8	50–60	12.4	69	0.45	0.35
	2.0	8.8	25–30	13.5	77	0.57	0.42
	2.4	8.8	<15	15.0	86	0.70	0.50
441	0	7.7	100	8.5	55	0.18	0.18
	0.7	11.9	100	9.1	55	0.18	0.18
	1.1	13.2	80–90	9.6	58	0.23	0.22
	1.5	13.2	70–80	10.2	64	0.30	0.27
	2.0	11.7	35–45	11.7	75	0.41	0.32
	2.4	11.0	<15	12.8	86	0.50	0.38
551	0	9.0	100	8.5	55	0.18	0.18
	0.7	14.1	100	8.9	55	0.18	0.18
	1.1	14.3	80–90	9.5	58	0.20	0.20
	1.5	12.8	55–65	10.5	72	0.29	0.26
	2.0	13.0	35–45	11.1	77	0.36	0.29
	2.4	14.3	20–25	11.4	80	0.38	0.31
	2.6	13.9	<15	11.9	86	0.43	0.33
661	0	10.4	100	8.6	55	0.18	0.18
	0.7	16.3	100	8.5	55	0.18	0.18
	1.1	16.3	80–90	9.2	61	0.19	0.19
	1.5	14.6	55–65	10.1	72	0.24	0.23
	2.0	15.0	35–45	10.4	77	0.28	0.25
	2.4	16.5	20–25	10.4	80	0.31	0.27
	2.6	15.9	<15	10.9	86	0.33	0.28
772	0	11.7	100	8.5	55	0.18	0.18
	0.7	18.1	100	8.5	55	0.18	0.18
	1.1	18.3	80–90	8.7	61	0.18	0.18
	1.5	17.4	55–65	9.2	69	0.19	0.19
	2.0	17.9	35–45	9.5	75	0.21	0.21
	2.4	18.3	20–25	9.9	80	0.24	0.23
	2.6	17.9	<15	10.0	86	0.26	0.25
882	0	13.0	100	8.5	55	0.18	0.18
	0.7	20.0	100	8.5	55	0.18	0.18
	1.1	18.7	70–80	8.8	64	0.18	0.18
	1.5	19.2	55–65	9.0	66	0.18	0.18
	2.0	18.5	20–25	9.4	77	0.20	0.20
	2.4	18.3	<15	9.7	86	0.23	0.22
992	0	14.1	100	8.5	55	0.18	0.18
	0.4	19.2	100	8.5	55	0.18	0.18
	1.1	20.5	70–80	8.6	64	0.18	0.18
	1.8	20.1	35–45	9.0	75	0.18	0.18
	2.2	18.7	<15	9.5	86	0.22	0.22

[a] Modified from "Nutrient Requirements of Beef Cattle," 5th rev. ed., National Academy of Sciences, Washington, D.C., 1976, Table 2A. With permission.

[b] The concentration of vitamin A in all diets for growing and finishing heifers is 2000 IU/lb dry diet.

TABLE AIV.6

Nutrient Requirements for Beef Cattle Breeding Herd (Nutrient Concentration in Diet Dry Matter)[a,b]

Weight (lb)	Daily gain (lb)	Dry matter consumption (lb)	Roughage (%)	Total protein (%)	TDN (%)	Ca (%)	P (%)
Pregnant yearling heifers—last third of pregnancy							
716	0.9	14.5	100	8.8	52	0.23	0.23
	1.3	18.7	100	8.8	52	0.21	0.21
	1.8	20.7	85–100	9.0	58	0.23	0.21
772	0.9	15.2	100	8.8	52	0.22	0.22
	1.3	19.6	100	8.8	52	0.21	0.21
	1.8	22.0	85–100	8.8	58	0.22	0.21
827	0.9	15.9	100	8.7	52	0.21	0.21
	1.3	20.5	100	8.7	52	0.20	0.20
	1.8	24.2	85–100	8.7	55	0.20	0.20
882	0.9	16.5	100	8.7	52	0.21	0.21
	1.3	21.4	100	8.7	52	0.20	0.20
	1.8	25.6	85–100	8.7	55	0.19	0.19
937	0.9	17.2	100	8.8	52	0.20	0.20
	1.3	22.3	100	8.7	52	0.19	0.19
	1.8	26.7	85–100	8.7	55	0.18	0.18
Dry pregnant mature cows—middle third of pregnancy							
772		12.2	100	5.9	52	0.18	0.18
882		13.4	100	5.9	52	0.18	0.18
992		14.8	100	5.9	52	0.18	0.18
1102		15.9	100	5.9	52	0.18	0.18
1213		17.0	100	5.9	52	0.18	0.18
1323		18.3	100	5.9	52	0.18	0.18
1433		19.4	100	5.9	52	0.18	0.18
Dry pregnant mature cows—last third of pregnancy							
772	0.9	13.9	100	5.9	52	0.18	0.18
882	0.9	15.4	100	5.9	52	0.18	0.18
992	0.9	17.9	100	5.9	52	0.18	0.18
1102	0.9	17.9	100	5.9	52	0.18	0.18
1213	0.9	19.0	100	5.9	52	0.18	0.18
1323	0.9	20.3	100	5.9	52	0.18	0.18
1433	0.9	22.4	100	5.9	52	0.18	0.18
Cows nursing calves—average milking ability, first 3–4 months postpartum							
772		18.1	100	9.2	52	0.29	0.29
882		19.4	100	9.2	52	0.28	0.28
992		20.5	100	9.2	52	0.28	0.28
1102		21.6	100	9.2	52	0.28	0.28
1213		23.1	100	9.2	52	0.27	0.27
1323		24.2	100	9.2	52	0.25	0.25
1433		25.1	100	9.2	52	0.25	0.25

TABLE AIV.6 (*continued*)

Weight (lb)	Daily gain (lb)	dry matter consumption (lb)	Roughage (%)	Total protein (lb)	TDN (lb)	Ca (gm)	P (gm)
Cows nursing calves—superior milking ability, first 3–4 months postpartum							
772		22.4	100	10.9	55	0.44	0.39
882		23.8	100	10.9	55	0.42	0.38
992		24.9	100	10.9	55	0.40	0.37
1102		26.0	100	10.9	55	0.39	0.36
1213		27.3	100	10.9	55	0.37	0.35
1323		28.4	100	10.9	55	0.36	0.34
1433		29.5	100	10.9	55	0.35	0.33
Bulls, growth and maintenance (moderate activity)							
661	2.2	19.4	70–75	10.2	64	0.31	0.26
882	2.0	24.2	70–75	9.4	64	0.21	0.21
1102	1.5	26.9	80–85	8.8	61	0.18	0.18
1323	1.1	26.4	80–85	8.8	61	0.18	0.18
1543	0.7	28.4	90–100	8.5	55	0.18	0.18
1764	0.0	23.1	100	8.5	55	0.18	0.18
1984	0.0	25.1	100	8.5	55	0.18	0.18
2205	0.0	27.3	100	8.5	55	0.18	0.18

[a] Modified from "Nutrient Requirements of Beef Cattle," 5th rev. ed., National Academy of Sciences, Washington, D.C., 1976, Table 2B. With permission.

[b] The concentration of vitamin A in all diets for pregnant heifers and cows is 1300 IU/lb dry diet; for lactating cows and breeding bulls, 1800 IU/lb.

TABLE AIV.7

Net Energy Requirements of Growing and Finishing Beef Cattle (Megacalories per Animal per Day)[a]

	Body weight (lb)								
Daily gain (lb)	220	330	440	550	660	770	880	990	1100
NE_g required	(2.43)[b]	(3.30)	(4.10)	(4.84)	(5.55)	(6.24)	(6.89)	(7.52)	(8.14)
Steers									
0.2	0.17	0.23	0.28	0.34	0.39	0.43	0.48	0.52	0.56
0.4	0.34	0.46	0.57	0.68	0.78	0.88	0.97	1.06	1.14
0.7	0.52	0.70	0.87	1.03	1.18	1.33	1.47	1.61	1.74
0.9	0.70	0.95	1.18	1.40	1.60	1.80	1.99	2.17	2.34
1.1	0.89	1.20	1.49	1.77	2.02	2.27	2.51	2.74	2.97
1.3	1.08	1.46	1.81	2.15	2.46	2.76	3.05	3.33	3.60
1.5	1.27	1.73	2.14	2.53	2.90	3.26	3.60	3.93	4.25
1.8	1.47	2.00	2.47	2.93	3.36	3.77	4.17	4.55	4.92
2.0	1.68	2.27	2.82	3.33	3.82	4.29	4.74	5.18	5.60
2.2	1.88	2.55	3.16	3.75	4.29	4.82	5.33	5.82	6.29
2.4	2.10	2.84	3.52	4.17	4.78	5.36	5.93	6.47	7.01
2.6	2.31	3.13	3.88	4.60	5.27	5.92	6.54	7.14	7.73
2.9	2.53	3.43	4.26	5.04	5.77	6.48	7.16	7.83	8.47
3.1	2.76	3.74	4.63	5.49	6.29	7.06	7.80	8.52	9.22
3.3	2.99	4.05	5.02	5.95	6.81	7.65	8.46	9.23	9.98
Heifers									
0.2	0.18	0.25	0.30	0.36	0.41	0.46	0.51	0.56	0.61
0.4	0.37	0.50	0.62	0.74	0.84	0.95	1.05	1.14	1.24
0.7	0.57	0.77	0.95	1.13	1.29	1.45	1.61	1.75	1.90
0.9	0.77	1.05	1.30	1.54	1.76	1.98	2.18	2.39	2.58
1.1	0.99	1.34	1.66	1.96	2.25	2.52	2.79	3.05	3.30
1.3	1.21	1.64	2.03	2.40	2.75	3.09	3.41	3.73	4.03
1.5	1.44	1.95	2.42	2.86	3.27	3.68	4.06	4.44	4.80
1.8	1.67	2.28	2.81	3.33	3.82	4.28	4.73	5.17	5.59
2.0	1.92	2.60	3.23	3.81	4.37	4.91	5.43	5.93	6.41
2.2	2.17	2.94	3.65	4.32	4.95	5.56	6.14	6.71	7.26
2.4	2.43	3.30	4.09	4.84	5.55	6.23	6.88	7.52	8.13
2.6	2.70	3.66	4.55	5.37	6.16	6.91	7.64	8.35	9.03
2.9	2.98	4.04	5.01	5.92	6.79	7.63	8.42	9.21	9.96
3.1	3.26	4.42	5.49	6.49	7.44	8.36	9.23	10.09	10.91
3.3	3.56	4.82	5.98	7.07	8.11	9.11	10.06	11.00	11.90

[a] Modified from "Nutrient Requirements of Beef Cattle," 5th rev. ed., National Academy of Sciences, Washington, D.C., 1976, Table 3. With permission.

[b] Numbers in parentheses are NE_m required to achieve body weight given.

TABLE AIV.8

Mineral and Vitamin Requirements of Beef Cattle (in Percentage of Diet Dry Matter or Amount per Pound of Dry Diet)[a]

Nutrient	Growing and finishing steers and heifers	Dry pregnant cows	Breeding bulls and lactating cows	Possible toxic levels (mg/lb diet)
Vitamin A activity (IU)	1000	1272	1770	
Vitamin D (IU)	125	125	125	
Vitamin E (IU)	7–27	—	7–27	
Minerals				
Sodium (%)	0.06	0.06	0.06	
Calcium (%)	0.18–1.04	0.18	0.18–0.44	
Phosphorus (%)	0.18–0.70	0.18	0.18–0.39	
Magnesium (%)	0.04–0.10	—[b]	0.18	
Potassium (%)	0.6–0.8	—[b]	—[b]	
Sulfur (%)	0.1	—[b]	—[b]	
Iodine (μg)	—[c]	23–45	23–45	23
Iron (mg)	4	—[b]	—[b]	180
Copper (mg)	2	—[b]	—[b]	50
Cobalt (mg)	0.02–0.04	0.02–0.04	0.02–0.04	4–6
Manganese (mg)	0.4–4.5	9	—[b]	68
Zinc (mg)	9.1–13.6	—[b]	—[b]	410
Selenium (mg)	0.04	0.02–0.04	0.04–0.04	2

[a] Modified from "Nutrient Requirements of Beef Cattle," 5th rev. ed., National Academy of Sciences, Washington, D.C., 1976, Table 4. With permission.

[b] Unknown.

[c] See Table 1.4 for more details.

Appendix V

Typical Composition of Feedstuff for Cattle

Typical Composition of Feeds for Cattle
(All Values Except Dry Matter Are Shown on a Dry Matter Basis)

				Fiber				Minerals					Energy			
FEEDSTUFFS	DM %	CP %	EE %	CF %	ADF %	NDF %	ASH %	Ca %	P %	K %	S %	Zn PPM	TDN %	DE Mcal/ Lb.	NEM Mcal/ Lb.	NEG Mcal/ Lb.
Alfalfa Cubes	91	18	2.0	29	34	45	11	1.3	0.23	1.9	0.35	18	57	1.14	0.56	0.25
Alfalfa Dehydrated 17% Protein	92	19	3.3	26	32	45	10	1.4	0.25	2.7	0.29	17	62	1.24	0.62	0.33
Alfalfa Fresh	26	21	2.2	27	32	44	8	1.6	0.32	2.3	0.34	21	57	1.14	0.56	0.25
Alfalfa Hay Early Bloom	90	18	2.2	29	35	47	8	1.4	0.25	2.3	0.30	18	57	1.14	0.56	0.25
Alfalfa Hay Mid-Bloom	89	17	2.0	30	38	50	10	1.4	0.23	1.8	0.30	17	56	1.12	0.55	0.23
Alfalfa Hay Full Bloom	88	16	1.8	34	41	56	8	1.3	0.20	1.7	0.29	17	53	1.06	0.52	0.18
Alfalfa Hay Mature	90	14	1.7	38	45	59	8	1.3	0.19	1.4	0.25	17	50	1.00	0.49	0.12
Alfalfa Silage Wilted	36	18	2.8	30	35	46	10	1.5	0.28	2.4	0.30	17	56	1.12	0.55	0.23
Alfalfa Seed Screenings	90	28	7.0	16	—	—	7	.03	0.67	—	—	—	76	1.52	0.80	0.52
Alfalfa Stems	89	11	1.3	44	51	68	6	0.9	0.18	2.5	—	—	47	0.94	0.46	0.07
Almond Hulls	90	3	3.0	11	28	30	10	0.4	0.06	0.5	0.11	—	67	1.34	0.68	0.40
Apple Pomace Wet	20	6	5.9	17	—	—	4	0.1	0.10	0.5	0.02	—	73	1.46	0.76	0.48
Apple Pomace Dried	89	4	3.4	15	—	—	2	0.1	0.12	0.5	0.02	—	69	1.38	0.70	0.43
Bagasse Sugar Cane	91	1	0.7	49	59	86	3	0.9	0.29	0.5	0.10	—	36	0.72	0.37	0.0
Bakery Product Dried	91	11	13.0	1	—	—	3	0.1	0.35	0.6	0.02	15	89	1.78	0.98	0.67
Barley Silage	32	10	4.0	34	—	—	10	0.3	0.30	1.6	0.17	—	50	1.00	0.49	0.12
Barley Silage Mature	40	9	4.0	34	—	—	10	0.2	0.30	1.5	0.15	—	67	1.34	0.68	0.40
Barley Straw	88	4	1.9	42	57	82	7	0.3	0.05	2.0	0.15	7	44	0.88	0.43	0.01
Barley Grain	89	12	2.0	6	7	19	3	0.1	0.40	0.5	0.16	20	82	1.64	0.88	0.59
Barley Grain Light Weight	88	13	2.3	9	—	—	4	0.0	—	—	—	—	79	1.58	0.84	0.55
Barley Grain Screenings	89	13	2.6	9	—	—	4	0.0	0.40	0.1	0.15	—	81	1.62	0.87	0.58
Beans Navy Cull	90	24	1.4	5	—	—	6	0.1	0.50	1.4	0.26	—	84	1.68	0.91	0.61
Beet Pulp Wet	11	10	2.0	20	34	59	5	0.9	0.10	0.2	0.22	1	68	1.36	0.69	0.41
Beet Pulp Dried	91	9	0.8	21	34	59	5	0.8	0.08	0.2	0.22	1	72	1.44	0.74	0.47
Beet Pulp Wet With Molasses	24	12	0.5	16	27	47	9	0.6	0.10	1.8	0.36	11	76	1.52	0.80	0.52
Beet Pulp Dry With Molasses	92	12	0.5	16	27	47	9	0.6	0.10	1.8	0.36	11	76	1.52	0.80	0.52
Beet Tops Sugar	20	13	1.4	9	—	—	25	0.7	0.24	4.8	0.45	20	58	1.16	0.57	0.26
Beet Top Silage	25	10	2.0	10	—	—	38	1.2	0.22	5.7	—	—	52	1.04	0.51	0.16
Bermudagrass Coastal Dehydrated	90	16	3.8	27	24	40	7	0.3	0.25	—	0.22	—	62	1.24	0.62	0.33
Bermudagrass Hay Coastal	91	9	2.0	30	33	75	5	0.4	0.18	1.5	0.14	—	49	0.98	0.48	0.11
Bermudagrass Hay	91	9	1.9	30	35	77	8	0.5	0.21	1.7	0.21	—	49	0.98	0.48	0.11
Birdsfoot Trefoil Fresh	22	21	4.7	21	—	—	7	1.8	0.25	2.3	0.25	—	70	1.40	0.72	0.44
Birdsfoot Trefoil Hay	89	16	2.2	30	34	44	8	1.8	0.22	1.8	0.25	—	57	1.14	0.56	0.25

Bluegrass Kentucky Fresh Early Bloom	36	14	3.9	27	—	—	7	0.5	0.39	2.3	0.12	25	72	1.44	0.74	0.47
Blue Grass Straw	93	6	1.1	40	50	78	6	0.2	0.10	—	—	—	45	0.90	0.44	0.03
Bluestem Fresh Mature	61	6	2.5	34	—	—	4	0.3	0.14	1.0	0.05	28	46	0.92	0.45	0.05
Bone Meal Steamed	95	13	11.6	1	—	—	79	30.7	12.86	0.2	0.25	130	16	0.32	0.34	0.0
Brewers Grains Wet	24	26	6.5	16	21	34	5	0.3	0.57	0.1	0.36	100	81	1.62	0.87	0.58
Brewers Dried Grains	92	28	7.5	15	19	32	4	0.3	0.57	0.1	0.36	100	81	1.62	0.87	0.58
Brewers Yeast Dried	94	48	1.0	3	—	—	7	0.1	1.56	1.8	0.41	41	79	1.58	0.84	0.55
Brome Grass Fresh Immature	32	15	4.1	28	33	54	10	0.4	0.39	2.7	0.20	—	64	1.28	0.64	0.36
Brome Grass Hay	89	10	2.5	32	39	69	9	0.5	0.23	2.5	0.16	—	52	1.04	0.51	0.16
Buckwheat Grain	88	12	2.8	11	—	—	2	0.1	0.36	0.5	0.16	10	79	1.58	0.84	0.55
Buttermilk Dried	94	34	5.6	0	0	0	10	1.5	1.03	1.1	0.09	44	88	1.76	0.97	0.65
Cactus	32	6	2.1	27	23	29	16	2.9	—	—	—	—	65	1.30	0.65	0.37
Calcium Carbonate	99	0	0.0	0	0	0	99	39.0	0.04	—	0.09	—	0	0.0	0.0	0.0
Cattle Manure Dried	92	15	2.6	34	37	57	10	1.3	1.00	0.5	—	240	41	0.82	0.41	0.0
Cheatgrass Fresh Immature	21	16	2.7	23	—	—	10	0.6	0.28	—	—	—	68	1.36	0.69	0.41
Citrus Pulp Dried	90	7	4.1	13	22	23	6	1.8	0.12	1.0	0.20	14	77	1.54	0.81	0.53
Clover Ladino Fresh	19	25	4.8	14	—	—	11	1.3	0.42	2.2	0.20	39	69	1.38	0.70	0.43
Clover Ladino Hay	90	21	2.0	24	—	—	9	1.7	0.32	2.4	0.22	17	61	1.22	0.61	0.31
Clover Red Fresh	24	18	4.0	24	33	44	9	1.7	0.26	2.0	0.17	23	64	1.28	0.64	0.36
Clover Red Hay	88	15	2.9	30	41	56	8	1.4	0.22	1.9	0.17	17	59	1.18	0.58	0.28
Coffee Grounds	88	13	22.3	41	68	80	1	0.1	0.08	—	—	—	13	0.26	0.20	0.0
Corn Whole Plant Pellets	91	9	2.4	21	—	—	6	0.5	0.24	1.0	0.14	—	63	1.26	0.63	0.34
Corn Fodder	80	9	2.4	27	29	48	7	0.3	0.18	1.0	0.14	—	65	1.30	0.65	0.37
Corn Stover Mature	80	5	1.3	35	41	71	7	0.5	0.09	1.6	0.17	—	59	1.18	0.58	0.28
Corn Silage Milk Stage	26	8	2.8	26	31	—	6	0.3	0.24	1.6	0.12	25	67	1.34	0.68	0.40
Corn Silage Mature Well Eared	36	8	2.7	23	—	—	7	0.3	0.20	1.0	0.11	24	69	1.38	0.70	0.43
Corn Grain Dent Yellow	88	10	4.0	3	3	10	2	0.0	0.30	0.4	0.14	14	90	1.80	0.99	0.68
Corn Grain High Lysine	92	12	4.4	4	—	—	2	0.0	0.24	0.3	0.11	—	84	1.68	0.91	0.61
Corn and Cob Meal	87	9	3.7	9	10	25	2	0.1	0.26	0.5	0.21	10	82	1.64	0.88	0.59
Corn Cobs	90	3	0.5	36	43	88	2	0.1	0.04	0.8	0.47	5	48	0.96	0.47	0.09
Corn Gluten Feed	90	27	2.9	8	—	—	7	0.5	0.86	0.6	0.24	100	82	1.64	0.88	0.59
Corn Gluten Meal	91	43	2.5	5	—	—	4	0.2	0.51	0.0	0.85	45	84	1.68	0.91	0.61
Corn Cannery Waste	29	8	3.0	28	36	59	5	0.1	0.29	1.0	0.13	25	68	1.36	0.69	0.41
Cotton Gin Trash	90	7	1.7	37	—	—	9	0.3	0.16	—	—	—	44	0.88	0.43	0.01
Cottonseed Hulls	90	4	1.5	48	67	86	3	0.1	0.07	1.0	0.09	22	44	0.88	0.43	0.01
Cottonseed Meal Screw Press 41%Protein	93	45	6.0	13	22	30	7	0.2	1.18	1.4	0.33	63	78	1.56	0.82	0.54
Cottonseed Meal Solvent 41% Protein	92	46	2.2	13	22	30	7	0.2	1.16	1.4	0.36	66	75	1.50	0.78	0.50
Crambe Meal Solvent	92	34	1.1	25	—	—	6	1.2	1.30	—	—	—	70	1.40	0.72	0.44
Cranberry Pulp Dried	88	7	15.7	26	47	54	2	—	—	—	—	—	49	0.98	0.48	0.11
Curaco Phosphate	99	0	0.0	0	0	0	95	34.0	15.00	—	—	—	0	0.0	0.0	0.0
Defluorinated Phosphate	99	0	0.0	0	0	0	95	32.6	18.07	1.0	—	100	0	0.0	0.0	0.0
Diammonium Phosphate	98	115	0.0	0	0	0	35	0.5	20.41	—	—	—	0	0.0	0.0	0.0
Dicalcium Phosphate	96	0	0.0	0	0	0	94	21.0	18.65	0.1	—	70	0	0.0	0.0	0.0

(continued)

Typical Composition of Feeds for Cattle *(continued)*

Feedstuffs	DM %	CP %	EE %	CF %	ADF %	NDF %	ASH %	Ca %	P %	K %	S %	Zn PPM	TDN %	DE Mcal/ Lb.	NE_m Mcal/ Lb.	NE_g Mcal/ Lb.
				Fiber				Minerals					Energy			
Distillers Grains Corn	92	30	8.2	14	16	39	2	0.1	0.43	0.2	0.46	35	84	1.68	0.91	0.61
Distillers Grains Corn with Solubles	92	29	9.8	10	—	—	5	0.2	0.85	0.7	0.32	90	87	1.74	0.95	0.64
Distillers Dried Solubles	92	30	9.5	4	—	—	8	0.4	1.48	1.9	0.40	91	88	1.76	0.97	0.65
Fat Animal-Poultry	99	0	99.0	0	0	0	0	0.0	0.0	0.0	0.0	0	198	3.96	2.43	1.84
Feather Meal Hydrolyzed	94	91	3.3	2	20	20	4	0.2	0.78	0.3	2.00	53	68	1.36	0.69	0.41
Fescue Kentucky 31 Fresh	29	15	5.5	25	—	—	10	0.4	0.35	2.6	—	22	66	1.32	0.67	0.38
Fescue Kentucky 31 Hay Early Bloom	88	18	6.6	25	—	—	9	0.5	0.37	—	—	—	64	1.28	0.64	0.36
Fescue Kentucky 31 Hay Mature	89	14	5.2	28	38	65	7	0.4	0.25	—	—	—	60	1.20	0.59	0.30
Fescue Straw (Red)	94	4	1.1	41	—	—	6	0.0	0.06	—	—	—	43	0.86	0.43	0.0
Garbage Municipal Cooked	23	16	23.3	8	50	59	11	1.6	0.45	—	—	—	75	1.50	0.78	0.50
Grain Screenings	90	15	5.5	14	—	—	9	0.5	0.43	—	—	17	63	1.26	0.63	0.34
Grain Dust	91	11	3.2	15	—	—	12	0.3	0.18	—	—	42	72	1.44	0.74	0.47
Grape Pomace Stemless	91	12	7.5	32	50	53	9	0.6	0.06	0.6	—	24	30	0.60	0.37	0.0
Grass Silage	26	12	4.6	34	38	66	9	0.8	0.22	2.0	—	29	61	1.22	0.61	0.31
Hominy Feed	90	12	7.7	6	12	56	3	0.1	0.58	0.7	0.06	3	95	1.90	1.06	0.73
Hop Leaves	37	15	3.6	15	—	—	35	2.8	0.64	—	—	—	49	0.98	0.48	0.11
Hop Vine Silage	30	15	3.1	21	—	—	20	3.3	0.37	1.8	0.22	44	53	1.06	0.52	0.18
Hops Spent	89	22	4.0	28	—	—	7	1.6	0.60	—	—	—	39	0.78	0.40	0.0
Lespedeza Fresh Early Bloom	25	16	2.0	32	—	—	10	1.4	0.21	1.1	0.21	—	67	1.34	0.68	0.40
Limestone Ground	98	0	0.0	0	0	0	96	33.8	0.02	—	—	—	0	0.0	0.0	0.0
Linseed Meal Solvent	91	40	1.9	10	17	25	6	0.4	0.91	1.5	0.40	35	76	1.52	0.80	0.52
Meadow Hay	92	8	2.5	33	—	—	9	0.6	0.17	—	—	—	46	0.92	0.45	0.05
Milo Grain	89	11	3.2	3	8	17	2	0.0	0.32	0.4	0.18	15	80	1.60	0.85	0.56
Molasses Beet	77	9	0.2	0	0	0	11	0.2	0.04	6.1	0.60	18	79	1.58	0.84	0.55
Molasses Cane	76	5	0.0	0	0	0	10	1.0	0.10	4.0	0.46	30	81	1.62	0.87	0.58
Molasses Cane Dried	94	11	0.6	3	0	0	14	1.2	0.15	4.0	0.46	30	81	1.62	0.87	0.58
Molasses Citrus	65	11	0.3	0	0	0	9	2.0	0.25	0.2	0.23	137	77	1.54	0.81	0.53
Molasses Wood	59	1	0.5	1	—	—	7	1.9	0.04	0.1	0.05	—	77	1.54	0.81	0.53
Monoammonium Phosphate	98	74	0.0	0	0	0	24	0.5	24.77	—	0.71	81	0	0.0	0.0	0.0
Mono-Dicalcium Phosphate	97	0	0.0	0	0	0	94	16.7	21.10	0.1	—	70	0	0.0	0.0	0.0
Oat Hay	87	9	2.1	30	38	63	9	0.2	0.22	1.0	0.30	39	59	1.18	0.58	0.28

Oat Silage	34	11	3.8	30	—	—	10	0.4	0.25	3.4	0.32	35	58	1.16	0.57	0.26
Oat Straw	90	4	2.3	41	46	70	8	0.3	0.10	2.2	0.22	6	52	1.04	0.51	0.16
Oats Grain	89	14	4.0	12	17	31	4	0.1	0.38	0.5	0.19	18	74	1.48	0.77	0.49
Oat Groats	91	18	6.6	3	—	—	2	0.1	0.47	0.4	0.22	—	93	1.86	1.03	0.71
Oat Meal Feeding	90	17	6.0	4	—	—	3	0.1	0.51	0.5	0.29	—	94	1.88	1.05	0.72
Oat Mill Byproduct	90	7	2.5	23	—	—	6	0.1	0.24	0.6	—	—	33	0.66	0.37	0.0
Oat Hulls	93	5	1.5	32	44	81	7	0.2	0.15	0.6	0.15	—	37	0.74	0.39	0.0
Orange Pulp Dried	89	9	1.8	9	—	—	4	0.7	0.11	—	—	—	87	1.74	0.95	0.64
Orchardgrass Fresh Immature	24	18	5.0	24	27	45	11	0.4	0.40	2.7	0.22	20	65	1.30	0.65	0.37
Orchardgrass Hay	88	11	3.3	34	40	70	7	0.3	0.28	2.8	0.26	18	59	1.18	0.58	028
Pea Vine Hay	89	10	1.8	32	—	—	7	1.2	0.21	1.8	0.17	15	60	1.20	0.59	0.30
Pea Vine Silage	24	13	3.3	31	—	—	8	1.3	0.24	—	0.29	—	57	1.14	0.56	0.25
Pea Straw	89	7	1.3	45	—	—	7	—	0.11	1.1	0.20	—	56	1.12	0.55	0.23
Peas Cull	89	25	1.5	8	—	—	5	0.2	0.43	1.1	0.26	30	79	1.58	0.84	0.55
Peanut Hulls	92	7	1.3	65	65	74	5	0.2	0.07	0.9	—	—	22	0.44	0.35	0.0
Peanut Meal Solvent	92	52	1.3	14	—	—	5	0.2	0.71	1.2	0.30	22	77	1.54	0.81	0.53
Pineapple Green Chop	17	9	2.6	23	35	64	8	—	—	—	—	—	45	0.90	0.44	0.03
Pineapple Bran	89	5	1.2	20	28	59	3	0.3	0.10	—	—	—	72	1.44	0.74	0.47
Pineapple Presscake	21	5	1.1	20	36	69	3	0.2	0.12	—	—	—	72	1.44	0.74	0.47
Potatoes Cull	21	10	0.4	2	—	—	5	0.0	0.24	2.2	0.09	—	80	1.60	0.85	0.56
Potato Waste Wet	14	7	1.5	9	—	—	3	0.2	0.26	1.3	0.11	12	82	1.64	0.88	0.59
Potato Waste Dried	89	8	0.5	7	—	—	5	0.1	0.13	1.2	—	—	85	1.70	0.92	0.62
Potato Waste Wet With Lime	17	5	0.3	10	—	—	9	4.2	0.18	—	—	—	80	1.60	0.85	0.56
Poultry Litter Dried	86	30	2.8	20	—	—	16	2.7	1.85	1.8	—	235	64	1.28	0.64	0.36
Poultry Manure Dried	89	30	2.1	15	16	33	28	8.6	2.30	1.7	—	400	53	1.06	0.52	0.18
Prairie Hay	91	7	2.0	35	—	—	8	0.4	0.13	1.1	—	34	50	1.00	0.49	0.12
Rapeseed Meal Solvent	91	41	2.2	14	—	—	8	0.7	1.10	1.4	0.28	66	70	1.40	0.72	0.44
Rice Straw Ammoniated	87	9	1.3	39	53	68	6	0.3	0.10	1.0	0.11	—	45	0.90	0.44	0.03
Rice Polishings	90	14	13.1	4	—	—	8	0.1	1.37	1.1	0.19	28	89	1.78	0.98	0.67
Rice Bran	91	13	15.2	13	20	26	13	0.1	1.68	1.9	0.20	33	66	1.32	0.67	0.38
Rice Hulls	92	3	0.9	44	70	80	20	0.1	0.08	0.3	0.09	—	13	0.26	0.33	0.0
Rice Mill Feed	91	7	5.4	33	—	—	—	0.4	0.62	—	—	—	42	0.84	0.42	0.0
Rye Straw	89	4	1.5	44	55	71	6	0.3	0.10	1.0	0.11	—	44	0.88	0.43	0.01
Rye Grain	89	13	1.7	2	—	—	2	0.1	0.38	0.5	0.17	34	81	1.62	0.87	0.58
Safflower Meal Solvent	91	22	1.0	33	41	59	6	0.3	0.80	0.8	0.20	44	55	1.10	0.54	0.21
Safflower Meal Dehulled Solvent	91	49	0.6	9	—	—	7	0.3	1.83	1.3	0.22	36	76	1.52	0.80	0.52
Sagebrush Fresh	50	13	9.2	25	—	—	10	1.0	0.25	—	0.22	—	50	1.00	0.49	0.12
Sodium Tripolyphosphate	96	0	0.0	0	0	0	96	0.0	25.98	—	—	—	0	0.0	0.0	0.0
Sorghum Stover	85	5	2.1	33	—	—	10	0.4	0.11	1.5	—	—	57	1.14	0.56	0.25
Sorghum Silage	31	8	2.8	24	21	39	7	0.4	0.11	1.5	0.11	32	57	1.14	0.56	0.25
Soybean Hay	89	15	2.2	37	—	—	8	1.3	0.32	1.0	0.24	24	52	1.04	0.51	0.16

(continued)

Typical Composition of Feeds for Cattle (*continued*)

Feedstuffs	DM %	CP %	EE %	CF %	ADF %	NDF %	ASH %	Ca %	P %	K %	S %	Zn PPM	TDN %	DE Mcal/ Lb.	NE_m Mcal/ Lb.	NE_g Mcal/ Lb.
				Fiber				Minerals					Energy			
Soybean Straw	88	5	1.4	44	54	70	6	1.6	0.06	0.6	0.26	—	42	0.84	0.42	0.0
Soybeans Whole	91	42	19.2	6	—	—	5	0.3	0.63	1.8	0.24	60	92	1.84	1.02	0.70
Soybean Meal Solvent 44% Protein	89	52	1.3	6	10	12	7	0.3	0.73	2.1	0.48	48	82	1.64	0.88	0.59
Soybean Meal Solvent 49% Protein	90	56	1.2	3	—	—	6	0.3	0.71	2.2	0.48	61	84	1.68	0.91	0.61
Soybran Flakes (Hulls)	91	12	2.8	39	44	60	4	0.6	0.17	1.0	0.09	24	65	1.30	0.65	0.37
Sudangrass Fresh Immature	18	17	3.9	31	—	—	9	0.5	0.31	2.0	0.04	—	70	1.40	0.72	0.44
Sudangrass Hay	89	10	1.8	31	43	68	10	0.4	0.30	2.1	0.06	—	59	1.18	0.58	0.28
Sudangrass Silage	23	10	3.1	34	—	—	10	0.4	0.25	3.5	0.05	—	57	1.14	0.56	0.25
Sunflower Meal Solvent	93	50	3.1	12	—	—	8	0.6	0.54	1.1	—	—	65	1.30	0.65	0.37
Sunflower Meal With Hulls	90	32	1.4	27	—	—	7	0.4	1.04	0.9	0.33	—	57	1.14	0.56	0.25
Sunflower Hulls	90	5	2.2	25	63	—	3	0.0	0.11	—	—	—	40	0.80	0.41	0.0
Timothy Fresh Pre-bloom	26	11	3.8	32	—	—	7	0.4	0.28	2.1	0.21	24	61	1.22	0.61	0.31
Timothy Hay Early Bloom	88	8	2.6	33	43	68	6	0.5	0.25	0.9	0.21	—	59	1.18	0.58	0.28
Timothy Hay Full Bloom	88	7	2.5	34	45	70	5	0.4	0.20	1.6	0.13	17	57	1.14	0.56	0.25
Tomato Pomace Dried	92	23	10.6	26	50	55	6	0.4	0.59	3.6	—	—	67	1.34	0.68	0.40
Triticale Silage	38	12	—	—	—	—	—	—	—	—	—	—	—	—	—	—
Triticale	90	16	4.6	4	—	—	2	0.1	0.34	0.4	0.17	—	86	1.72	0.94	0.63
Turnip Tops (Purple)	17	16	2.6	10	—	—	14	3.2	0.34	3.0	0.27	—	69	1.38	0.70	0.43
Turnip Roots	9	12	1.5	11	—	—	8	0.8	0.40	3.4	0.43	40	86	1.72	0.94	0.63
Urea 45% N	98	287	0.0	0	0	0	0	0.0	0.0	0.0	0.0	0	0	0.0	0.0	0.0
Wheat Fresh (Pasture)	21	28	4.0	18	—	—	14	0.4	0.40	3.5	—	—	69	1.38	0.70	0.43
Wheat Silage	28	10	3.2	28	—	—	8	0.3	0.27	1.2	0.23	—	63	1.26	0.63	0.34
Wheat Straw	88	3	1.5	42	56	85	7	0.2	0.08	1.2	0.14	7	44	0.88	0.43	0.01
Wheat Grain Hard	89	14	2.0	3	—	—	2	0.1	0.45	0.5	0.17	16	89	1.78	0.98	0.67
Wheat Grain Soft	89	12	2.0	3	—	—	2	0.1	0.35	0.4	0.17	16	89	1.78	0.98	0.67
Wheat Bran	89	18	4.8	11	12	44	7	0.2	1.32	1.4	0.25	105	70	1.40	0.72	0.44
Wheat Middlings	88	18	3.9	3	—	—	3	0.1	0.57	0.6	0.22	70	90	1.80	0.99	0.68
Wheat Mill Run	90	17	4.7	9	—	—	6	0.1	1.15	1.4	0.28	—	75	1.50	0.78	0.50
Wheat Shorts	89	20	5.4	7	—	—	5	0.1	0.99	1.1	0.19	118	80	1.60	0.85	0.56
Wheatgrass Crested Fresh Early Bloom	37	11	1.6	30	—	—	7	0.3	0.30	—	—	—	58	1.16	0.57	0.26
Wheatgrass Crested Fresh Full Bloom	50	10	1.6	33	—	—	7	0.4	0.28	—	—	—	55	1.10	0.54	0.21
Wheatgrass Crested Hay	92	10	2.4	33	—	—	7	0.3	0.15	—	—	32	54	1.08	0.53	0.20
Whey Dried	94	16	0.9	0	0	0	10	1.0	0.81	1.6	1.10	3	84	1.68	0.91	0.61

Index

D

R

S

T